oekom

Climate neutral
Publisher
ClimatePartner.com/53585-1805-1001

Commitment to sustainable publishing
The oekom verlag takes a consistent stand for sustainability as a publisher and also as a company. Our facilities and production of publications are subject to the highest ecological criteria. This book was printed on 100% recycled paper, certified with the FSC® stamp. The paper chosen for the coating of the cover is also made out of 100% recycled material with FSC® approval. All the CO_2 emissions caused by this publication are compensated by investments in a Gold Standard project. The additional costs for this are covered by the publisher.

Bibliographic information of the German National Library:
The German National Library has registered this publication in the German National Bibliography; detailed bibliographic information can be found online at: http://dnb.d-nb.de.

Gesellschaft für ökologische Kommunikation mbH
Waltherstraße 29, 80337 München

Translated by Ray Cunningham (http://www.raycunningham.eu/).
Translation Note: except where an English language source is given, all quotations are translated by the translator and lead editor.

Layout and typesetting: Reihs Satzstudio, Lohmar
Proofreading: Ashley Curtis
Cover design: Elisabeth Fürnstein, oekom verlag
Image credits for the book cover: top and middle – © Pierre Ibisch;
below © tuomaslehtinen – Fotolia.com
Print: Friedrich Pustet GmbH & Co. KG, Regensburg

ISBN 978-3-96238-117-2

Humans in the Global Ecosystem

An Introduction to Sustainable Development

Translated by
Ray Cunningham

Contents

Part 4
The transformation to sustainability

Appendix

Preface I

I would like to begin with a quotation from the German author Roger Willemsen, from his last piece of writing, 'Who We Are' (2016), the publication of which Willemsen himself did not live to see. 'We were the ones who knew but did not understand; full of information but without insight; brimful of knowledge, but poor in experience; and so we went on, not checking or constraining ourselves'—a grim preview of things to come and a shocking retrospective recognition of human ineptitude! Today I meet more people with similar forebodings because of their experience and perception of what is currently happening on and to our earth. The ongoing impacts of the various forms of interference into the ecosystems that sustain us have reached genuinely frightening proportions. The influence of our human civilization on the wonderfully ecologically constructed house of the earth is becoming, from a global perspective especially, ever more destructive and ever less reparable within human timescales. The geological epoch of the Anthropocene has begun. This raises the inescapable question: will *Homo sapiens*, endowed with reason and elevated above the animal kingdom, succeed in becoming once again a part of the earth's ecosystem, in self-subjugation to the laws of nature? This does not seem to be the case at the moment: our unique biosphere—this thin, vital skin—is being depleted like a stone quarry.

As a result, the Earth is ageing prematurely and rapidly, and apparently unstoppably—no longer naturally, but in response to the actions of humans and our 'advanced civilization'. The dramatic changes to the climate which can no longer be denied, the decline in the natural fertility of our soils, the decline in biodiversity are all compelling us to re-think and re-formulate our dealings with nature—which will remain the foundation for our subsistence in the future. This requires knowledge, combined with responsibility and practical application: knowledge regarding the diversity and functioning of the ecosystems on our earth and their interactions; knowledge, too, regarding how nature—driven by the principles of evolution—is able to continue to develop, to optimize and thus to grow without collapsing; knowledge regarding nature's ability to adapt and to regenerate after so-called 'catastrophes'.

I am very grateful to the authors of this book for the fact that — 25 years after the foundation of the Eberswalde University for Sustainable Development, and the beginning of its own evolution — they are addressing the topic of sustainable development in all its complexity. They begin with an analysis of the challenges as they confront us today, and this analysis results in a concept, a synthesis for durable environmental interaction between human beings and nature — the only sustainable path for human civilization. This requires an investigation into complex philosophical, ethical and psychological issues concerning the relationship between human beings and nature. Following on from this, the authors endeavour to devise approaches and solutions for sustainable action across the entire spectrum of our social system. This is followed by proposals for urgent transformations in all areas of society towards the goal of sustainability. This is a courageous project, taking forward a process that will — that must — have a greater importance for us in the future. The fact that such a young university, in its teaching and research, is facing up to the challenges posed by sustainability in all its complexity should serve, and should be acknowledged, as an example. The students, the graduates, will have reason to be grateful. More than ever, our university graduates need this kind of comprehensive understanding of nature, the basis for our existence. Because in nature, as in society, everything is interconnected. We understand more clearly now that the 'miracle' of evolution (of creation) is not essentially predatory, cut-throat competition; it is not 'grow or die', as is often proclaimed — alongside dubious invocations of Darwin — in our governing social system. Rather, in its essence, evolution is about the interplay, the interaction of the individual components, to which we also belong. This applies both to individual organisms and to the ecosystems which they together make up. Such complex interactions speak to the success of evolution, which has repeatedly shown itself capable of overcoming all manner of catastrophes so as to recreate itself in the wake of the great species extinctions of past geological epochs.

This book can — must — help to clarify the function and functionality of the natural economy, to show its capacity and its limits, to understand its vulnerability. In a time of ever more intensive use of nature's capital stock — whether through consumption of natural resources, land use or infrastructure — nature must not be the loser. For let us be clear: the project of nature will go on in any event. My concern is for the project of humanity, which will ultimately prove to be weaker.

This book is a first start, an attempt—already highly persuasive in parts—not only to gather knowledge and to compress it but to combine knowledge with sustainable action, which means with responsibility. It is about the future of our human civilization, which will only succeed if we come to see ourselves once more as part of our biosphere, which in turn is no more and no less than sustainability; or, to put it in other words: our and the planet's future viability and future capacity to support our grandchildren.

Greifswald, in the early spring of 2018

Prof. em. Dr. Michael Succow
Michael Succow Foundation for the Protection of Nature

Preface II

Everything we do affects our social environment — our colleagues, our children and friends, their friends, and the friends and colleagues of our friends' friends. And what they do affects us. We are embedded in social networks. And I'm not just referring to networks like Facebook, but more especially to networks and systems not dependent on technological media. Many thinkers — including Max Weber and Niklas Luhmann among others — have studied how these systems emerge and function. They note that the communication of ideas and projects can itself contribute effectively to their realization. So conviction often produces conviction.

I would go even further. It is often the case that simply asking the right question is enough to reveal that our culture has no answer, no idea. This is currently the situation regarding finite resources, climate change and many aspects of the energy transition. It is also certainly the case with regard to the fragmentation of the social centre of our societies and to political decisions on pensions policy — which, given the facts of demographic change, are moving in precisely the wrong direction.

Here, asking the right questions is the real added value that communication can bring. It is a necessary condition for change. A sufficient condition is finding answers. However, answers can only be anticipated from interdisciplinary and transdisciplinary discursive approaches in science and society. Unfortunately, the proper coordination and structural embedding for such discursive approaches are still lacking in the German science and research system. We cannot afford to wait for these optimal conditions to emerge. It is therefore all the more important that individuals take the lead and dare to change the system. How does that work in a context in which knowledge is created and communicated? The two components — knowledge and change — are not dependent on each other in a linear fashion. Knowledge alone does not change reality, nor will a changed world automatically result in the generation of new knowledge. Their mutual connection is weak and contingent. The point at which a society becomes ready, on the basis of knowledge, to undergo fundamental change, or when social change sets new demands and questions for science and research, is an open question — and therefore subject to influence. It becomes altogether different — and this is the goal of education for sustainable development — when a third element is added to the mix, one that binds the two components together and aligns them. Change and knowledge reciprocally strengthen one another when understanding and a guiding vision come into play. For this we need individual and shared experiences, discursive processes and transformation. In this context, transformation is an apt term for what is less accurately called the 'mediation' or — perhaps too robustly — the 'application' of knowledge.

Because knowledge changes in the course of its application and implementation. The transformation of techno-scientific, ecological, social and economic knowledge into certitudes on which societal action can be based is a demanding task. It releases us from a nit-picking attachment to factual knowledge and enables us to develop benchmarks for the assessment and categorization of facts. This is why the link to practice in education and research for sustainability is so indispensable. So I am putting my faith not least in the activities of the Eberswalde University for Sustainable Development and its close links with politics and the wider society.

Sustainable action in politics, business and society can be made concrete through clear communication, coordination and structural innovations. We must trust ourselves, and there must be sufficient incentive, to set off on new pathways and to create the framework conditions for a sustainable society — one that contributes to the preservation of our natural existential foundations instead of excluding future generations from access to those very foundations.

This book begins with a systemic perspective on the global challenges of our time. Its structured presentation of current systems and instruments provides a good overview and establishes clarity. In addition, the book provides answers on how to design systems to enable and implement a transformation towards sustainable development.

9 March 2018

Marlehn Thieme
Chair of the German Council for Sustainable Development

Foreword by the Editors

'Sustainability' is an imprecise term, and by no means a word on everyone's lips. Nonetheless, the term has achieved a certain prominence over the past three decades — with all the associated advantages and disadvantages. For hardly has 'sustainability' become more prominent than already its regular use leads to confusion and scepticism — and of course criticism. 'Carlowitz's once noble word has become a shallow advertising slogan. [...] The term "sustainability" leads all too easily to dishonest marketing, something that is certainly common in environmental policy, even if usually not meant to be fraudulent. [...] The term "sustainability" obscures the complex relationships within nature and between environment and society. It provides no answers, but raises questions. The term is harmful. Let's leave it to sly salespeople' (Bojanowski 2014). 'It is remarkable how quickly the value of words can evaporate. For the conscientious consumer, terms such as "ecological", "sustainable" or simply "responsible" are as worn out as "quality" on packet soups: the more it is emphasized, the less it delivers' (Hartmann 2015). The concept of sustainable development is for some critics even more problematic, as it seems to be a contradiction in terms, an oxymoron (Patten 2013).

It remains the case, however, that something does not become bad simply through misuse. And if important terms that once stood for a paradigm shift have become blunted through prolonged and varied use, nevertheless it may not be a good idea to give them up lightly. For example, the aim that as many companies as possible should have a sustainability strategy is not bad simply because many such strategies achieve too little 'genuine sustainability', and are used principally for 'greenwashing'. Rather, this makes it all the more important to hold on to the power to determine exactly what a word means — on the basis of knowledge and scientific facts, and taking its historical development into account. And it needs to be made clear to everyone that they bear a certain responsibility when using the term 'sustainability'. This introductory textbook on sustainable development aims to do precisely that. It has been conceived as a dispassionate treatment of the topic, as an overview of relevant discourses, as food for thought, with new proposals for defending and elaborating the concept of sustainable development. Designed as an interdisciplinary course book for students and prospective students, it is also suitable for interested laypeople and those already engaged in supporting sustainable development.

The book covers a wide range of fundamental questions—such as where the idea of sustainable development comes from, and why we need it—right up to current approaches to political and economic transformation. It addresses the position of humans in the global (eco) system. Systems theory is the preferred approach, and is deployed as a basis for understanding sustainability across disciplines. The analysis of practical problems is just as much at the centre of the book as solutions and the necessary conditions for sustainable development. The four sections of the book deal with discrete and important thematic complexes. After an introductory part (Part 1) comes a study of humans as the agents of sustainability and of related anthropological and psychological aspects (Part 2). There then follows an account of selected systems and of whether or not their development can be classified as sustainable (Part 3). The final part comprises a description of important fields within, and existing approaches to, sustainable transformation (Part 4).

The chapters all follow a similar structure, with recurring elements:

- **Marginal notes** use key words to lead the reader through the chapters and to make it easier for them to find their way or to search for specific topics.

- **Important terms** are highlighted in the text and explained at the end of the paragraph in 'definition boxes' (also highlighted).

- Technical digressions or **in-depth explanations** are presented in highlighted **boxes** marked with a large **magnifying glass**. References to these boxes are sometimes indicated in the text with a small magnifying glass.

- Particularly **important text sections** are highlighted throughout.

- Photos, charts and graphs illustrate and visualize the contents of the chapters and provide valuable additional information.

References facilitate further reading and the appropriate classification of the knowledge presented. All the literature cited can be found in the list of references at the end of the book.

The index can be used to search for specific topics and terms. This book has its origins in an inter-departmental foundational lecture on sustainable development at the Eberswalde University for Sustainable Development, first put together by a large professorial team and now continually revised and developed. The lecture is part of the process that accompanied the renaming of the Eberswalde University (in 2010) and the unprecedented exercise in strategic refocusing and reorientation that followed. As such, the book also summarizes the preliminary results of an ongoing research process, which seeks to understand and elucidate the significance of sustainable development today.

The distinctive character of the book arises not only from the joint conception and interdisciplinary coordination of its contents and the repeated collaborative editing of the draft chapters by the editors, with their diverse (inter-)disciplinary backgrounds in ecology, economics, political science, cultural studies and physics. In addition, the various specialist fields represented at the university (such as nature conservation, environmental education, education for sustainable development, transformation governance and public administration) and the very different professional backgrounds of the editors also contributed to the breadth of the core concept and the detailed chapters. Other colleagues at the university brought additional expertise to the authorial collective for a number of chapters. Moreover, all the chapters were reviewed by other colleagues and students at the university. This was not only in the interests of academic rigour, but also to ensure the suitability of the texts for a significant target market by involving representative readers in the process. The book is thus based on an intense creative process characterized by considerable exchange and communication — a unique experience. What remains, among other things, is a strong impression of how much the concept of sustainable development can help people with very diverse backgrounds to work together for a great common cause. The process also afforded us all a greater sense of the intensity needed to continue the struggle for unbounded knowledge, mutual understanding and the development of effective and sustainable solutions.

The editors would like to thank their co-authors: Prof. Dr. Norbert Jung, Prof. Dr. Benjamin Nölting, Prof. Dr. Hermann Ott, Prof. Dr. Martin Welp, Prof. Dr. Wilhelm-Günther Vahrson, Prof. Dr. Jan König, Prof. Dr. Hans-Peter Benedikt, Kerstin Kräusche, Prof. Dr. Vera Luthardt, Prof. Dr. Peter Spathelf,

Prof. Dr. Martin Guericke and Hannah Mundry for their valuable support. Likewise, our great thanks go to the reviewers, all of whom work at, or are closely associated with, the University for Sustainable Development: Prof. Dr. Hartmut Ihne (Philosophy, Ethics), Prof. Dr. Horst Luley (Social Science, Regional Development), Prof. Dr. Carsten Mann (Forest Resource Economics), Dr. Dörte Martens (Psychology), Prof. Dr. Udo Simonis (Political Science, Psychology, esp. Environmental Psychology), Prof. Dr. Martin Welp (Socio-economics) and the student reviewers Alina Conrady, Sebastian Döring, Petra Heilig, Laura Koller, Stefanie Logge, Friedrich Mauel, Hannah Mundry and Fabian Rösch. We are particularly pleased to have prefaces written by Prof. Dr. Michael Succow and Marlehn Thieme. We would like to thank the Eberswalde University for Sustainable Development and its President Prof. Dr. Wilhelm-Günther Vahrson for their support, without which this project could not have been realized. For additional financial support we would like to thank the Science Foundation of the Sparkasse Barnim.

The editors

Eberswalde, April 2018

1
Understanding the challenge: the systems question

The biosphere: a short report on the state of the Earth

1.1

Pierre L. Ibisch

As far as we know, within our solar system the Earth is a unique planet: its size, its orbit around and distance from the sun, its rotation around its own axis, its atmosphere, surface temperature and other interrelated characteristics mean that water is present in liquid form on its surface, thus meeting an essential condition for life — at least, for life as we know it and are capable of imagining it. We humans have become a determinant factor of the biosphere, and exercise as much influence on it as previously only cosmic, climatological and geological drivers did. We transform the biosphere — in part deliberately to meet our needs and desires, but in considerable part also unconsciously and unwillingly through our sheer numbers, through everything we do, on an accelerating scale. Environmental and societal change are now both of enormous magnitude and are exerting reciprocal influence on each other. So how great, and how urgent, is the need for sustainable development?

Human life and its impact on the biosphere: the Anthropocene and the 'Tachycene'

A thin-skinned planet

The Earth has entered a new epoch — the human age, or Anthropocene (Crutzen 2002; Bonneuil & Fressoz 2015). Man-made changes on Earth affect not only the biosphere but also the atmosphere and increasingly even the geosphere (the geological ground below). The biosphere — enclosed by the gaseous atmosphere — surrounds the Earth like a thin skin; it is a kind of macroscopic biofilm:

'Our globe, which is 12,700 km "thick" in diameter, is composed of matter that is dead and predominantly hostile to life. Two-thirds of its surface is washed over by a thin layer of water, at its maximum barely thicker than 10 km (the mean sea depth is only 3.8 km, or 0.003 % of the Earth's diameter). This water reservoir itself is for the most part devoid of life; living things float in it in low concentrations — this "living" epidermis of the ocean is bioproductive to a significant degree only to depths of about 200 m. […] It is porous, and in many regions, where it is still formed solely by microbes, only a few millimetres thick. Even in the areas of the mightiest forests, the vegetation height represents less than 0.0005 % of the Earth's diameter. This fragmented and delicate film captures a small portion of the solar energy reaching the Earth and transforms it into chemical energy, which is either used directly for activity and work or else stored. It also exerts considerable influence on the composition of the atmosphere and on the so-called greenhouse effect. It is our home, the source of our food and our existence — our everything' (Ibisch 2016a, p. 91 f.).

The sudden appearance of humans

For hundreds of thousands of years, *Homo sapiens* has been active in this precariously thin, film-like biosphere — a period which in turn represents a vanishingly small fragment of the 4 billion-year history of life. Only in the last five centuries have people learned that this earth is a sphere and that there are other continents on which people live. Only around three centuries ago did humans embark on the journey of systematic exploration and surveying of planet Earth, thus heralding the beginning of an age which would later go down in history as the Enlightenment. At that time, just over half a billion people were living on Earth. It had taken about 700 years for the medieval population to double in number. The doubling time until the first billion (in 1803) was about 260 years. This was when the rapid growth of the species really took off. By 1928, 125 years later, it was already 2 billion, and after a further 32 years, it reached 3 billion in 1960. The shortest doubling time came in the phase up to 1987 — by which time there were 6 billion (Roser & Ortiz-Ospina 2018). The highest annual growth rate was 2.1% in 1962 (Roser & Ortiz-Ospina 2018). And in 2019 we will surpass 7.7 billion Citizens of the Earth (see Fig. 1). Currently, another billion is added every 15 years; the United Nations expects that by 2030 human beings will number 8.5 billion, by 2050 9.7 billion and by 2100 11.2 billion (DESA 2015).

Demographic change is not the same everywhere

Demographic change varies over both time and space, and this leads to a further geographical shift in geopolitical power relations as well as to new challenges. Between 2015 and 2050, population growth will essentially be concentrated in nine countries: India, Nigeria, Pakistan, the Democratic Republic of the Congo, Ethiopia, Tanzania, the USA, Indonesia and Uganda. In ten sub Saharan African countries, the population is expected to increase five-fold by 2100 (Angola, Burundi, the Democratic Republic of the Congo, Malawi, Mali, Niger, Somalia, Uganda, Tanzania, Zambia; DESA 2015).

Population density

Global population density (for all state territories, excluding oceans and large bodies of water) in 1960 was less than 24 people per km². By 2016, it had risen to over 56, with certain heavily-populated regions experiencing above-average increases (The World Bank 2017c). The increase has been particularly concentrated in the urban centres of the Global South, where 1 billion city dwellers in 1960 have now grown to over 4 billion, and the rate of urbanization continues to rise (The World

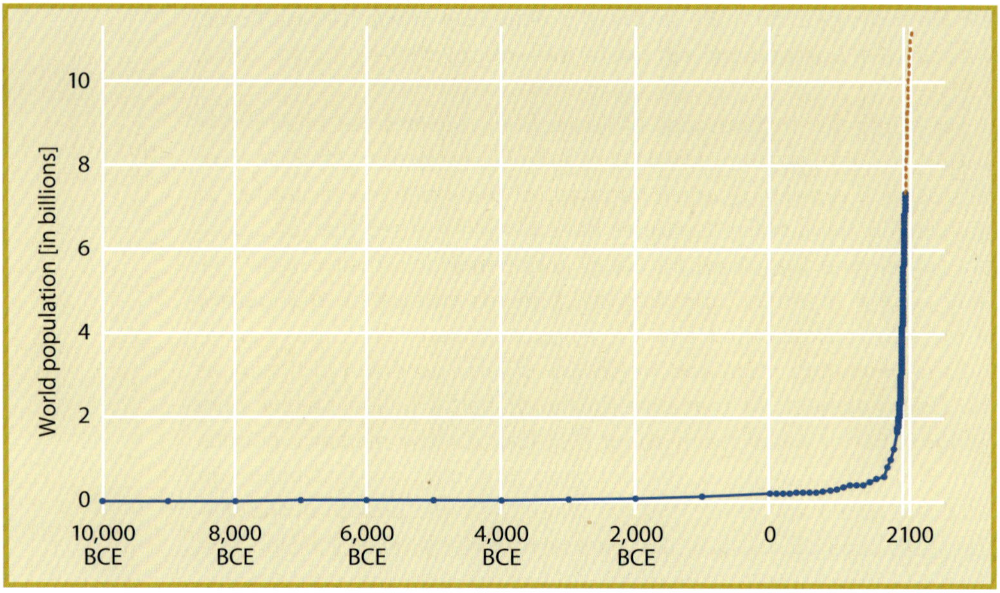

Figure 1: Human population growth over the last 12,000 years. The development after 2018 is speculative, and takes into account a slight decline in the growth rate. (source: Roser & Ortiz-Ospina 2018)

Bank 2017e). In 1960, just under 14 % of people lived in cities with a population of more than 1 million; by 2016, it was already 23 % (The World Bank 2017d). For many people, this meant a major change to their entire lifestyles: smallholders often became slum dwellers, people who had been self-sufficient became dependent service providers, and discrete rural communities were replaced by anonymous, sprawling urban settlements. In South Sudan in 2014, 96 % of the urban population lived in slums; in the Central African Republic, 93 %; in Sudan, 92 %; 82 % in Guinea-Bissau; 80 % in Mauritania and Mozambique; and over 60 in many other African countries (United Nations Statistics Division 2015).

Increasing inequality

An important facet of global social change is the increasingly unequal distribution of resources. This is described in the World Inequality Report (Alvaredo et al. 2018). For example, the richest 10 % of European citizens earn 37 % of all income; in sub-Saharan Africa, the figure is 55 %, and in the Middle East it is no less than 61 %. In the 1980s, the equivalent global figure was well below 40 %. Since then, socio-economic inequality has been increasing in virtually all regions of the world, albeit at different rates. In the USA, the growth in income inequality has been especially marked — there, the richest 1 % of citizens earn 20 % of all income (1980: 12 %). The poorest 50 % take only 13 % of the income (1980: 20 %). A key driver of the growth of inequality is the concentration of private capital. Another aspect of this process is the shrinking of public capital and the corresponding reduction in the capacity for investment. In addition, the concentration of capital and income and the shrinking of public budgets have a significant impact on poverty.

Poverty and malnutrition

In the last few decades, many people across the world have been able to escape poverty and become integrated in the global economic system, and problems of global health and nutrition have been reduced. This applies above all to China and India. However, a global average of 60 % of young people between the ages of 15 and 24 are currently unemployed — half of all the people in this age group live in South Asia and sub-Saharan Africa (Barne & Khokhar 2017). Worldwide, the number of undernourished children has also sunk; only in sub-Saharan Africa has it risen, from 45 to 57 million (from 1990 to 2015). In 2017, 83 million people in 45 countries needed food aid, an increase of 70 % over the previous year (Barne & Khokhar 2017).

Hyper-exponential growth of the human population and of economic activity

Human population growth is hyperexponential, meaning that the growth rate is not constant, but accelerates (although in recent years the growth rate has actually declined). In consequence, not only have population densities in many parts of the world grown enormously, but so too has every form of human interaction. Ever more 'condensed' groups of people, with better access to education, have created knowledge ever faster, have exchanged this knowledge with a growing number of their fellow humans and generated new ideas at an accelerating rate, ideas which entail a large number of activities. Ever more new inventions and technologies have swollen, among other things, our infrastructure, our energy throughput and our mobility, so that ever more people are able to interact with ever more other people, ever faster — to learn, trade, build, research…: a self-reinforcing process. So it is not surprising that for about 100 years now the scale of economic output has also been rising sharply. Especially since the end of World War II, humanity has been experiencing something called 'the Great Acceleration'. Global gross domestic product, which includes all the economic output from all countries, rose from an estimated 643 billion international dollars in 1700 to 108 trillion international dollars in 2015 (Roser 2018a, see Fig. 2). This means an approximately 168-fold increase in economic output over a period when the human population grew by a factor of 'only' 12.6. The curves in graphic illustrations of this growth have long had the shape of a hockey stick, which is why it is often referred to as 'hockey stick growth' (e.g. McCloskey 2013). However, in some of these curves it is possible to see the shape of a boomerang beginning to emerge.

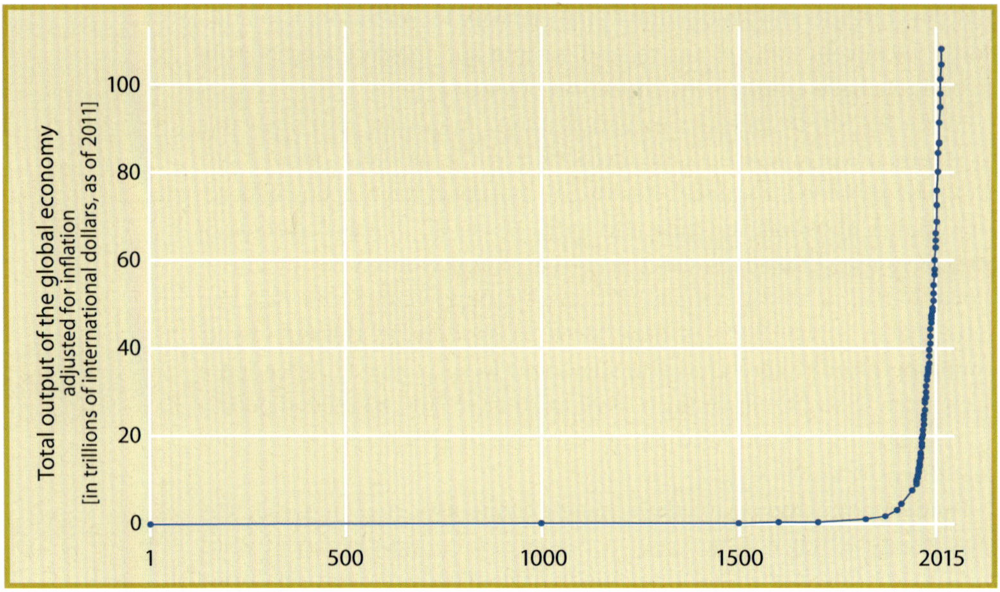

Figure 2: Global Gross National Product over the last two millennia – abrupt onset of global economic growth, in a phase of hyperexponential growth since 1945 (source: Roser 2018a)

Many indicators of economic development show that global economic growth has indeed persisted for some time in some regions, but also that some regions take part only after a delay, but then often show very strong growth rates. For example, the tourism industry—which reveals a great deal about the economic and political status of a country or region—has long been dominated by European destinations. However, since the American continents have become more strongly involved, the Asian travel market has more recently been experiencing a boom (Roser 2018b, see Fig. 3).

The boost to innovation, networking and globalization from the Internet

A similar trend can be seen in technology markets. The invention of the Internet and of mobile communications networks and the progress of digitalization have inspired a worldwide leap forward, particularly since the 2000s, in innovation, networking and globalization. Those regions that had already been economically strong for some time, such as Europe and North America, have long since been outstripped by the Asia-Pacific region. Sub-Saharan Africa was the last region to show substantial growth (Murphy & Roser 2017, see Fig. 4). The Internet and information technology are also good examples of how the hyperexponential growth of technology and the economy constantly creates new markets. Not only is the consumption of traditional products—such as food, clothing and cosmetics—growing, but new goods are constantly being developed to further fuel and accelerate growth.

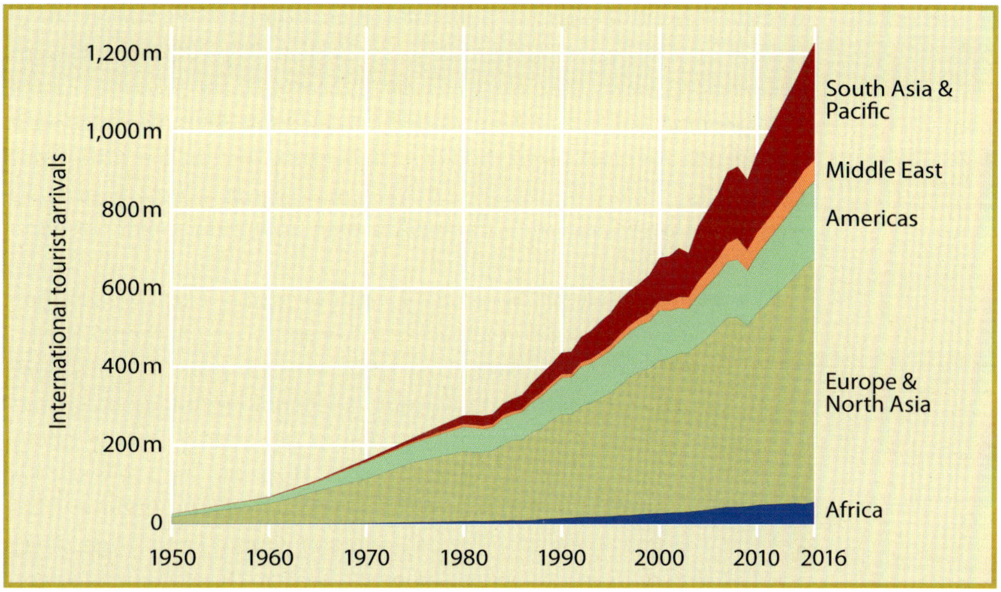

Figure 3: Development of global tourism since 1950 measured by arrivals by region (Source: Roser 2018b)

The beginning of the epoch of the 'Great Acceleration' marks a new turning point in the history of humankind and of the planet: it is the 'Tachycene' era (Greek *tachys* = fast, accelerated). This acceleration is not just something that is subjectively perceived, brought on by the media and faster transportation. Many hyperexponential changes are objectively measurable. Information technology, with the Internet and the technologies it has enabled, constitutes a catalyst and accelerator sector, one which provides new growth opportunities for other sectors. The Internet, e-mail, Skype and social media link economic agents and companies that previously could not trade or work together at all, or only at very high cost.

The Tachycene – the age of acceleration

The very sudden onset of the revolution in personal communications, bringing with it complex social consequences that remain only partially understood (e.g. changes in the presentation of self and in self-perception, changes in the use of information and in opinion formation) (Li & Sakamoto 2014), was due in part to so-called 'social media'. In the third quarter of 2008, there were 100 million active Facebook users; ten years later (Q3 2017) there were over 2 billion (Statista 2018). The volume of mobile data is currently growing at around 60 % per year (Andrae & Edler 2015). In theory, the explosive growth in 'common-sense knowledge' and ever easier access to information should bring with them new developmental and educational opportunities. Whether this will lead

The transformation of information, communications, education and politics by digitalization

to a greater assumption of political responsibility and increased participation on the part of citizens remains unclear. The *2017 World Governance Report* shows that elections are increasingly perceived as unfair. Voter turnout has been falling worldwide since the 1940s. Globally, political parties are considered the least trustworthy institutions (The World Bank 2017 f.). 2017 was the twelfth year in a row in which more countries saw a deterioration in democracy than an improvement (Freedom House 2017).

The risks of digitalization

The Internet itself has become a major driver of energy consumption. By 2010, data centres alone already consumed about 1% of global electricity; in 2030, this will have risen to between 3% and 13%. By 2030, under certain scenarios, communications technology could be consuming as much as 51% of electricity and thus generating 23% of greenhouse gas emissions (Andrae & Edler 2015). However, the rapid pace of digitalization not only increases material and energy throughput in the economy; it also accelerates the degrading of current knowledge, which is called into question or diluted by new research. This, in turn, increases the insecurity of those people who are not part of the information elite, or heightens their inclination to remain within seemingly secure information parameters. Control of the information filters surrounding so-called 'echo chambers'—where people prefer to expose themselves only to information that fits with their existing world view (see, inter

Figure 4: Internet users by world region since 1995 (Source: Murphy & Roser 2017)

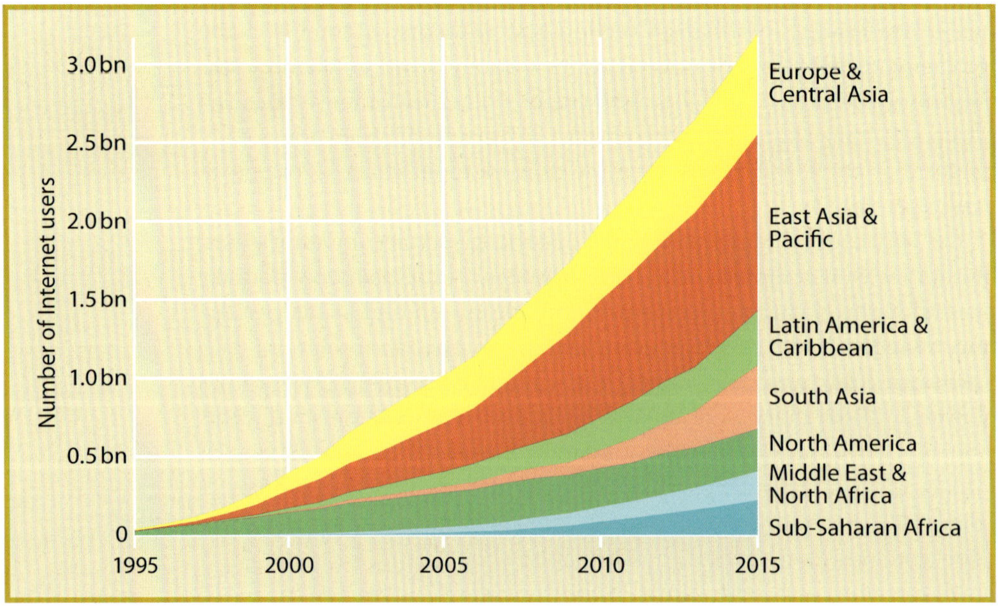

alia, Williams et al. 2015; Flaxman et al. 2016) — is becoming a new site of power. This new instrument of power is mainly in the hands of private corporations (especially internet companies and platforms). Other processes of societal change driven by digitalization are related to the transformation and loss of jobs through the rapid spread of robotization and artificial intelligence. There is already talk of 'human obsolescence', and more and more people are becoming superfluous to the workforce; by 2030 already, a large proportion of the workforce in logistics and sales may no longer be needed (Cross 2017). Some of the impacts on education systems and other social structures, on inequality, the economy and politics, can already be foreseen; but they are also likely to hold many surprises.

The growth of interacting environmental problems

The escalation of environmental problems

The hyperexponential growth of economic activity produces not only goods and services that people want, but also new, unintended and escalating environmental problems ('added damage' ▹ Chapter 3.1). To date, there are very few examples of the decoupling of economic growth from resource consumption; moreover, the consequences of virtual globalization are decidedly real, and they manifest themselves physically, for example in the form of increasing transportation and travel and of the corresponding infrastructures, and above all in increased material and energy throughput. The global road network connects people and the goods they trade, but it slices through the natural world, with significant impacts on the functioning of ecosystems (Ibisch et al. 2016; Alamgir et al. 2017). Towards the end of 2017, more than 72 million kilometres of road worldwide had been uploaded onto the *OpenStreetMap* data platform.[1] One of the largest development projects ever planned is China's 'One Belt, One Road' initiative, which aims to cover Eurasia and Africa with new transport routes, not least to make natural resources more accessible and marketable and to meet growing demand and need (Lechner et al. 2018).

The material requirements and throughput of construction projects is reaching hitherto unimagined proportions. For example, from 2011 to 2013, China used more cement than the USA used over the entire

1 ESRI Germany; personal communication with the author. In 2013, there were only just under half this number of road kilometres (Ibisch et al. 2016); quite a few of these streets were new, but most had only recently been digitalized and therefore become visible online — an example of how the digital surveying of the Earth is growing rapidly.

Construction projects consume land, energy and water

20th century (Swanson 2015, see Fig. 5). Following a 23-fold increase in the consumption of building materials from 1900 to 2010, there is now even a shortage of sand (Torres et al. 2017). The construction industry is not just intensive in its use of materials, space and energy, it also consumes vast amounts of water: 9% of global industrial water consumption is now accounted for by concrete production; water consumption is growing at twice the rate of the global population, and by 2050, 75% of the water needed for concrete will be in areas of acute water scarcity (Miller et al. 2018).

Fresh water crisis

In many parts of the world, the supply of fresh water is likely to determine the future for megacities and the viability of the agricultural industry. In 1999/2000, the first 'water war' took place in the city of Cochabamba, in a dry valley in the Bolivian Andes, when water prices rose as a result of the privatization of the water supply (Global Water Partnership 2017).

Figure 5: New buildings in Weichang, a growing metropolis near Chengde. In China, new megacities are being built every few years. In 2009, China already had 221 cities with a population of over one million (compared to 35 in Europe, Swanson 2015). The environmental impacts include the destruction of productive ecosystems and soil sealing, but also massive and energy-intensive consumption of materials (photo: P. Ibisch)

Figure 6: Dam and depleted reservoir in the winter following several years of drought (Lake Kaweah near Three Rivers, USA, photo: P. Ibisch)

Currently, concern is growing in California, where in a 'once in a millennium drought' nearly 2,000 wells have dried up since 2011 (Stevens 2015, see Fig. 6). In South Africa, Cape Town's water supply threatened to break down in the spring of 2018 (Onishi & Senguptajan 2018). From 2001 to 2009, Australia suffered from a 'millennium drought' that had far-reaching socio-economic and environmental consequences — for example, rice and cotton production decreased by 99% and 84% respectively, and many social and health problems were also reported (van Dijk et al. 2013). Once-major lakes such as Lake Chad and the Aral Sea have now almost dried up due to drought and unsustainable water abstraction (from tributary zones). Meanwhile, much of the world's population (80%) is severely threatened by inadequate water supply (Vörösmarty et al. 2010). A worldwide boom in dam construction is endangering river ecosystems and bringing with it various socio-political and economic risks; the number of free-flowing rivers is expected to decrease by 21% worldwide (Zarfl et al. 2015).

The release of substances harmful to nature and human health

An important aspect of global change is the release of substances which would not occur in nature without human activity, or at least not in such concentrations. For example, it has recently been discovered that the quantity of mercury released into the environment has tripled since the beginning of the Industrial Revolution. Prior to that, 100,000

Figure 7: Air pollution from exhaust fumes in Beijing, China (photo: P. Ibisch)

tons of mercury had already been emitted through silver mining (Stokstad 2014). The production of significant quantities of plastic began only about 70 years ago. By 2017, however, 6.3 billion tonnes of plastic waste had already been generated, of which almost 80 % was released into the environment (Geyer et al. 2017b). Virtually all commercially used plastic products release hormonally active compounds (mainly with estrogenic effects) as they degrade — resulting in insidious dangers to the health of animals and humans (e.g. Wagner & Oehlmann 2009; Yang et al. 2011). On a worldwide average, there are 44 µg of pollutants per m^3 in the air (in Egypt the figure is 105; in South Asia, on average, 74; in OECD states, 15; Brauer et al. 2016). In 2013, 87 % of people lived in areas where air pollution limits were being exceeded (Brauer et al. 2016). In cities like Beijing, industrial production has had to be cut back several times in order to reduce supercritical pollutant levels.

Record atmospheric levels of carbon dioxide

As is well known, emissions which do not directly affect human health have also become a problem. First and foremost among them is the otherwise innocuous gas carbon dioxide, which, as a greenhouse gas, has begun to indirectly impact on the condition of the earth and humanity. The CO_2 concentration in the atmosphere is currently reaching pre-

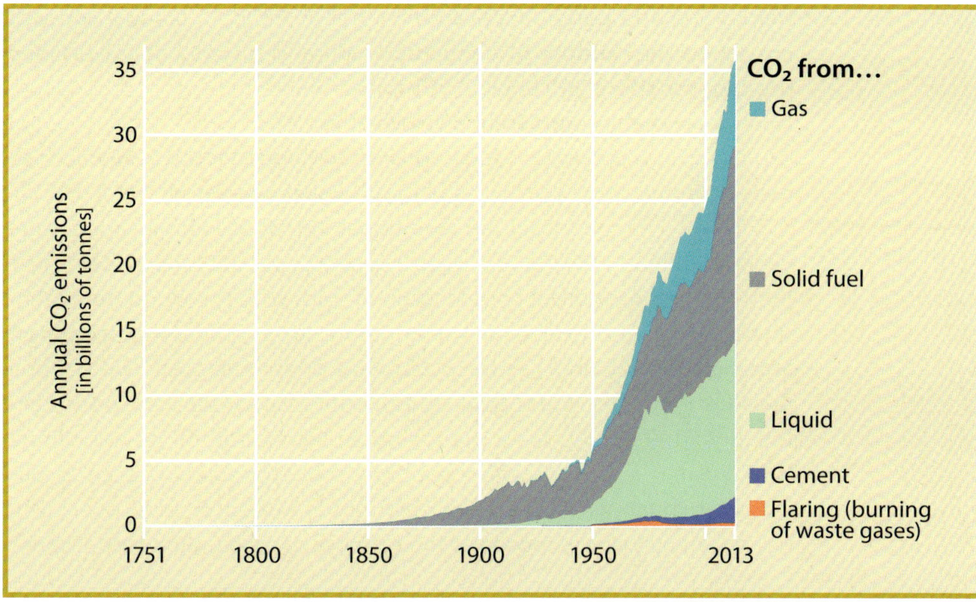

Figure 8: Growth in carbon dioxide emissions from energy use, showing a sharp increase since 1950 from solid fuel (e.g. coal), liquid fuel (e.g. oil), gas (e.g. natural gas), cement production and flaring (burning of waste gases) (Source: Ritchie & Roser 2017)

viously unrecorded levels (at least in the last 800,000 years) every year; between 1990 and 2014, emissions grew by 60 % (Barne & Khokhar 2017). The overwhelming proportion comes from the burning of fossil fuels (Fig. 8).

Climate change

The increasing accumulation of greenhouse gases in the atmosphere leads, with a time lag, to global warming and to many other changes in climate. In 2017, the global mean temperature was already 1.1 to 1.2 degrees above pre-industrial levels. It was the 41st year in succession with a temperature above the 20th century average. After 2015 and 2016, 2017 was the third-warmest year since weather records began. The consequences for the earth's ecosystems are incalculable. For example, there has been a significant reduction in Arctic and Antarctic sea ice in their respective summer months (and a new record low in the Antarctic in 2017), and record droughts in many regions (e.g. Portugal); great damage has also been caused by heavy rainfalls and severe storms in many areas (NOAA 2018). Climate models predict another decade-long increase in global temperature, even if there were to be a sharp decline in greenhouse gas emissions; the risks of flooding alone mean that significant adaptation measures are needed, above all in North America, Central Europe and parts of Africa (Willner et al. 2018). In comparison with the 1960s, the number of dangerous natural disasters has quadrupled; the biggest rise is in climate- or weather-related events. If the Antarctic land ice melts completely, a sea level rise of 3 cm per century can be expected for a thousand years—this could happen if the available fossil fuels continue to be burned as before (Winkelmann et al. 2015).

Farming and food

Climate change poses an additional threat to global agriculture, which is already struggling with the effects of the spread of deserts (desertification) and the deterioration of soils. With the world population skyrocketing, and despite ongoing deforestation, the amount of available agricultural land has been stagnant or shrinking for two to three decades now (The World Bank 2017a). Both erosion due to overuse and the spread of settled areas are taking an increasing toll on agricultural land. Even in a country with a relatively stable population, such as Germany, as much as 100 ha have been either sealed or built over every day for many years—mostly at the expense of agricultural land. There is

Figure 9: Intensive forestry in the form of a plantation and clear-cutting system: eucalyptus plantation after the harvest, in a formerly very biodiverse Atlantic rainforest ecosystem in Bahia, Brazil (photo: P. Ibisch)

an increasing shift towards more intensive land-use practices as well as higher-yielding and energy-rich crops. The cultivation of grain, for example, has increased (The World Bank 2017b). Based on current consumption habits and trends, by 2050 global demand for starchy foods will increase by 40 % and for meat by 60 %; in large parts of Africa, demand is growing by 100 % to 200 % and more (IFPRI 2017). Intensification is also happening in other systems of natural resource use, such as forestry (Fig. 9). The intensification of land use entails serious impacts on ecosystems (▹ chapter 3.1).

Loss of biodiversity

Data on biodiversity on all continents clearly indicate an escalating crisis. Between 2000 and 2013, for example, 919,000 km^2 of intact forests disappeared (a loss of 7.2 %; Potapov et al. 2017). At least 36 % and perhaps up to 57 % of the more than 40,000 Amazonian tree species must now be considered endangered. The populations of large herbivores are plummeting worldwide — about 60 % of species are threatened with extinction (Ripple et al. 2015). Similarly, 60 % of primate species — our closest relatives among mammals — are threatened; 75 % of species show a decline in population sizes (Estrada et al. 2017). Currently, the dramatic decline in insect numbers is causing great concern. For example, it has been established that in protected areas in Germany, 75 % of insects have disappeared within the last 27 years (Hallmann et al. 2017).

A warning for humanity

It would be difficult not to link the many negative trends in the environment with the spread of humans and their activities across the planet. The scientific findings are unequivocal, and the available information is more than sufficient — humanity has been warned (see Ripple et al. 2017: World Scientists' Warning to Humanity: A Second Notice). The various ecological and social problems are interlocking and mutually reinforcing. Their scale is historically unprecedented, and no one can seriously claim that the situation is under control. The steeply rising curves — plotting growth, consumption and destruction — signal disaster. When the famous study on the limits to growth was published in 1972 (Meadows et al. 1972) (▹ Chapter 1.2), it was very difficult to imagine that events would snowball in quite this fashion. Nevertheless, humanity then set off in search of a 'pathway to sustainable development'. For around 50 years, humanity has been reacting with increasing attentiveness to the global crisis. The problems may be growing, but so too are the insight, knowledge and strategies for their containment. Sustainability as a reaction… (▹ Chapter 1.2).

1.2

Sustainability as a reaction: what has been happening so far

Heike Molitor and Pierre L. Ibisch

The concepts of sustainability and sustainable development have often emerged in reaction to pressing problems: a shortage of timber or some other resource scarcity, pollution, poverty, then climate change, global change—always new challenges. Humanity has been chasing answers to self-made problems. There is no sign of any about-turn as yet, even though the warning signs have been there since at least the middle of the last century. Europe, a densely populated continent with ecosystems already subject to change for millennia, is the birthplace of the modern concept of sustainability. But other regions of the world, too, have developed strategies for creating a good life and a better future. What is the state of play? Has sustainability discourse now established itself across all societies?

The origin of sustainability – as a reaction to an increasing scarcity of wood in Europe

Historical origins

The basis for the term and the concept of sustainability was originated by Hans (Hannß) Carl von Carlowitz, Director of the Inspectorate of Mines for Saxony, in his 1713 publication *Sylvicultura oeconomica, or economic information and instructions for the growing of wild trees in accordance with nature*. The text was written in response to a timber shortage in Germany. Carlowitz had travelled in England and France, where the idea of re-stocking forests had been discussed and written about much earlier already.

In 1664, the Englishman John Evelyn wrote a treatise for the Royal Society *(Sylva, or a Discourse of Forest Trees and the Propagation of Timber in His Majesty's Dominions)* in which he complained of the destruction of '...all those many goodly Woods and Forests, which our more Prudent Ancestors left standing for the Ornament and Service of their Country'. Evelyn painted a scenario in which the nation would be severely weakened and its survival threatened if the loss of the wood that was so desperately needed for the construction of ships were to continue. He not only identified the causes of deforestation (shipbuilding, glassworks, ironmaking, the expansion of farming), he also recommended the planting of trees:

» Truly the Waste and Destruction of our Woods has been so universal, that I conceive nothing less than a universal Plantation of all sorts of Trees will supply, and well encounter the Defect … (Evelyn 1664, p. 3).

Under the direction of the French Minister of State, Jean-Baptiste Colbert, regulations issued by King Louis XIV ('Ordonnance sur le fait des Eaux et Forets' = Ordinance on the Facts of Waters and Forests) were published in France in 1669 which paved the way for Carlowitz and for forestry in much the same way as Evelyn had in England. Carlowitz looked at the wood crisis in Saxony from an economic perspective:

» Once the wood and the forest have been ruined / so then the income will be lost for an infinite number of years / and the treasury will be entirely exhausted, so that beneath continuing apparent profit, irreparable damage lies hidden (von Carlowitz 1713, p. 87).

Here Carlowitz clearly distinguishes between short-term profit and long-term costs. His forward-looking approach entailed the careful management of resources.

» Old clothes should not be discarded / until one has new ones / and just so, the stock of fully-grown timber should not be cut down / until one sees / that there is sufficient re-growth. […] Therefore the greatest art / science / diligence and institution of this country will consist in / how such a Conservation and cultivation of the wood can be arranged / as to make possible a continuous, steady and sustaining use / because it is an indispensable thing / without which the country cannot maintain its Being (von Carlowitz 1713, p. 150).

Strictly speaking, then, at this point Carlowitz did not speak of sustainability per se; rather, he combined the idea of 'continuous and sustaining use' with preserving resources for the future. This was not for environmental or ethical reasons, but out of concern for the public finances and the supply of important resources — as had been the case in England and France. The old meaning of the German term 'nachhalten' is 'to hold back' or 'to last, or persist' (cf. Duden 2014, p. 582). 'Sustainability' was spoken of only in later centuries. What is important to note is that sustainability was born as a reaction to scarcity and crisis — and this should be remembered as the concept is developed and elaborated in modern times.

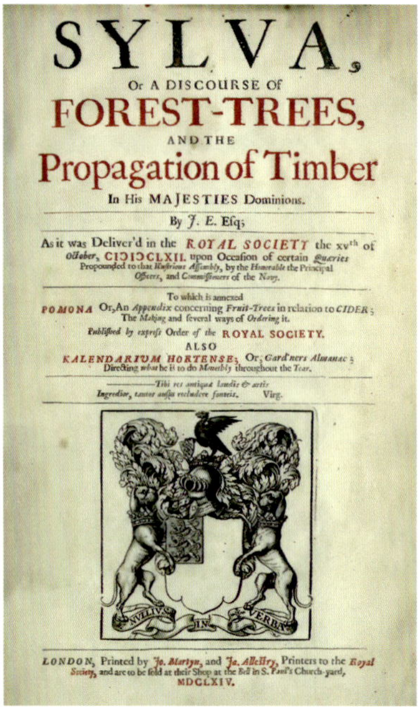

SYLVA,

Or A DISCOURSE Of

FOREST-TREES,

AND THE

Propagation of Timber

In His MAJESTIES Dominions.

By *J. E.* Esq;

As it was Deliver'd in the ROYAL SOCIETY the xv^th^ of *October*, CIↃIↃCLXII. upon Occasion of certain *Quæries* Propounded to that *Illustrious Assembly*, by the *Honorable* the Principal *Officers*, and *Commissioners* of the *Navy*.

To which is annexed

POMONA Or, An *Appendix* concerning *Fruit-Trees* in relation to *CIDER*; The *Making* and several ways of *Ordering* it.

Published by express Order *of the* ROYAL SOCIETY.

ALSO

KALENDARIUM HORTENSE; Or, *Gard'ners Almanac*; Directing *what* he is to do *Monethly* throughout the *Year*.

——*Tibi res antiquæ laudis & artis*
Ingredior, tantos ausus recludere fonteis. Virg.

LONDON, Printed by *Jo. Martyn*, and *Ja. Allestry*, Printers to the *Royal Society*, and are to be sold at their Shop at the Bell in S. *Paul's* Church-yard, MDCLXIV.

Mit GOtt!

SYLVICULTURA OECONOMICA,

Oder

Haußwirthliche Nachricht und Naturmäßige

Anweisung

Zur

Wilden Baum-Zucht,

Nebst

Gründlicher Darstellung/

Wie zu förderst durch Göttliches Benedeyen dem allenthalben und insgemein einreissenden

Grossen Holtz-Mangel/

Vermittelst Säe- Pflantz- und Versetzung vielerhand Bäume zu prospiciren/

Alles zu nothdürfftiger Versorgung des Hauß- Bau- Brau- Berg- und Schmeltz-Wesens/ und wie eine immerwährende Holtz-Nutzung, Land und Leuten/ auch jedem Hauß-Wirthe zu unschätzbaren grossen Auffnehmen/ pfleglich und füglich zu erzielen und einzuführen.

Worbey zugleich eine gründliche Nachricht von den in Churfl. Sächß. Landen

Gefundenen Turff

Dessen Natürliche Beschaffenheit/ grossen Nutzen/ Gebrauch und nützlichen Verkohlung.

Aus Liebe zu Beförderung des algemeinen Bestens beschrieben

Von

Hannß Carl von Carlowitz/

Königl. Pohl. und Churfl. Sächß. Cammer-Rath/ und Ober-Berg-Hauptmann.

Mit Königl. Pohl. und Churfürstl. Sächß. allergnädigsten PRIVILEGIO.

LEIPZIG/

verlegts Johann Friedrich Braun 1713.

Figure 1: Historical documents and authors linked to the origin of sustainability in Europe. Top left: *Sylva or a Discourse of Forest Trees and the Propagation of Timber in His Majesties Dominions*, Evelyn (1664) (source: openlibrary.org); top right: *Sylvicultura oeconomica, oder haußwirthliche Nachricht und Naturmaßige Anweisung zur wilden Baum-Zucht*, von Carlowitz (1713) (source: scan of Exemplar II-7c in university library USD Eberswalde); bottom left: John Evelyn (source: Wellcome Collection; CC BY); bottom right: Hans Carl von Carlowitz (source: SLUB Dresden/Deutsche Fotothek/ Hans Loos, Foto vor 1945, Johann Martin Bernigeroth: 'Bildnis Hans Carl von Carlowitz', Inv. Nr. A 26171 im Kupferstich-Kabinett, Staatliche Kunstsammlungen Dresden)

Evolution of the term 'sustainable'

In Latin, the word *sustinere* means 'to hold out/endure/withstand'. In English, the word sustainable has been known since the 1610s, though initially it also included the sense of something being 'tolerable/bearable'. From 1845 onwards, the meaning changed to 'acceptable'. Beginning in the 1950s and 1960s, economists, too, increasingly talked about sustainable growth, self-sustaining growth (Rostow 1956).

The birth of environmental consciousness and awareness of the limits to growth

Thomas Robert Malthus: the theory of overpopulation

Since the early modern period, a deeper understanding of natural boundaries has gradually evolved. An early and provocative critical work in this area was published in 1798 by the economist and social philosopher Thomas Robert Malthus. In his 'Essay on the Principle of Population', Malthus opposed the view that the Earth's carrying capacity could continually grow. He was the first critic of growth to articulate the fear that overpopulation would become a problem for society and the economy. His inhumane and morally questionable formulations — that there was simply not room for everyone at 'the great banquet of Nature' — and the consequences of his findings led to centuries of criticism. In the end, the idea prevailed that human ingenuity and technical progress, as well as economic mechanisms, made it possible to feed an ever-expanding human population. The continuing exponential growth of the global population seemed for a long time to support this view.

Industrialization

Two centuries later, with the onset of industrialization, the problems grew worse. Resource exploitation and throughput, as well as the pollution of the human environment with waste and new toxins, were again hugely accelerated in the decades following the Second World War. In many places, it became clear that the blessings of life in industrial society came at a price. In the 1960s and 1970s, water, soil and air were severely polluted, and not just in the industrial centres of Europe and North America. Rivers turned into sewers, while poor air quality led to public health disasters. December 1952 saw the Great Smog of London, which resulted in the deaths of thousands of people.

In 1962, the biologist Rachel Carson (1907–1964) published *Silent Spring*, which described the effects of the pesticide dichlorodiphenyl-

Rachel Carson: *Silent Spring*

trichloroethane (DDT) on ecosystems and its accumulation in the food chain (Carson 2007). On account of its impact at the time, the book is still seen today as the starting point of the worldwide environmental movement. Shortly after its publication, DDT was banned in the USA. In 1970 and 1972, the Clear Air Act and the Clean Water Act were passed (von Weizsäcker 1994). Six years later, the moon landing enabled the global community to see for the first time images of the earth from space. This view of the earth as a 'vulnerable, tiny little planet' made it clear that there is only one place where human beings can live — this earth.

The Club of Rome: *The Limits to Growth*

In 1968, the Club of Rome — today limited to 100 people — was founded as a non-commercial organization for the development of a sustainable future for humanity. In 1972, Donella and Dennis Meadows, with two other editors and a total of 17 authors from six countries, published the first Report to the Club of Rome, *The Limits to Growth* (Meadows et al. 1972). In this study, they projected a number of threatening scenarios of resource scarcity and population decline based on a systemic world model. With the aid of what was for that time a powerful computer, the projections took into account, amongst other things, birth rates, industrial production, food, stocks of raw materials and environmental pollution. Although the study has been criticized in many respects, especially by the countries of the South, it prompted the global community to think about whether resources were endless or finite, and about the relationship between lifestyles and economic growth (Grober 2012; Grunwald & Kopfmüller 2012).

Planetary boundaries

The concept of planetary boundaries follows on from the limits to growth (Rockström 2009; Steffen et al. 2015). However, it not only attempts to identify ecological limits, but also examines to what extent there can be a safe space for human development in relation to various man-made environmental problems such as climate change and loss of biodiversity.

Development: a discourse becomes sustainable

Progress and modernization

For centuries, human beings have striven for progress and an easier life. This pursuit was the driving force behind the steady progress of technology. Especially since the Age of Enlightenment, science has become increasingly systematic, giving many societies unprecedented opportunities for intellectual development and new means of modernizing all areas of life on the basis of new knowledge and technology. Progress, modernization, growth and development became practically synonymous.

Beginning in the second half of the twentieth century, the concept of human development was reinvented, subjected to scientific analysis, and politicized.[1] After 1945, following the collapse of the world order after the two world wars, the desire for or belief in modernization and progress quickly regained momentum. The USA did its utmost to stabilize and rebuild the ruined European continent from which the wars had sprung. Development aid became a vehicle for peace as well as an instrument for the pursuit of geopolitical and economic interests.

Development aid

The fight against hunger and poverty in other parts of the world became the focus. The process of decolonization[2] and the creation of a new political world order were also important concerns. It came to an antagonistic confrontation between the Western bloc of capitalist countries (the First World) and the socialist states (the Second World). Less industrialized countries — which, as part of the so-called Third World, were not clearly associated with either of the two blocs — were increasingly regarded as developing countries.

Three worlds

The *United Nations* was established as a higher-level political entity. It proclaimed the 1960s to be the first decade of development in order to 'accelerate progress towards self-sustaining growth of the economy' (United Nations 1962, pp. iii and 7). This was seen as a way to satisfy basic needs and to raise the material standard of living. Economic growth was the binding leitmotif for global development. At the same time, the institutions established by the Bretton Woods Conference of 1944 set the economic course for globalization under US leadership: the World Bank, the International Monetary Fund and the General Agreement on Tariffs and Trade (GATT).

The United Nations and the first UN Development Decade

During the second United Nations Development Decade (1970–1980), development discourse became significantly more complex. The rapid growth of world trade and the internationalization of markets challenged the ability of states to regulate. Critical and pessimistic voices tempered the earlier optimism, and the so-called Third World grew in importance as it improved its organization and cooperation. The exploitation of poor countries and their descent into new forms of dependency following independence became critical issues; there were increasing

The second UN Development Decade

1 The content of the following sections follows Benjamin Knutsson (2009), who has written a condensed history of development.

2 The process of the abolition of the colonies, which, from the 15th century onwards, had been occupied mainly by European countries and treated as part of their domains, and of the achievement of (national) self-determination by these areas; the foundation of new, independent states.

demands for more social justice, participation and respect for the environment.

The Stockholm UN Environment Conference (1972)

The 1970s also saw the start of a global environmental movement. Against the background of the events and studies above, as well as other socially and politically relevant changes, the United Nations Conference on the Human Environment was organized in Stockholm in 1972 at the initiative of the United States and Sweden. However, the Western countries and Japan stood almost alone in their concern for the environment.[3] The Global South (still referred to then as the 'developing countries') attacked any environmental restrictions as 'neo-colonialist and completely unacceptable' (von Weizsäcker 1994, p. 17). Instead they urged action to combat poverty. The *United Nations Environment Programme (UNEP)* was launched. Nairobi was chosen as its headquarters. Many countries set up national environment ministries as a result of the conference (ibid.; Grunwald & Kopfmüller 2012).

Famines and economic crisis

Between 1972 and 1975, severe famines that were not due to war struck for the first time in the modern era. Millions of people died, mainly in West Africa and Ethiopia, but also in parts of India and Bangladesh. After a protracted economic boom, there followed a worldwide economic crisis (as also happened later from 2008–2011) which resulted in a tripling in the price of internationally traded grain within 20 months. This led to turmoil in the global grain market: attempts were made to capitalize agriculture in countries of the Global South, and there was a turn away from traditional food aid (Gerlach 2015).

The Brandt Report (1980)

In 1977, prompted by stagnating global development and new problems such as environmental crises, the World Bank President suggested establishing an Independent Commission for International Development Issues, made up of 20 states. In 1980, under the chairmanship of former German Chancellor Willy Brandt, it produced the so-called Brandt Report, subtitled 'A Programme for Survival' (ICIDI 1980). Hunger, poverty, inequality, war and rearmament were identified as major threats to human well-being. Population growth and overpopulation, as well as migration and refugee crises, were all on the agenda, alongside the growth of the industrialized economies. The notion of the limits to growth influenced the report: deforestation and even climate change featured prom-

3 Although a comprehensive environmental framework law was introduced in the GDR (the second country in Europe to do so) in 1970, and the protection of nature was even introduced into the constitution, the real condition of the natural world was very different. Environmental degradation was considerable and visible everywhere by the 1980s if not earlier (BMUB 2015).

inently. There was a strong call for alternative energy sources, and nuclear energy was for the first time identified as a concern.

The Brundtland Report (1987)

Against the background of further escalating ecological, social and economic problems (worldwide recession, international debt crisis, unemployment, etc.), the UN Commission on Environment and Development (the so-called *Brundtland Commission*) began work in 1983 under the leadership of the former Norwegian Prime Minister Gro Harlem Brundtland. In 1987, its final report, 'Our Common Future', was published, a milestone in the evolution of sustainability discourse.

For the first time, the term 'sustainable development' was defined from an **anthropocentric perspective**. This definition is still recognized worldwide to this day. It states that development is sustainable if it 'meets the needs of the present without compromising the ability of future generations to meet their own needs' (World Commission on Environment and Development 1987, p. 41).

The anthropocentric approach

Anthropocentric ethical approaches to the environment place human beings at the centre of moral action. Only human beings are attributed intrinsic value. This is the approach behind nature conservation, where nature is to be conserved *for* human beings.

Pathocentric approaches attribute intrinsic value to all sentient beings (humans and higher animals).

Biocentric approaches assume that all living beings have a moral status.

Physiocentric or holistic or ecocentric approaches take the broadest point of view. Everything that exists has intrinsic value and potentially moral status. 'Nothing in nature exists purely as an instrument for anything else' (Gorke 2000, p. 85).

Sustainable development: stable, durable or sustainable?

Chapter 2 of the *Brundtland Report* deals with sustainable development. Interestingly, the term *sustainable development* was not consistently used in German. The German translation used by the GDR state publishing house uses *'stable development'* ('stabile Entwicklung') (Weltkommission für Umwelt und Entwicklung 1988, p. 57). In 2017, the Federal German Development Ministry was still translating sustainable as 'durable' ('dauerhaft'), following the (West) German translation of Brundtland (Hauff 1987, p. 46).

This semantic confusion reflects a more widespread struggle to understand what exactly sustainability stands for. In French, *développement durable* is used more frequently than *développement soutenable*, but there is a view that *durable* does not convey the spirit of the English 'sustainability', a concept that implies both longevity and capacity within certain limits. In Spanish, there has been an ongoing debate as to whether *sustentable* or *sostenible* is more appropriate. It has been argued that *sostenible* implies a process in space and time which is effective, efficient, and self-sustained, whilst *sustentable* refers only to a more immediate effectiveness, to resources that can be sustained. In different Spanish-speaking countries, the preference varies. In Italian, it is *sviluppo sostenibile* and in Portuguese, *desenvolvimento sustentável*.

The report outlined the relationship between environmental and development issues and considered how these relate to the task of securing inter- and intragenerational equity, from a global perspective. Based on the analysis of the situation at that time, the Brundtland Commission concluded that sustainable development required fundamental changes and new orientations in politics and institutions (Hauff 1987). The Brundtland Report has sometimes been criticized for leaving too broad a scope for policy implementation and for containing few concrete recommendations for action. But this also enabled it to find wide acceptance. Ultimately, the Brundtland Commission succeeded in combining a problem analysis with basic demands, and this triggered a broad international discussion about appropriate ways to implement sustainable development (Grunwald & Kopfmüller 2012). It should be emphasized that a number of concepts directly addressed in the report would only later gain currency in the wider discussion; indeed, many of the central concepts remain topical almost 30 years after the report's publication. These include, for example, the importance of qualitative growth, consideration of the value of 'natural capital' stocks, and environmental pollution as a source of conflict. In this regard, the Brundtland Report marks the end of a decade dominated by the arms race and economic liberalization, one which, from a development policy perspective, is often looked back on as the 'lost decade' (Knutsson 2009).

The Earth Summit – Rio de Janeiro (1992)

On the recommendation of the Brundtland Commission, in 1992 the United Nations hosted the largest ever UN Conference on Environment and Development (the so-called *Earth Summit*) in Rio de Janeiro. The conference was attended by around 10,000 delegates from 178 states. Thanks to the 'Rio spirit' that characterized the negotiations, the summit succeeded in tackling the still-pressing problems of global development in an environmental context. Epoch-making documents — such as the Framework Convention on Climate Change and the Conventions on Biodiversity and Desertification, Agenda 21, the Forest Principles and the Rio Declaration on Environment and Development — were adopted at the conference. For the first time, a comprehensive commitment to sustainable development was clearly set out and adopted by almost all countries in the world.

Agenda 21 Education

Agenda 21 established that states should implement sustainable development at the national level in the form of strategies or national action plans. Chapter 36 called for a reorientation of education and educators to integrate sustainable development (United Nations 1992a). However, none of the documents adopted in Rio contained verifiable obligations on the contracting states. The Rio Declaration, the Forest Principles and Agenda 21 remained entirely without international legal obligation. Nevertheless, the conference generated a worldwide momentum which endures to this day (Grunwald & Kopfmüller 2012).

The Millennium Development Goals

In the years that followed, the guiding principles of the Rio Conference were taken forward in international politics and further elaborated. In the year 2000, the Millennium Declaration of the United Nations appeared (*Millennium Declaration*, United Nations 2000). Eight Millennium Development Goals (MDGs) set the agenda for the coming decade and became the basis for a new development cooperation architecture (Knutsson 2009). The seventh MDG was environmental sustainability. Ten years after Rio (2002), the world community met in Johannesburg for the *World Summit on Sustainable Development* in order to clarify and realise the plans made at the first Earth Summit (Rio + 10). The result was an action plan aimed at solving humanity's fundamental problems and addressing ongoing environmental risks.

UN Decade of Education for Sustainable Development

In the course of this conference, the United Nations General Assembly decided to give a special role to education in the sustainability agenda and to launch the *UN Decade of Education for Sustainable Development* (▹ Chapter 4.4) for the period 2005–2014. The concrete goal was to implement Education for Sustainable Development in all areas of education, including universities.

The Earth Summit Rio+20 – Rio de Janeiro, 2012

In 2012 (Rio +20), political representatives met again at a World Summit in Rio de Janeiro — once more with the aim of finally achieving the urgent change of course needed for sustainable development. The outcome was the adoption of the document 'The Future We Want' (United Nations 2012). It remained difficult to reconcile the interests of the countries of the Global South with those of the Global North. Even before the summit began, the growing calls for a so-called *Green Economy* (▹ Chapter 4.3) were criticized by some of the representatives from the South. They feared that they would suffer as the rich countries blocked access to their markets under the pretext of meeting sustainability goals by raising environmental standards. In contrast, for many industrialized nations, including Germany, the outcomes did not go far enough in view of the global situation. Nevertheless, the political commitment to sustainable development was reaffirmed.

The 2030 Agenda – the Sustainable Development Goals, SDGs

Another outcome of the summit was the adoption by all member states of a catalogue of goals, by 2015, replacing the eight Millennium Development Goals (MDGs) with 17 Sustainable Development Goals (SDGs). The SDGs were published under the title 'Transforming Our World: The 2030 Agenda for Sustainable Development' (United Nations 2015) (see Fig. 2) with a total of 169 sub-goals to be implemented by 2030 (Bundesregierung 2017). In the end, the MDGs had not been achieved. The 2030 Agenda was not merely a postponement of the original targets, but turned into something more comprehensive and ambitious than the MDGs. The most important goal, the first, is to bring an end to extreme poverty, succinctly expressed as 'no poverty'.

The United Nations SDGs were the result of an unprecedented global process, one in which civil society was involved. Not least for that reason, they are all-encompassing, and represent an incoherent 'wish list' of goals which sometimes contradict one another. There is no attempt to prioritize — all the goals seem to be equally important. In the final analysis, they reflect a sustainability model that has its roots in European

Figure 2: 17 Sustainable Development Goals were adopted in 2015
(United Nations, https://www.un.org/sustainabledevelopment/news/communications-material/)

thinking, according to which people and the environment are, at best, independent actors with equal rights. Notwithstanding this, the SDGs have so far had a considerable impact on the international debate.

The development discourse becomes more pluralistic

Over time, the international development discourse has become much more pluralistic. The centuries-long dominance of the Western, originally European development model also shaped development aid and sustainability policy. But resource conservation on other continents has its own traditions; moreover, in some other cultures, a careful use of nature is more firmly engrained than in the West. In parallel with the policy of the United Nations as described above, different discourse strands have emerged.

Sustainable development in the present-day constitutions of Bolivia and Ecuador

'The good life': *Vivir bien/ Buen vivir*

An important example is the (central) Andean culture, which recently has been the source of a number of ideas that have gained international attention. In the Andean cosmovision,[4] the earth plays an important role in space and time *(pacha)*, and Mother Earth *(pachamama)* is revered. The socialist governments of Bolivia and Ecuador especially have in recent years linked elements of traditional cultures with the modern discourse on sustainability. The new constitutions contain far-reaching innovations (Gudynas 2009). The new constitution of the Plurinational

4 Worldview: culturally influenced way of looking at the world, and explanatory approach to causal relationships and phenomena.

Has the sustainable development discourse spread to the general population?

The discourse on sustainable development has evolved over decades now, and has become an integral part of the international political agenda. However, the temptation to overestimate what has been achieved should be resisted. In the general population, serious consideration of the far-reaching changes needed for sustainable development plays a very minor role, if any. According to Google Trends (based on changes in Google search queries over time), the general trend in public interest is downward (Fig. 3); the UN's 2030 Agenda has not reversed this trend, at least not in German-speaking countries. In the English-speaking world (using Google searches for 'sustainable development'), a slight trend reversal may be discernible from 2015 onwards. Of course, the decline in searches may be explained by the fact that the terms are better known and understood, and, over time, have been replaced by more specific searches, or that other new terms are receiving more attention. This is the case, for instance, with the relatively new Latin American concept of 'the good life'—often in the Spanish original form *buen vivir*—which has seen a definite rise in interest recently.

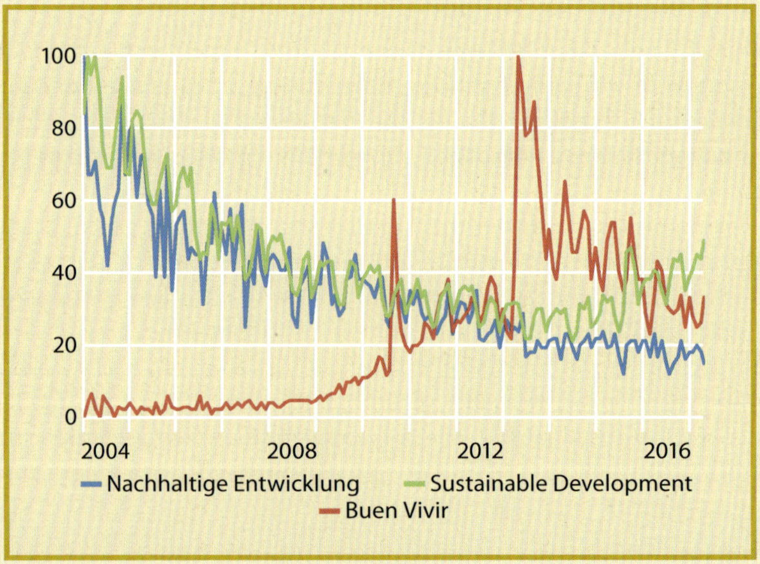

Figure 3: Google Trends results for 'Sustainable development', German *'Nachhaltige Entwicklung'*, and the alternative concept of *'Buen vivir'* from South America. Based on Google searches for these terms since 2004. (www.trends.google.de)

State of Bolivia[5] (2009) is not only aimed at improving the quality of life and the Vivir Bien ('good living' or 'the good life') for all Bolivians, but is also explicitly committed to sustainability.

The Ecuadorian constitution of 2008 also combines these elements. It even suggests that Nature or *Pachamama* has 'the right to integral respect for its existence and for the maintenance and regeneration of its life cycles, structure, functions and evolutionary processes' (Article 71). *Pachamama* also has the 'right to a complete restoration' (Article 72). The constitution also describes a new development regime *(Régimen de desarrollo)*. The general principles (Article 275) include the description of the developmental regime as an organized, sustainable and dynamic set of economic, political, socio-cultural and environmental systems that guarantee the 'good life' (*sumak kawsay* in the Quechua language). This is a novel, comprehensive, dynamic and systemic (▹Chapter 1.3) approach to sustainable development.

Pioneers and the need for development

This illustrates that in the further development of what used to be called the developing countries, it is possible for them not only to skip developmental stages — for example, in the course of the spread of telecommunications and digitalization — but also, in certain respects, to become conceptual pioneers from whom industrialized countries can learn. The global adoption of the Sustainable Development Goals is also a recognition on the part of those states which previously considered themselves 'developed' that all countries have significant developmental needs, at least with respect to some of the 17 goals.

Spiritual-religious approaches

As sustainability thinking embeds itself more strongly in society, spiritual-religious aspects are acquiring an important role. Bolivia was one of the states that advocated for the United Nations General Assembly to address the importance of living in harmony with nature in order to secure sustainable development. Certain regions of the world are regarded as placing more importance on the idea that people should recognize themselves as an integral part of nature — this applies, for example, to Buddhism (Grunwald & Kopfmüller 2012, p. 227). At the Earth Summit in Johannesburg in 2002, an Islamic Declaration on Sustainable Development (United Nations 2002) was adopted which makes it clear that the environment should be understood as a gift from Allah. Pope Francis attracted considerable attention in 2015 with the encyclical *Laudato si'*, which constitutes a manifesto for sustainability 'out of concern for our common home':

5 Official name of the newly established state: *Estado Plurinacional de Bolivia*.

» If we approach nature and the environment without this openness to awe and wonder, if we no longer speak the language of fraternity and beauty in our relationship with the world, our attitude will be that of masters, consumers, ruthless exploiters, unable to set limits on their immediate needs. By contrast, if we feel intimately united with all that exists, then sobriety and care will well up spontaneously. The poverty and austerity of Saint Francis were no mere veneer of asceticism, but something much more radical: a refusal to turn reality into an object simply to be used and controlled. (Pope Francis, in Libreria Editrice Vaticana 2015, p. 14).

The Dalai Lama: ethics are more important than religion

In the same year, the Dalai Lama went a step further in his appeal to the world, subtitled 'Ethics are more important than religion':

» Given the problems of our time, it is no longer enough to base ethics only on the values of religions. Rather, it is high time to open a new path beyond religions for our understanding of spirituality and ethics in the globalized world. (Dalai-Lama XIV 2015, p. 35).

Figure 4: Spirituality and nature: a centuries-old statue of Buddha carved in rock in a temple forest near the Bongam-sa monastery in South Korea (photo: P. Ibisch)

Models of sustainability

Three dimensions of sustainability

Several models have been proposed to help make sustainable development more concrete and practical, some of which are presented here. Often, three dimensions of sustainability (ecology, economy, society) are identified, which are regarded as integrated but (potentially) different in their relative importance. The ecological dimension refers to the ecological carrying capacity of the planet; the social dimension covers social justice; and the economic dimension relates to economic efficacy and efficiency. Sustainable development results when these three dimensions are all taken into account.

The 'three pillars' model and its critics

In the context of sustainable development, the three pillars model represents a simultaneous, equal and 'harmonious' implementation of environmental, social and economic goals for the continuing viability of a society. This model does not describe the potential for irresolvable conflict arising from the fact that social systems are dependent components of the global ecosystem (▷ Chapters 1.3, 3.1, 3.2, 4.1). Incompatibilities are not addressed, and therefore no steps towards integration are proposed (SRU 2008). The model assumes that strengthening individual pillars can compensate for the weakening of others. It does not address the fact that, even from an anthropocentric point of view, the

Figure 5: Schematic illustrations of the 'three pillars' model of sustainability (left) and the sustainability triangle (right)

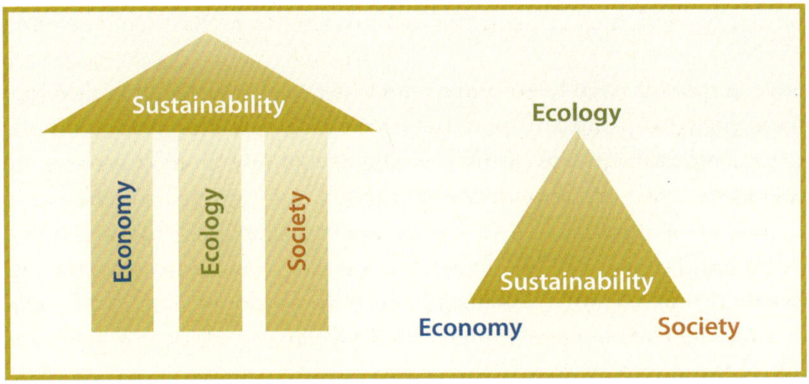

long-term viability of the ecological foundations for human life must be more important than the needs of individual people or of particular social systems.

» Sustainability is and remains a regulatory idea for the long-term use of natural capital. The 'three pillars' concept, which has been politically influential in Germany, initially led to a greater appreciation for environmental concerns, as it postulates the equal importance of economic, ecological and social development. However, the results of research projects and political approaches which applied this concept make it clear that the three pillars model collapses into a kind of wish list into which each actor inputs whatever seems important to them. The model thus enables increasingly arbitrary outcomes (SRU 2002, p. 21).

The sustainability triangle

A slightly different perspective is provided by the more integrative approach of the three dimensions of the so-called sustainability triangle. The Enquete Study Commission on 'The Protection of Man and the Environment' of the 13th German Bundestag preferred this approach, and characterized it at the time not as 'bringing together three pillars standing side by side, but (as) developing a three-dimensional perspective based on real experience' (Deutscher Bundestag 1998, p. 13). The three dimensions are to be seen as equally important and granted equal consideration in terms of public policy.

Guard rail model

The 'guard rail model' used by the German Advisory Council on Global Change (WBGU), an advisory body to the Federal Government, expresses the idea that the future development of humankind has to take place within a defined 'development corridor'. This development corridor is given by ecological parameters that ensure long-term stable living conditions. Only within this corridor is it possible to realize economic and social goals. If its external walls (or guard rails) are transgressed, development is no longer sustainable. The identification of these guard rails for sustainable development is an important responsibility for those giving policy advice on global change. The task of politics then is to ensure that such advice is followed (WBGU 1993).

A systemic approach

The systemic approach to explaining sustainable development describes and visualizes how different dimensions are dependent on each other. The global ecosystems provide the foundation on which social systems (society) develop. Embedded within society is the economy, which serves the needs of society (Fig. 6). This model is explained in detail in ▹ Chapter 1.3 and serves as the basis for this textbook.

Strong and weak sustainability

In the discourse on sustainability, a distinction is sometimes drawn between strong and weak sustainability concepts as a way of assessing the potential for substituting natural capital (▹ Section 4.3). Weak sus-

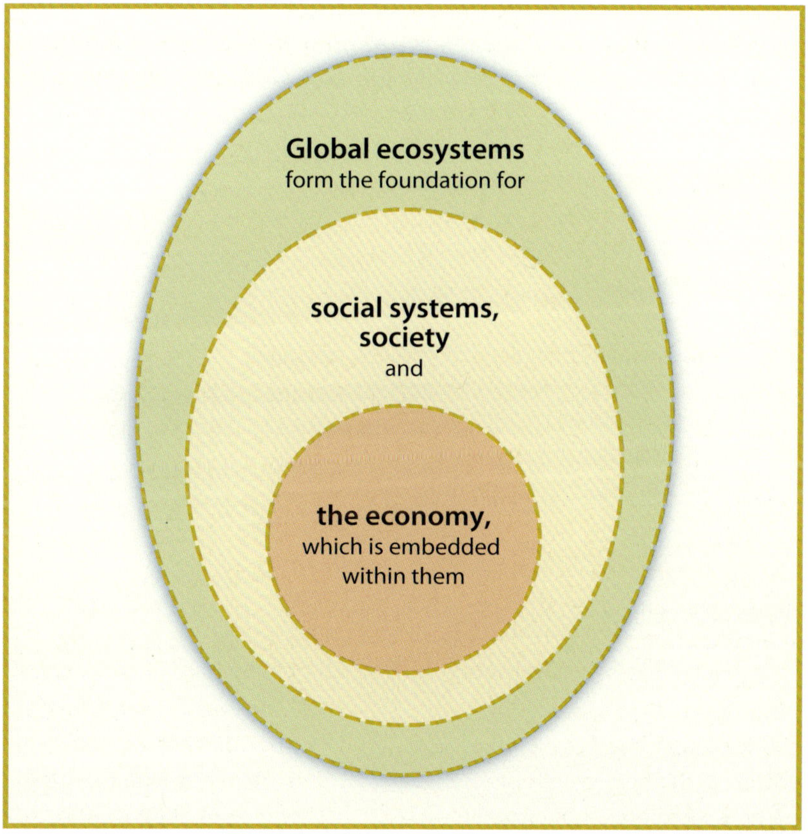

Figure 6: Schematic illustration of the relationship between the economy, society and the ecosystems in the systemic approach to sustainable development

tainability concepts assume that elements of the natural environment are generally substitutable in principle. Thus, a forest can be replaced by a landscape park, as the park will provide the same recreational service as the forest (Ott & Döring 2001; SRU 2002; Grunwald & Kopfmüller 2012); if necessary, the service could also be provided by nature films or by virtual nature. Strong sustainability concepts are based on the assumption that nature (or natural capital) should be kept constant over time and that it can only be substituted (temporarily) to a very limited degree. A forest which is cleared for a construction project cannot simply be replanted elsewhere; at best, it would take decades, if not centuries, for it to acquire a comparable value or to fulfil the same (ecological) function.

The sustainability models outlined above can be arranged on a spectrum between the poles of strong and weak sustainability. The three pillars model, which places the dimensions of sustainability on a flat plane, is closer to the idea of weak sustainability, while the more integrative sustainability triangle tends toward strong sustainability. The guard rail model and the systemic approach prioritize the ecological dimension as the basis for human subsistence and represent strong sustainability (Fig. 7).

Figure 7: Positions of different sustainability models on a spectrum from weak to strong sustainability

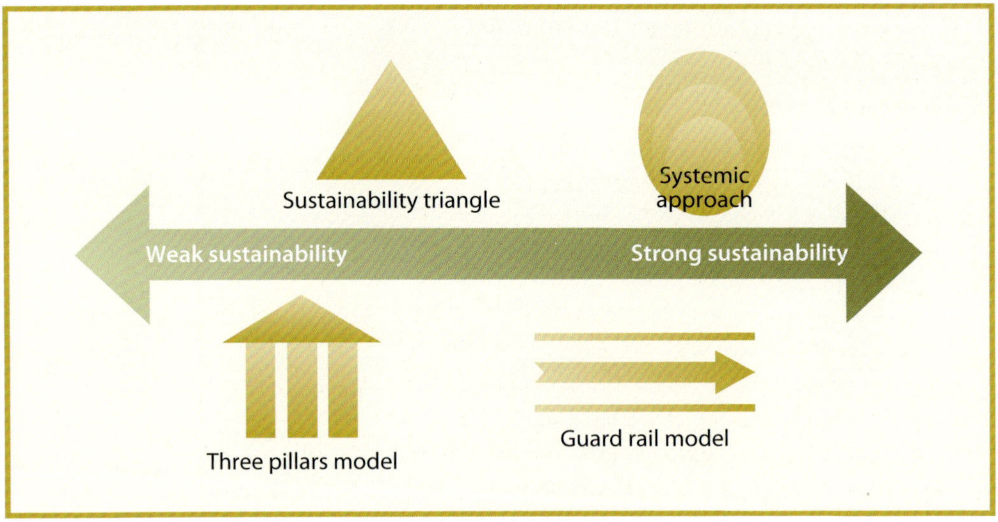

Fundamental principles of sustainable development

The implementation of the concept of sustainable development can be pursued via different approaches, which are often referred to in the relevant literature as 'strategies', but it is more accurate to say that they represent basic principles. Strategies based respectively on efficiency, sufficiency and consistency were established quite early on in the sustainability debate, with the education strategy being added later. The Brundtland Report already set out perspectives on how sustainable development can be implemented both on a technical and on a societal level (World Commission on Environment and Development 1987).

Efficiency

The ▹ **efficiency** strategy pursues the goal of reducing material and energy used per unit of output (minimizing input) while maintaining or increasing production (maximizing output). There is an emphasis on technical solutions (technologies), seeking high potential reductions through efficient resource management, which can also bring significant cost savings. However, there is a risk that the reductions will be offset by volume effects (rebound effect; ▹ Chapter 3.4, 4.3). Potential reductions in resource consumption are sometimes as high as 90 % (efficiency revolution). The efficiency strategy stands firmly in the tradition of market economy thinking, and for that reason it is the preferred strategy of the business community (Huber 1995; Sachs 1997; Schmidt 200; Grunwald & Kopfmüller 2012).

> **Efficiency: a reduction in resource consumption while maintaining production levels.**

At the heart of the ▹ **sufficiency strategy** is the cultural issue of whether a change in human needs or wants is necessarily implied by sustainable lifestyles and other alternative models of prosperity. The starting point is a reduction in consumption. This is often criticized as a call for self-denial, which is considered unrealistic and impractical. This pointed debate restates a question that was already implicit in the Brundtland Report, which can perhaps be described as the basis of any needs approach: 'How much is enough?' In 1996, a study by BUND and Misereor entitled 'Sustainable Germany' devised a set of concepts — such as 'living well rather than having much' — which placed questions about sufficiency at the heart of sustainability action and anticipated the 'good

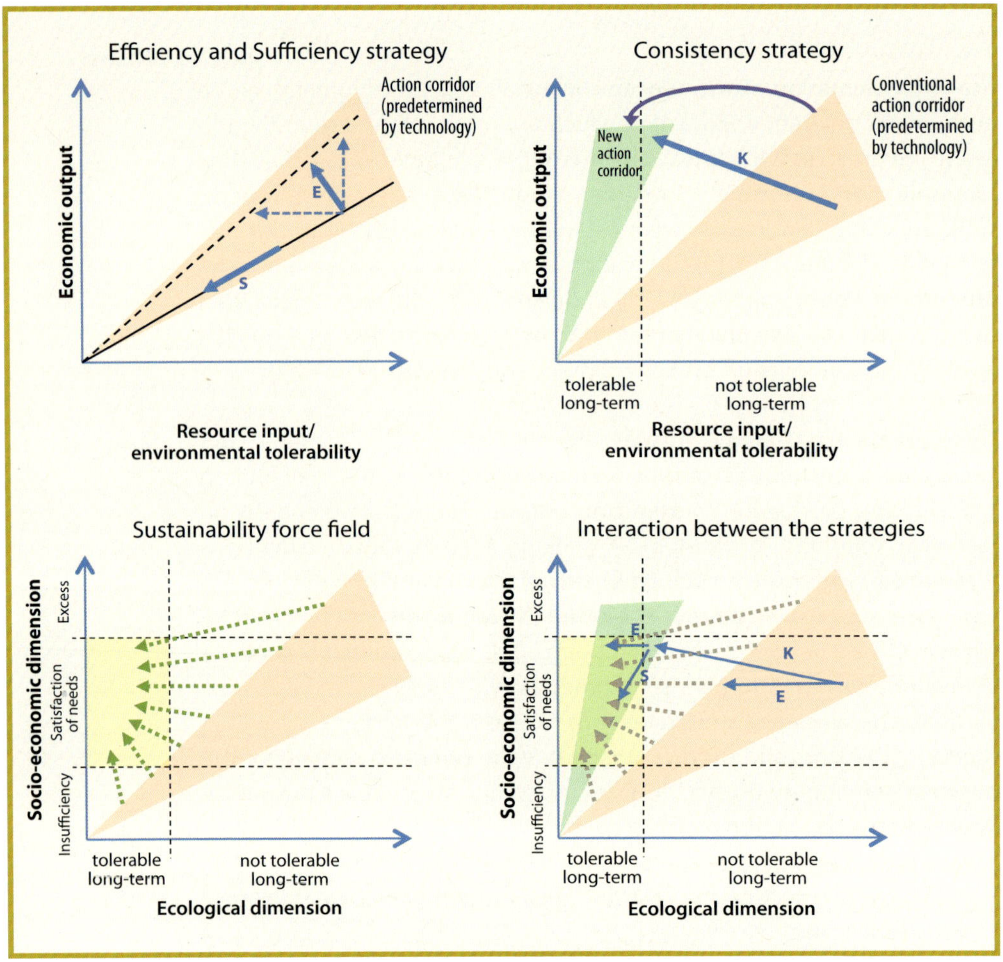

Figure 8: Strategies or principles of sustainable development and their interaction (following Schmidt 2008). Efficiency (E) and sufficiency (S) alone can reduce resource consumption only within a given action corridor (top left). The consistency strategy (C) creates a new action corridor through alternative products and technologies with lower resource consumption (top right). Sustainability (yellow field) is characterized by long-term (ecological) tolerability and a balanced (fair) satisfaction of human needs (bottom left). Only through the interaction of the three strategies (C, S and E) can sustainability (yellow field) be achieved (bottom right).

life' approach (BUND & Misereor 1996). The sufficiency strategy seems to be at odds with today's market economy, which has made economic growth into an economic maxim (▷ Chapter 3.2, 4.3).

> **Sufficiency: Consideration and limitation of one's own (superfluous) consumption, together with respect for the absolute limits (to growth).**

Consistency

The ▷ **consistency strategy** broadly means the substitution of environmentally harmful products, substances and technologies. This substitution must be compatible with the natural cycles of matter and 'the adaptation of the flows of material produced by human activity to natural metabolic processes' (Grunwald & Kopfmüller 2012, p. 93). Examples include the use of renewable raw materials and the circular economy (Huber 1995).

> **Consistency: The replacement of environmentally damaging substances and technologies with others which are environmentally friendly.**

The interaction of the principles

The interaction of the three basic principles of efficiency, sufficiency and consistency is shown in Fig. 8 and underlines the importance of the consistency strategy. While efficiency and sufficiency strategies lead to lower resource consumption, the consistency strategy has the potential to enable permanently low and therefore sustainable resource consumption and environmental impact.

Education

At the beginning of this century, a more intensive discussion opened up over the role of education in the context of the wider debate. Sustainable development as a learning, search and design process includes learning for sustainability within the overall process (Stoltenberg 2009). This means that ideas such as the reorientation of education itself towards 'education for sustainable development' acquire the status of an essential principle (or strategy) for achieving a sustainable society, and thus become indispensable (▷ Chapter 4.4).

Concluding remarks

The 'invention' of sustainability is not the result of a sudden inspiration. Its roots can be traced back through many countries over hundreds of years. The search for a form of development that is fair to all, and to both humans and nature, has been a feature of human history for a long time, and has grown in intensity since the middle of the last century. Efforts to implement it are far from over, and require the participation of all — both now and in the future. In the chapters to follow, systemic issues and implementation strategies will be discussed and reflected upon with the aid of in-depth investigations and case studies in key areas such as energy supplies and economic systems.

1.3

Systems theory: towards a holistic and interdisciplinary understanding of sustainable development

Pierre L. Ibisch

The concept of sustainability was invented by Hans Carl von Carlowitz in the pursuit of economic interests, and then became widely known centuries later as an ethical obligation towards future generations. The current socio-political understanding of sustainability, based on the idea of intergenerational equity, entails a lack of conceptual clarity and an ethical dilemma. Because of course it is not clear what future generations will really need or want. Perhaps they will solve what seem to us insoluble problems with the help of technologies that we cannot even imagine. It seems ethically unjustifiable to deprive today's generations of opportunities for development, or even to let them suffer, simply in order to favour generations that still have plenty of time to adapt to changing resource availability. Such considerations give a certain ideological charge to the political concept of sustainability. But can sustainability not perhaps be described more objectively? What might such a scientific theory of sustainability look like?

The search for a scientific theory of sustainability

Sustainability as the long-term existence and functionality of complex structures

What is sustainable? It could first be stipulated that sustainability means quite simply the possibility of the long-term existence and functionality of structures of varying complexity. In fact, many structures do persist in the world over relatively long periods of time — this applies to atoms and certain substances as well as to certain ecosystems, to all life on Earth, and to planetary systems and galaxies. Why are some structures relatively stable and exist 'sustainably', even though external disturbances regularly endanger their existence and functionality? A key question is whether there might be certain principles or mechanisms that explain the persistence of these wholly different structures. If so, could human society learn from such principles? Is it possible that human structures function according to the same principles?

The search for answers to these questions is nothing less than the search for a comprehensive scientific theory of sustainability.

'What holds the world together at its core?'

For centuries now, many scientists would have given their soul — like Goethe's Faust — to find the answer to this question. And in fact various scientific disciplines have made great progress, since Goethe's day, in capturing and describing all observable structures and processes. For example, we now understand fairly well what matter consists of: a confusing multiplicity of tiny elementary particles, which, in various combinations, form atoms and molecules, which in turn form the substance of everything. Physics and chemistry can also describe why certain elements interact with others, creating new structures with new properties. These are energetic interactions, meaning the transmission principally of power and energy.

Thanks to Albert Einstein, we also know that mass is ultimately nothing more than materialized energy — the famous energy formula shows the relationship between mass (m) and the speed of light (c): $E = mc^2$. It is becoming clear that everything must be described and understood above all in terms of energy. Without wishing to delve too deeply here into physics, the existence of certain atoms, molecules and substances can be linked to energetically favourable states. For example, with respect to molecules: energy in nature is stored or transferred by means of chemical compounds. Different atoms with their charged particles combine to form a complex, because this means the potential energy is minimized. In order to separate compounds of atoms, some energy must be expended — for example, by applying heat — but the transformation often also releases energy.

The cosmos: order based on nested building blocks

Elementary particles form the smallest systems. Through energetic interaction, particles of matter — leptons, gauge bosons, quarks, neutrons, protons, atoms, molecules and so on (see e.g. Fritzsch 2004) — form higher-order structures which can be observed and which have properties that do not arise directly from the properties of the individual particles. The interaction of the two gases hydrogen and oxygen, for example, produces water, a liquid substance with physical and chemical properties that differ dramatically from those of its building blocks. In this case, therefore, the whole is 'greater than the sum of its parts'.

Planets, galaxies and the cosmic network: moving up in scale from the building blocks of matter to the vast expanse of space, the findings are similar. Due to the spatial concentration of matter, various celestial bodies arose. These structures of matter, sometimes of enormous mass, exert influence on each other (by bending space-time) and thereby form new superordinate structures. Thus, planets

with their central stars form solar systems, and these in turn form galaxies. These are not chaotic in structure but are located at the nodes of a cosmic network of filaments (e.g. Bonjean et al. 2018). The individual parts together form ordered or structured larger systems. Using the example of our planetary system, we can recognize the inner dynamic which is a prerequisite for the temporary existence of the individual parts and of the whole.
For example, the movements of the celestial bodies around each other explain why, with a certain degree of probability, collisions are avoided, at least intermittently. The interaction of the components results in some degree of self-organization and in the fact that the larger whole is recognizable as an independent entity, even though it has open borders and is therefore subject to influence from the outside — for example, from incoming radiation or asteroids. In fact, collisions occur regularly in planetary systems, after which the entire system reorganizes itself (see, for example, Zuckerman et al. 2008).

The discovery of systems theory

Scientists recognized in the middle of the last century already that there are fundamental principles which connect everything in existence and which are responsible for the existing orders and dynamics. It was the Austrian Karl Ludwig von Bertalanffy (1901–1972) — in his reflections on the 'development of forms' in nature (Kritische Theorie der Formbildung, 1928), the 'fabric of life' (Das Gefüge des Lebens, 1937) and the connecting links 'from the molecule to the world of organisms' (Vom Molekül zur Organismenwelt, 1940) — who made the leap from a 'biological world view' (Das biologische Weltbild, 1949) to a 'theory of open systems in physics and biology' (1950). In 1953 (Biophysik des Fließgleichgewichts), he described the 'biophysics of equilibrium', which formed a further basis for 'Modern Theories of Development' (1962) and ultimately a 'General System Theory' (1968a). He considered organisms as open physical **systems** in an apparent state of equilibrium. This laid the foundation for a revolutionary worldview that not only promised to explain the development and function of so-called systems, but also integrated separate bodies of knowledge from different disciplines. Systems theory (or systemics) became a source of inspiration not only for ecology, but also for areas such as control engineering and psychology (see, inter alia, von Bertalanffy 1968b). In 1975, Ervin Laszlo distinguished seven different types of system: physico-chemical and biological systems, organ systems, socio-ecological and sociocultural systems, organizational and technical systems.

Systemics, or systems theory, is a scientific concept that not only builds bridges between different scientific disciplines *(interdisciplinary approach)* but also dissolves the boundaries between disciplines and between science and practice *(transdisciplinary approach)*.

Systems theory provides a bridge between disciplines and between research and practice

Systems theory is the basis for modelling the development of complex structures and for **cybernetics**, the study of the control of systems. Systems theory was of crucial importance in the design of the Club of Rome study on the limits to growth (Meadows et al. 1972), which analysed growth processes in the light of positive feedback loops. A central chapter of the study deals with the 'world system'. The co-author of the Club of Rome study, Donella Meadows, wrote a highly recommended introduction to systems thinking (Meadows 2010).

Systems and cybernetics – early roots

The word 'system' is derived from the Latin *systēma*, a loan word from the Ancient Greek *σύστημα* for 'composition' or 'structure', which describes a concept composed of different parts. The term was used in a philosophical context from the 17th century onwards. Initially, it was about cohesive, coordinated rules, methods and practices (François 1999). Cybernetics, which goes back to the Ancient Greek word for 'helmsman', was first used in 1843, both in a book on the art of government (Ampère, *Essai sur la Philosophie des Sciences*, 1843) and in a Polish book on management (Vallée 1993).

The basic insights of systems theory (see Laszlo & Krippner 1998) are as follows:

1. systems consist of elements (components);
2. these components do not exist in isolation, but interact with each other energetically, materially and/or informationally;[1]
3. these interactions bring about structuring, self-organization, and the formation of (open) borders, which enable the system to appear and act as a greater whole;
4. properties or qualities are thereby produced which cannot be explained solely by the properties of the component parts, but are only created by their interaction. These are called emergent properties, meaning that a new whole emerges and that the whole is greater than the sum of its parts.

Figure 1: Emergence: Certain system elements interact with each other, creating a new larger whole with properties that do not arise (only) from the properties of these elements, but above all from their interaction.

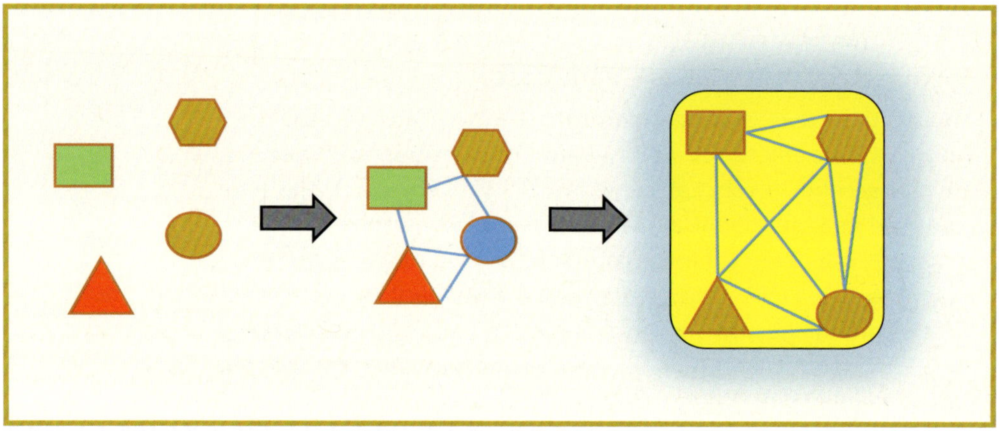

1 Both matter and information can be described in terms of their energy equivalence (Pagel 2013). In that sense, systems are ultimately always about energetic interactions.

Functions enabled by unpredictable interactions in the complex world system

The earth is a complex integrated system. The interaction of the constituent elements — such as the earth's core, the earth's mantle, the tectonic plates, the mountains, ocean and atmosphere, in combination with the earth's rotation — create an unpredictable dynamic, which is manifested for example in emergent properties such as plate tectonics, earthquakes and volcanism, ocean currents and weather and climate. The earth's rotation, involving its inner core of iron, leads to the creation of a magnetic field, which in turn protects the atmosphere from the harmful effects of the solar wind. On our neighbouring planet, Mars, there seems to have been at one time both an atmosphere and water — but both have been lost. Taking the earth system as an example, it is clear that the integration of the earth system into a superordinate system is of crucial importance. For example, its various interactions with the other planets and the sun combine to create a specific orbit and an energy input. This, in turn, means that water can be present in liquid form, which (together with the atmosphere) was a key factor in the evolution of a higher level of complex systems: life! The most complex systems we know are living beings, or those complex systems that are joined together by the interaction of living systems. The simplest organisms — bacteria — consist of millions upon millions of molecules. Through their interaction in aqueous solutions they organize themselves into a cell, in a form that can take in substances, water and energy in a controlled manner (metabolism), which means the cell can grow and multiply. The cell thus interacts with its surroundings — and can thereby also exchange information with other living things, which in turn brings about reactions in the cell.

Complex systems

The emergent properties of systems become all the more surprising and — in the truest sense of the word — incalculable, the more complex the systems are. The complexity of systems is a measure of the variety of interactions between system components. The complexity in the world system is increased by the fact that it is present in both horizontal and vertical forms. Horizontal complexity is about the interaction of system components within a system — for example, individuals within a population, or species within an ecosystem. Vertical complexity results from the fact that systems are stacked inside each other; these is also referred to as 'nested systems'. For example, the existence of an organism is on the one hand determined by the operation of its components, which in turn are also systems — such as organs or cells or molecules — and on the other hand by the proper functioning of the higher-order systems on which the system depends — for example, population or ecosystem. It is a peculiarity of vertical complexity that it is not a hierarchy in which the enclosed systems are 'rule takers' subordinate to the higher-order system. Rather, the existence and function of the higher-order system also depend on how the system components behave. The Austro-Hungarian writer and thinker Arthur Koestler (1905–1983) described this situation by saying that everything is a large whole and at the same time part of a larger whole. For this he coined the term holon (Koestler 1967; Koestler & Smythies 1969; Koestler 1970)[2]. In a system composed of such holons, order and function arise through holarchy — not through hierarchy (Fig. 2). No top-down commands are enforced, and a nested system of a particular order cannot break away from the holarchy — but it

Figure 2: Top-down: in a hierarchy, higher-ranking elements determine the lower ones. Top-down and bottom-up: in a holarchy, holons are nested within each other – a larger whole and at the same time a part of an even larger whole.

2 It is not widely known that the term was first used by a South African general (Smuts 1926); Koestler probably only rediscovered the concept (François 1999).

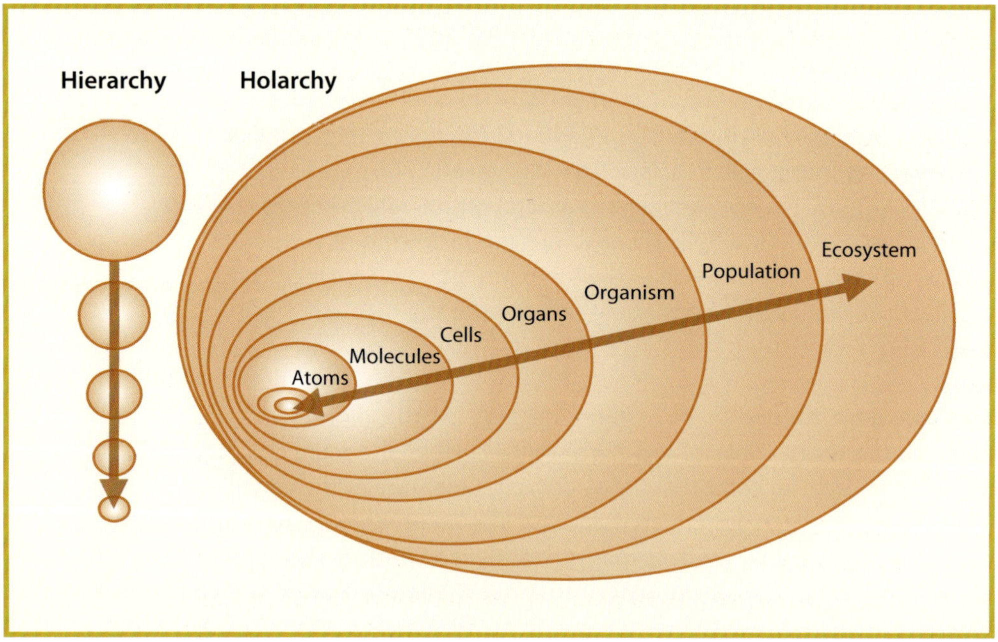

can affect the capacity of both lower-order and higher-order systems to function properly.

Life and evolution as emergent properties of complex systems

The basic principle of life is based on an interaction between a code stored in the genetic material and the proteins that organize all life operations. So life itself is an emergent property of a highly complex, self-organizing and self-regulating system. Modern genetic techniques enable the reconstruction of the genealogy of all living things, and show that all organisms on Earth are descended from a common ancestor. This primal being, called LUCA *(Last Universal Common Ancestor)*, is believed to have lived in areas of hot springs under oxygen-free conditions 3.5 to 3.8 billion years ago, and was capable of carbon dioxide and nitrogen fixation (Martin & Russell 2007). The biological evolution that, over the course of billions of years, has led to the formation of many millions of life forms can be understood as a systemic process. It led not only to the differentiation and diversification of single cell organisms, but also to temporary or permanent cooperation between them. From this emerged multicellular organisms and symbiotic double and multiple organisms. Thus, for example, α-proteobacteria and cyanobacteria were integrated into other cells as so-called endosymbionts and have become mitochondria or plastids (Thrash et al. 2011) — an important step in the development of complex organized life forms. It has been discovered that mitochondria, the power sources for the cells of all eukaryotes, are derived from α-proteobacteria which lived as plankton in the sea and repeatedly interacted with other cells, sometimes as parasites. When competition turned into coexistence, a very successful new system type emerged.

Innovation arising from the integration and cooperation of different systems

In the course of evolution and the multiplication of different life systems on the Earth's surface, integration and cooperation between systems have in fact become increasingly important, and have become a significant factor in innovations that have given systems advantages and secured them a relatively sustainable existence. Symbiosis between organisms of different species can bring significant benefits (mutualism). Cooperation is of particular relevance for social creatures — and a key attribute of human social systems, too (▹ Chapter 2.1). The evolution of life, with its continuous innovations, regularly led to significant impacts on the environment. The development of photosynthesis (in which carbon dioxide is fixed and oxygen released through the harnessing of the inexhaustible supply of solar energy on the Earth's surface) led to changes in the atmosphere, which became downright toxic for some living organisms that were well-adapted to previous conditions.

In addition, life thus began to influence the natural greenhouse effect and the global climate. The formation of multicellular organisms opened the way for the specialization and the formation of organs within complex organism systems. This included the development of mechanisms for the surveillance of the environment (sensory organs) and for communication. The result was so-called self-referential systems that interact with themselves. We human beings—the furthest point to date in the development of complex living beings—are systems that can recognize themselves and can even think about themselves and their own future. Ultimately, this involves information stored in the brain interacting self-reflexively and generating ideas and new knowledge (▹ Chapter 2.1).

»There are self-referential systems. In the first place, this means only in a very general sense that there are systems with the ability to establish relationships with themselves and to differentiate these relationships from relationships with their environment (Luhmann 1987, p. 31).

Systemic effects: escalation and feedback

The case of an organism that interacts with its environment, perceives it or even communicates with it, and actively or passively changes its own life processes on the basis of the information gathered, helps us to identify some important functional processes of systems. These are ▹ **positive or negative feedbacks**. This can be seen particularly clearly in natural selection in the course of evolution. For example, if a plant is protected from predation by the formation of spines, predators will eventually develop adaptations which will enable them to continue to eat the plant, which may result in turn in the spines becoming more numerous and/or longer. Such reciprocal reactions can lead to an evolutionary escalation and the formation of extreme features. However, simple but nevertheless significant feedbacks are present even in much less complex systems. One example is the accelerated melting of the Arctic sea ice, which is triggered by global warming and has surprised scientists by its scale and speed (Stroeve et al. 2007). In fact, the models had underestimated the melting because a feedback effect was overlooked: the loss of the white ice surface leads to less radiation being reflected back into space. Instead, it is absorbed by the dark surface of the ocean and leads to additional warming.

A positive or negative feedback occurs when a process amplifies or diminishes as its effects are turned back — via another system component — on itself.

Systemic effects: tipping points and non-linear change

Feedback and mutually reinforcing interactions between systems and their components enable systems to find an equilibrium in certain states. This is also the case with so-called attractors, which explain why systems seem to remain in a 'stable' state (e.g. global mean temperature, population sizes, the price of a commodity). The more complex and well-regulated the systems are, the greater their ability to cope with disturbances or environmental change. But if a system is 'pushed' or 'pulled' too strongly from the outside, and feedback processes work in a single direction such that so-called tipping points are exceeded, the probability increases that the system will very quickly shift to a very different state. A well-regulated, resistant system may not display any perceptible initial

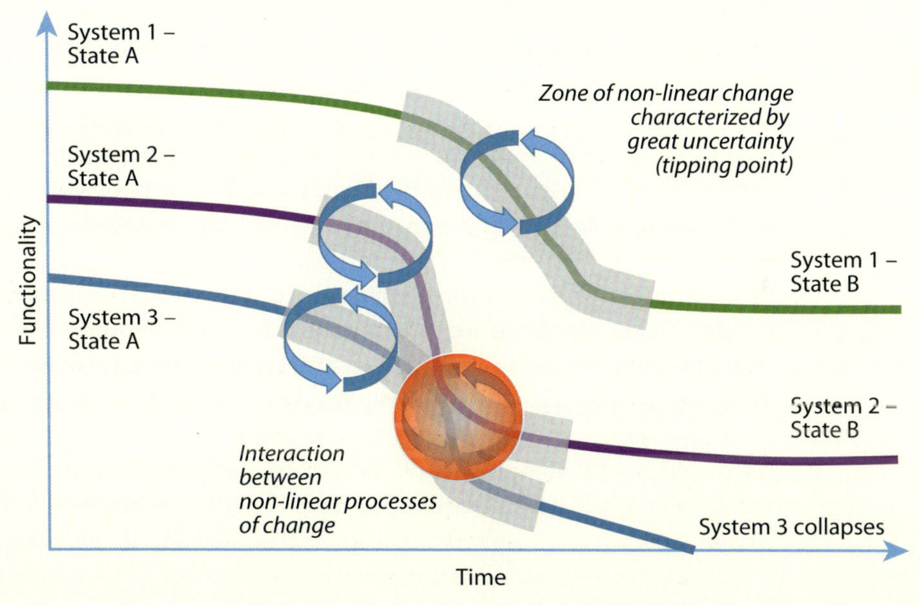

Figure 3: Schematic illustration of system change. After periods of (apparent) stability, pressure on systems can trigger non-linear change, which can lead to a new state but also to system collapse. Often, feedback effects play an important role. Systems changing in a non-linear way often interact, exerting pressure on each other — this can lead to unexpected change dynamics.

reaction to a disturbance, and will always react with a time lag, which can often be deceptive. In such systems, there is often only an 'either/or' reaction, with no 'in-between'. The phases of a change of state from one attractor to the next are often characterized by non-linearity. A typical example of non-linear change is exponential growth (▷ Chapter 1.1). For example, the capacity of a population to grow increases rapidly as the number of individuals capable of breeding grows. If this results in the depletion of important resources, it may lead to a self-regulated deceleration of growth (see negative feedback: inhibition of a component through its own effects). A well-known phenomenon illustrating such an event is the explosive algae bloom in a body of water. As soon as more and more dead algae decompose, the rapid reduction in the oxygen content can cause the water to 'tip over'.

Systems: from self-organization and self-regulation to husbandry and long-term functionality

Ecology, ecosystem, systems ecology

The simple multiplication of the first existing organisms through the copying of the genome and the division of all other elements of the cell created the components of a higher-order system: the first population of living things. Simultaneously, the first ecosystem emerged, as organisms started to use environmental resources and to create an energetic structure. The publication by Charles Darwin in 1859 of the theory of evolution, which first explained how the multiplicity and diversity of life came about, inspired the German zoologist Ernst Haeckel (1834–1919) to think about the interactions of organisms with their environment (Haeckel 1866). He coined the term '*eco*logy' for the study of the household [in nature] (*Oikos*, Ancient Greek *οἶκος*: household/house). The word 'ecosystem' itself was first introduced — in the 1930s — by the British biologist Arthur George Tansley (Tansley 1935). The American Eugene Odum was the founding father of systems ecology (see Odum 1964). For Odum, one of the core aspects of systems ecology was eco-energetics (Odum 1968).

No systems without energy

Ecosystems are complex systems that arise because their components, the individual living beings, interact with each other and with non-living resources. They develop emergent properties that safeguard or even promote their survival. For complex systems to exist and develop, energy is needed. As part of the interaction of system components, for example of organisms or ecosystems, energy is transmitted, converted, stored and above all used, for work. Energy is needed, for example, for the production of biomass and for growth, temperature regulation, mobility, the transportation of matter, and information processing by means of electrical signals. It cannot be created or destroyed (the first law of thermodynamics), but every 'use' of energy, and the corresponding performance of work, reduces its quality — the laws of physics also state that entropy increases (the second law of thermodynamics). All energy transformations end in heat, with which work can be done, but only with the help of additional energy: ▹ **exergy**.

The usable energy still available for work in a system is called exergy.

Ilya Prigogine was awarded the Nobel Prize in 1977 for his theory of the creation of order by so-called 'dissipative' structures (see, e.g. Nicolis & Prigogine 1977). Complex systems 'brace' themselves — by means of energy input — against the second law of thermodynamics. Self-organization is driven by the emergence of structures that keep the energy supplied in the system for longer, or slow down the degradation of energy.

Eco-exergy: ecosystem reserves of high-quality energy still useful for work

Life, in the sense of the creation of ever new complex systems that work and interact with each other, requires a continuous supply of energy. Almost all the energy for life on Earth is provided through photosynthesis. High-energy photons are used to produce, from simple inorganic compounds and elements, organic molecules of greater or lesser complexity and biomass, which fulfil a dual function. They form the structure of the complex living systems and at the same time a reservoir of high-quality energy in the ecosystem — eco-exergy. The eco-exergy increases in line with the biomass, but also with the diversity of the components of the ecosystem and the complex interactions. Biomass, information and network are therefore the three fundamental attributes of ecosystem functionality (see Jørgensen & Müller 2000; Jørgensen & Fath 2004; Jørgensen 2007). Their growth increases the system's ability to absorb more energy and to retain it for longer. Exergy increases with the lengthening of food chains and faster recycling due to higher decomposition rates in the ecosystem (Jørgensen & Fath 2006). In this context, ▹ **biodiversity**, the diversity of all living systems in the global ecosystem, must also be considered. For example, biodiversity has been shown to improve the productivity of forest ecosystems around the world (Liang et al. 2016). Biodiversity and hierarchical organization also mitigate disturbances (Jørgensen et al. 2016). The energetic-thermodynamic perspective represented an important advance in our understanding of the functioning of ecosystems and of sustainability based on an ecosystem approach (Kay & Schneider 1992; Schneider & Kay 1994; Kay 2008; Kay & Boyle 2008) (▹ Chapter 4.1).

> **Biodiversity or 'biological diversity', as defined by the UN Convention on Biological Diversity, means 'the variability among living organisms from all sources including, inter alia, terrestrial, marine and other aquatic ecosystems and the ecological complexes of which they are part; this includes diversity within species, between species and of ecosystems' (United Nations 1992b).**

Efficiency

Working systems can exist only as long as they have exergy available. Together with water and so-called nutrients for the construction of molecules, exergy is therefore a fundamental resource for living systems. Living beings and ecosystems can only function or grow or expand if they take up energy in an ▹ **efficient** way, retain it as long as possible in the system, and degrade it. An efficient ecosystem uses light energy for photosynthesis, via diverse plant forms with the largest possible surface area. Moreover, the energy taken in is not 'burned in a straw fire', but efficiently passed on in small steps along food chains and used for work. The imperative to efficiently manage not only energy but also water and scarce material resources such as nitrogen, phosphorus and trace elements is a main driver of biological evolution. Every living system is therefore based on husbandry, and has to balance input, output, and resources (Ripl & Wolter 2002). Systemic husbandry of the household economy occurs along the entire holarchy, from individual cells via individuals and species to the global ecosystem. In the same way, it plays a crucial role in social systems of all kinds (▹ Chapter 3.1).

> **Efficiency is the ability of a system to husband resources (e.g. energy, substances and water); that is, to minimize the resource output (loss) compared with input, and to build up and/or maintain a reservoir of resources.**

Resistance and (adaptive) resilience

However, living systems are not just selected for efficiency; they also have to deal with disturbances and environmental change. These disturbances can occur in the ecosystem itself, but can also come from outer space or from the Earth's interior (e.g. changes in radiation conditions, meteorite impacts, volcanic eruptions). The ability to continue functioning despite disturbances is called ▹ **resistance**. For example, the strong trunk of a palm tree can resist even very strong storms without breaking. ▹ **Resilience**, on the other hand, enables systems to recover or improve their functionality after it has been degraded by disturbances. An example would be the reforestation of a burnt area of land (Fig. 5), or the recovery of a population drastically reduced by an epidemic. If, however, the disturbance is persistent and consistent in effect, systems will adapt. Resilience does not mean that a system returns to its original state, but rather that it regains most of its emergent properties and functions with-

Figure 4: When a dune is colonized, it is possible to observe a change in the functionality of the ecosystem mediated by positive feedback. The initially very open system of the uncolonized sand dune is hardly able to husband resources such as radiant energy, rainwater or nutrients. The very first pioneer plants, such as beach grass, serve to hold in place some soil, nutrients and water, and use light to build up biomass. The degradation of plant residues leads to soil formation and to the improvement of the water retention capacity. The system starts to efficiently husband resources: the input-output ratio gets smaller — the system starts to store and enrich energy (eco-exergy) and nutrients. As a result, more and more plants are able to thrive on the site, helping to maintain and grow their own habitat. The vegetation becomes three-dimensional and complex (pine forest); effective microclimatic regulation begins (shading, cooling, water storage); more and more plant biomass is available as food for consumers; life escalates. (Amrum Island, photo: P. Ibisch)

out tipping over into a completely different functional state. Ultimately, from the perspective of the higher-order system, tipping over into a different state is also a kind of adaptation. However, if a forest collapses through climate change-induced drier and hotter conditions leading to fire, and is then replaced by a steppe inhabited by entirely different species, this is not an adaptation but a change of system. At the level of the global ecosystem, this means that important functions such as carbon storage and the regulation of greenhouse gas concentrations are maintained, even if at a different scale.

> **Resistance and (adaptive) resilience describe the ability of a system to withstand disturbances and to adapt to changes in (frame or environmental) conditions without collapsing and undergoing a loss of emergent properties equivalent to a system change.**

Diversity and redundancy (the presence of multiple instances of system components that can to some extent replace one another) play a vital role in ecosystems, enabling collapsing and failing system components to be replaced. Through constant experimentation, biological evolution via innovation not only ensures ever more efficient systems, but also creates 'substitutes' and new components able to withstand changed conditions. In the course of the emergence of complex systems, structural and functional constraints on further development always arise — so-called path dependencies. Retrospectively, therefore, system breakdowns always also create the potential for a creative new start. The emergence, growth and persistence of complex structures successively alternate within systems in countless adaptive cycles involving decay and reorganization (Gunderson & Holling 2002; see also Fig. 3, ▹ Chapter 4.1).

Sufficiency

As long as resources are available, living systems that use and husband those resources grow, and their growth is usually accompanied by increasing functionality. In larger populations, for example, specialization offers potential for increasing efficiency — this can be observed in the case of social insects, which form colonies. Larger populations also have greater genetic diversity, which can potentially provide a life insurance against unexpected disturbances or diseases. At the ecosystem level, growth occurs both quantitatively, through an increase in biomass, and qualitatively, through the information held and the networking between all the subcomponents. However, because of general limitations on resources, living systems have not been able to rely on sustained

Figure 5: After a forest fire on Tenerife, it transpires that most tree heather individuals *(Erica arborea)* have in fact survived underground and emerge again. This is an example of ecosystem resilience, in this case a kind of self-repair, based on the fact that biological information can be retrieved to restore the superficially destroyed vegetation relatively quickly, without the ecosystem having to go through all the lengthy succession phases. (Parque Rural Teno, Tenerife, Spain; photo: P. Ibisch)

growth, and instead have always had to be able to remain functional at zero growth, or even to shrink without thereby collapsing. The relevant abilities that enable this are what provide ▹ **sufficiency**.

> **Sufficiency: The capacity of a system to operate within limits, or even at zero growth or while shrinking.**

Consistency The increase in information held by, and in networks within, complex systems is the primary reason for the ever-increasing integration and cooperation of the system components. The coordination and interlocking of processes leads to ever greater self-organization, functionality and regulation.

> **Consistency or coherence: the coordinated functioning of system components, which leads to improved self-organization and self-regulation.**

The emergent properties of complex living systems described above—efficiency, resistance and resilience, sufficiency and coherence or consistency—enable and reinforce their continued ▷ **functionality** even under changing environmental conditions.

> **Functionality: For systems, functionality can be described as a state of the system characterized by inherent structures, functions and dynamics that provide it with both the necessary efficiency and resilience to develop without abrupt changes in emergent system properties or geographic distribution, and to be able to respond flexibly to external change (according to Freudenberger et al. 2012). Functionality therefore means, in summary: higher energy intake, more exergy storage, more self-regulation and more (regulatory) influence on the environment, and finally greater contributions to self-preservation.**

Efficiency, resilience, sufficiency and consistency are also key terms for the description and analysis of sustainability strategies (▷ Chapter 1.2). All the more important, then, that they should be explained in the context of a systems-theoretical analysis of their implications and effects.

Humans, social systems and sustainable development

Humans as a biological and cultural system

Humans originated as animals in the savanna ecosystems of East Africa and were subject to the same basic physical laws and systemic conditions as all other living systems (▹Chapter 2.1). What is unusual about them is that the human system — or anthroposystem — consists of not only a biological system, but also a cultural system. It is not just a population of breeding organisms (exchanging DNA), but of intelligent, self-reflective and intensively communicating individuals who have organized themselves into countless social systems (see, inter alia, Luhmann 1987).[3]

The anthroposystem is therefore much more complex than the systems (interacting individuals) of other species. Thanks to information

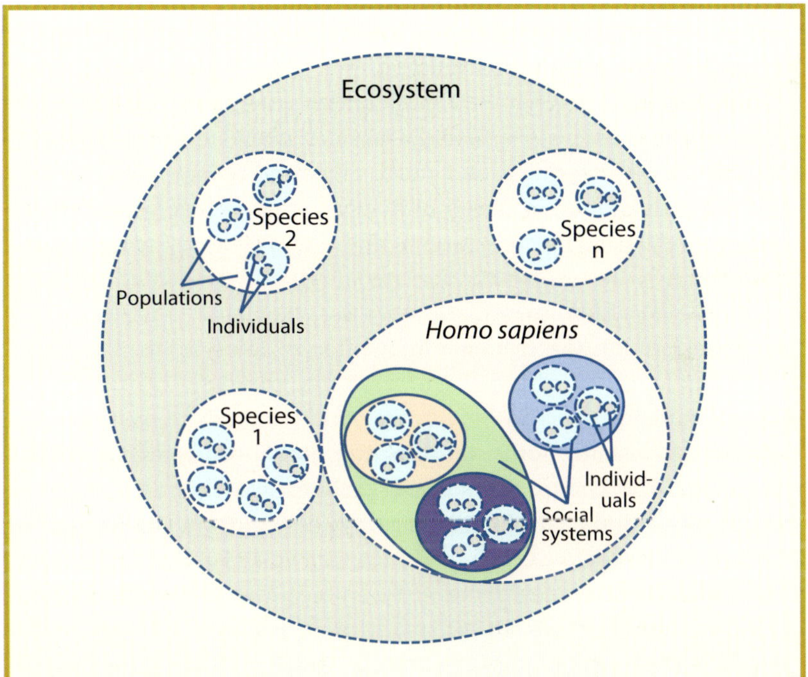

Figure 6: The species *Homo sapiens*, part of the global ecosystem, consists of a biological system and complex, nested social systems based on the interaction of human individuals.

3 The sociologist Niklas Luhmann has had a major influence on the understanding of social systems. Communication between people is at the heart of his thinking. 'A social system comes about whenever an autopoietic communication context arises and distinguishes itself from its environment by restricting the appropriate communication. Accordingly, social systems do not consist of people, nor of actions, but of communications.' (Luhmann 1986, p. 269). This definition is not followed here. Instead, here communication is understood as a process of interaction between socially active people as system components.

Social systems: extremely complex, but dependent on ecosystems

technology, there is now a cross-generational exchange of information; global interaction between physically separated individuals and social systems; and interaction between and with 'artificial systems' such as computers. Multiple memberships of human individuals in sometimes short-lived social systems (e.g. firms, associations) dramatically increase complexity. The human brain enables people to be aware of the future. As a result, projections of the future interact with current decision-making. This means that people act according to plans, which gives rise to another peculiarity of social systems. Social systems arise emergently out of interactions, but are also often deliberately formed to serve a particular purpose. Despite all these peculiarities, the anthroposystem is completely embedded in the ecological system by dint of its being composed of biological individuals. All human cultural achievements and technologies are based on the cognitive abilities that arose through biological evolution. Moreover, all intellectual and cultural achievements exist only thanks to the satisfaction of individual human beings' needs by the ecosystems (▷ Chapter 2.1). All social systems are subject to the same principles as living or ecological ones: they are open systems in which energy is degraded and must therefore be constantly 'replenished'. Social systems are based primarily on the exchange of information, but they also channel material and energy flows. Their emergent properties may be completely dematerialized and concerned only with the processing of information. Nonetheless, they arise — as always with systems — through the dynamic interaction of subsystems. Even man-made, purely virtual systems — for example, computer-controlled systems for trading in virtual financial products — produce real effects in higher-order systems, for instance in the global ecosystem, by directing material and energy flows. Social systems operating in ecosystems are also referred to as 'socio-ecological systems' and are closely studied, especially with regard to their resilience (e.g. Folke 2006).

Social system groups: state, market, civil society and criminal

In the course of its development, the anthroposystem has continued to diversify. Currently, there are essentially three groups of subsystems that determine the functioning of the anthroposystem — that is, the changes in mass (number of people), information and networks (between individuals and social systems) in the ecosystem. These three groups are state systems, market systems and civil society systems, which sometimes cooperate and sometimes compete in shaping how the anthroposystem develops (▷ Chapters 3.2, 3.4, 3.5). In addition, there is a fourth group, which affects the other three and their interactions and therefore often has a decisive influence on their development — antisocial and criminal

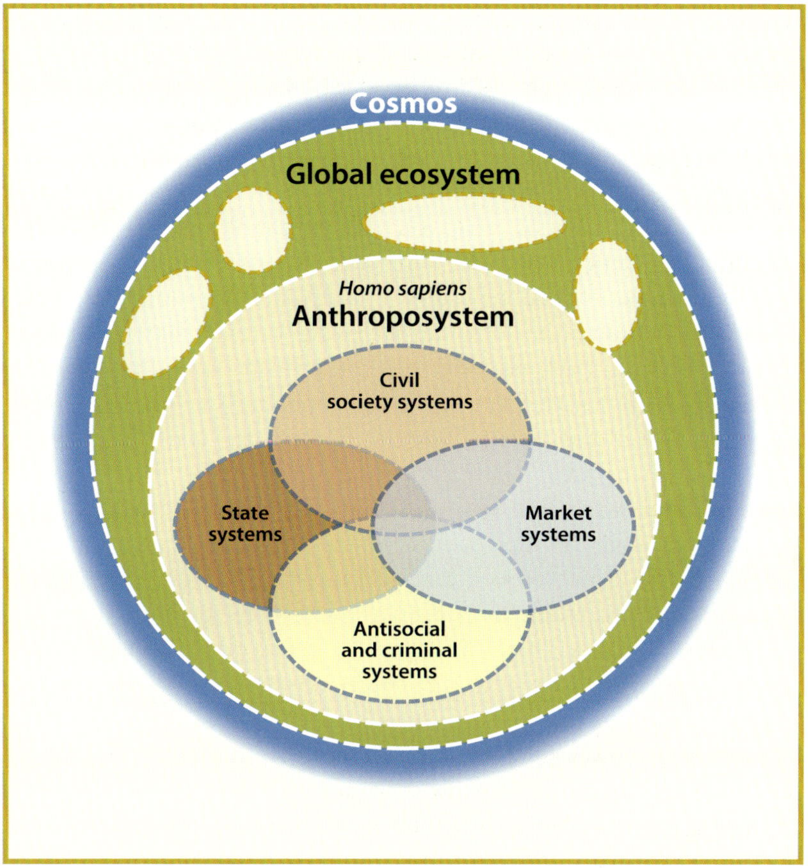

Figure 7: Systemic embedding and subdivision of the current anthroposystem. The development of human societies currently consists mainly of a struggle between four fundamental systems — state, market, civil society, organized crime — with different relative proportions and local characteristics in the different nation states. Depending on prevailing ideologies and forms of statehood, one of these four system types is usually dominant. The various institutional systems such as intergovernmental organizations, parastatal institutions, NGOs, etc., are situated at the interfaces between the respective fundamental systems. Human individuals, but also lower-order systems such as companies or clubs, often have multiple memberships in different system types, which significantly increases overall complexity.

systems. This last group may sometimes present itself as a state, market or civil society system, but may also generate entirely new forms of organization (e.g. organized crime, international terrorism).

Systems theory analysis of sustainable development

To sum up, a systemic theory of sustainable development can be proposed. Complex, work-performing and energy-transforming systems consist of interacting components. The interaction gives rise to emergent properties, which serve to increase self-organization and self-regulation. As a result, efficiency, resilience, and coherence (consistency) increase in long-term viable (functional) systems. Sustainable development, which is essentially based on maintaining the existence or increasing the functionality of the system, also includes the conversion and replacement of system components. Sustainable development is in no way equivalent to growth of the system (whether of mass, expansion or throughput). A sustainably developing system can grow — often at the expense of other, less sustainable systems — but only within certain limits of resource availability. It represents a constant qualitative change, which means a better protected and continuing existence, even if the system's environmental conditions change. Sustainable development is not a contradiction in terms; it is not — as Patten (2014) and others claim — an oxymoron, because development is of course not the same as growth.

Sustainability as a transient achievement

Indeed, as all complex, living systems are constantly changing, adapting, and dynamically husbanding their resource budgets, they can be said to be developing sustainably only until such time as the opposite is proven by system collapse. Sustainability can only ever be achieved temporarily. It requires — following the laws of thermodynamics — the continuous supply of high-quality energy. If the energy supply to a system causes functional components to be impaired or destroyed, the system is consuming itself, as it were, and cannot be sustainable.

The most sustainable of all known systems

The most sustainable system we know of — which has maintained this kind of change, as well as growth within clearly defined borders, for almost 4 billion years — is life itself, and thus the global ecosystem. During this time period, the exergy in the system has grown over the long term (with short-term downswings); this is due in large part to the qualitative development achieved through the growth in information, networking and intelligence in the system. The energy captured and circulated in the energetic web of this global system is distributed to ever more numerous components over ever longer and more complex paths of material and energy transformation.

One very important insight is that, as the global ecosystem sustainably develops, all its components are replaced and changed. The deaths of individuals, the extinction of life forms and the reorganization of existing information and relationships are a necessary condition for maintaining mutability and adaptability. The most sustainable components of the global ecosystem are not necessarily the most complex ones. Biological evolution has produced extremely complex life forms, such as vertebrates; but these have so far participated in the evolutionary process for a much shorter time than, for example, molluscs or insects. Among the vertebrates, there are forms that have existed for a long time, even hundreds of millions of years, such as fish or lizards; the human species is young, and as a life form has yet to stand the test of time. There are currently many indications that the hypercomplexity of the anthroposystem, which has developed over a mere few centuries, is leading to an abrupt change in global environmental conditions that calls into question the sustainable existence of many tried and tested life forms.

To date, the most sustainable forms of life have included first and foremost the simplest ones, such as bacteria and cyanobacteria. Since their emergence, they have been present in large numbers in all habitats (including extreme ones), have hardly changed structurally or functionally over many hundreds of millions of years, and have survived drastic changes in the atmosphere, the global climate and the biosphere. Although highly complex and structured organisms may flourish very rapidly, they also show less versatility (mobility, mutability) and considerable systemic vulnerability — which can be fatal, especially in rapidly changing environmental conditions. In this context, humanity should be listening attentively. Humans would be well advised to study the prescriptions for the ▷ **sustainable development** of ecosystems and to apply them to their own social systems (▷ Chapter 4.1).

Sustainable development: an emergent property of complex systems that process an indispensable external supply of energy and that continue to function and exist while changing (usually) by increasing efficiency, resilience and coherence (consistency). Sustainable system change also includes the conversion and replacement of system components. It may be temporarily associated with growth within given limits, but sustainable development also involves stagnation and shrinkage while maintaining the functionality of the total system.

2
We humans: the agents of sustainable development

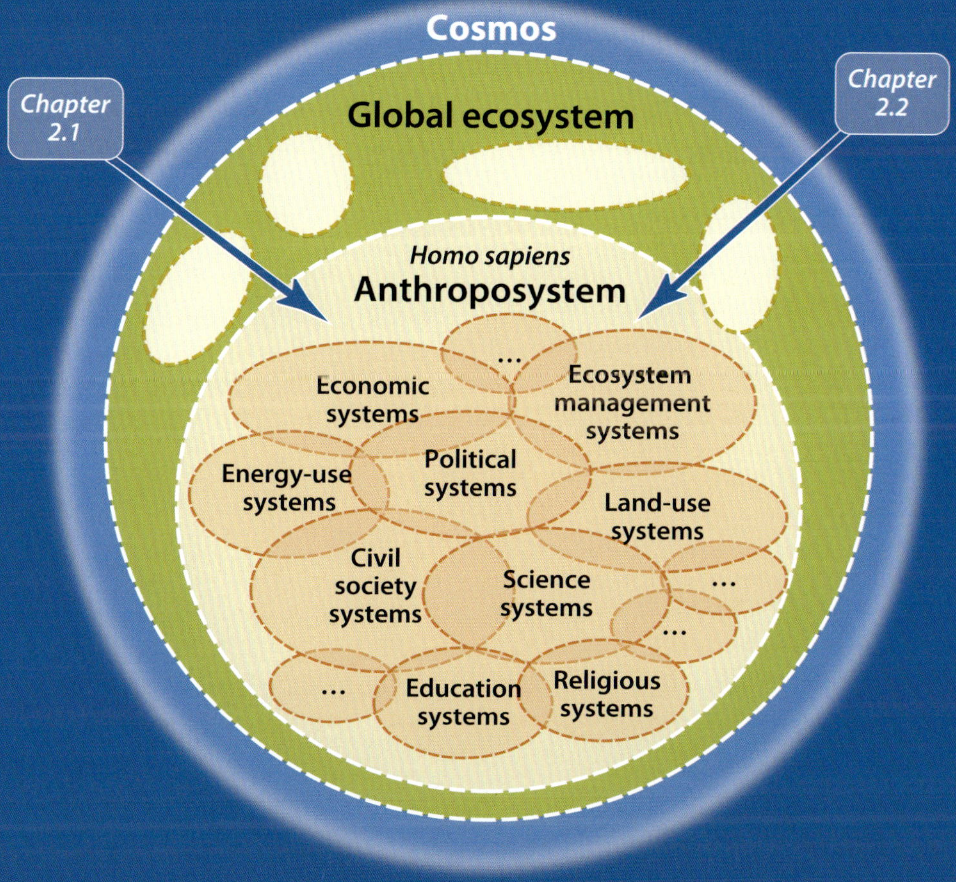

Good or evil – are we capable of sustainability?

2.1

Pierre L. Ibisch and Norbert Jung

We are living beyond our means; we are ruining the foundation for our subsistence and that of other species — humanity has become a force that is shaping the planet. At the same time, we are making increasing efforts to reverse the trend and to change course towards sustainable development. It is clear that we are capable of both creating and solving problems. This raises the question of how these capabilities came about. In order to develop effective solutions for a shift to sustainability, we have to deal with the agent of (un)sustainability — the human being. How did we become a global power capable of shaping nature? Were we born to dominate the Earth and transform the global ecosystem? Many people are inclined to believe that humankind is an aberration of evolution, destroying what has been produced by millions of years of development. Are we good enough to bring the forces we ourselves have unleashed back under control, or are we by nature too evil to live in a truly sustainable way? And do these moral categories even apply, given that they always refer only to human behaviour?

Introduction: our relatives

Humans and human culture as a product of biological evolution – humankind as part of nature

To understand how and what we are and what path we may take in the future, we need to understand where we come from. A first basic insight is that we humans and our culture have developed in a biological way in the course of evolution. We are 'by nature cultural beings' (Gehlen 2004, p. 80; see also Voland 2007; Wessel 2015). In that sense, even if the human mind can open up additional dimensions, everything man-made and all culture can be considered a part of nature. We and our culture are the result of our engagement with the ecosystem that has brought us into being, and with ourselves and our ideas. Even our ability to influence our own habitat does not elevate us above nature (Ibisch et al. 2010); it is a basic principle of all living things. This is also a key principle of the so-called ecosystem approach (▷ Chapter 4.1).

Our nearest relatives: chimpanzees and bonobos

Today, we know for certain that we belong to the group of primates, and that our closest living relatives are the African chimpanzee *(Pan troglodytes)* and the bonobo *(Pan paniscus)*, with whom we share 99.6 % of our genes. Our last common ancestors lived more than 4 million years ago (Prüfer et al. 2012). This finding is of great importance because while the way of life and the character of the chimpanzee and the bonobo differ very clearly from one another, they each have properties that are not foreign to us.

Figure 1: Chimpanzees (above) and bonobos (below) are humans' closest living relatives. (Zoologischer Garten Berlin, photos: P. Ibisch)

Chimpanzee males are aggressive, fighting for rank in the community and for access to females. Chimpanzees cooperate with each other to defend their territory: attacks on other groups quite often result in fatalities (Boesch et al. 2002). According to a study in Gombe National Park, 20 % of chimpanzee deaths were due to chimpanzee violence (Williams et al. 2008). Perpetrators and victims are mostly males (Wilson et al. 2014). Infanticide carried out by males is relatively frequent (Furuichi 2011).

Among bonobos, on the other hand, evolution has evidently led to the suppression of aggression: the males subordinate themselves to the females and do not fight over rank. Bonobos do not cooperate in groups, and they do not attack others. They tend instead to be playful, and maintain an intense sexual behaviour, which does not always include a direct reproductive function, often with the same partners (Hare et al. 2012). Females maintain intensive social relationships with each other, which has been linked to the much more peaceful coexistence among bonobos (Furuichi 2011). Chimpanzees regularly hunt other animals, including primates. Similarly, organized hunting has been observed among bonobos, with females involved too (Hohmann & Fruth 2008) — which does not occur in chimpanzees.

Such, then, are our relatives, the evolutionary environment from which we are descended. It raises the question of what it is that makes us human, for all that we have in common with African anthropoid apes. Which path have human beings taken? Are our social behaviours, the propensity for violence towards our fellow humans and our environment, something we have in common with chimpanzees, or do we have the potential for a character that is more akin to the bonobos, more peaceful and 'loving'? This question is not to suggest that bonobos are the good relatives and chimpanzees the bad ones — a moral evaluation of animals using human categories is always a mistake. In any event, we are not descended from them; we merely have common ancestors. The point is to develop a greater understanding of ourselves and our evolutionary roots, and especially for those peculiarities of our behaviour that can be considered 'typically human'.

The emergence of modern hominids through environmental change

As with all other species, a systemic event in the course of the struggle of the great apes, including humans, with their changing surroundings and fluctuating living conditions — as well as with themselves and the consequences of their own actions — led to an open-ended but nevertheless directional evolution. The following is a plausible description of a possible history of interactions and consequences. Roughly 6 million years ago, a long-term trend towards dryer conditions and increasing environmental variability seems to have begun on the African continent (reinforced in the last 3 million years, Potts 2013). As a result, the rainforests, which are believed to have previously been more extensive, retreated to the western part of the continent, while the east saw the spread of dry forests and savannas. At the same time, sedimentary evidence suggests that in the east of the continent, above all in the region of the African Rift Valley, wetter and drier conditions alternated relatively quickly. These ecological conditions caused a separation of the African great apes into two groups: the rainforest taxa of chimpanzees and bonobos in the tropical west and the hominids in the eastern savannas.

The emergence of human beings was only possible through cooperation and fairness

Evolutionary drivers lead to intelligence and a social way of life

Life in the open savanna favoured an upright gait (McHenry 2004). This made new uses for arms and hands possible; motor skills improved, the use of tools such as sticks and stones increased, food became more varied, and energy-rich meat — probably carrion at first, or prey killed by other animals (Aiello & Wells 2002; Stiner 2002) — played an increasingly significant part in the diet. The increased energy supply enabled the growth of the brain, the organ that accounts for up to 20 % of energy use in modern humans (while accounting for only 2 % of body weight, Engl & Attwell 2015). The availability of energy limits the capacity to process information. This is a classic case of positive feedback: increasing intellectual ability facilitated strategic food procurement, and the additional energy from better nutrition could then be invested in the development of the brain and intellect.

Helpless newborns and intensive family life require high levels of cooperativity

At some point, coordinated hunting in groups and the use of fire and weapons became more and more important. The increase in brain size led to the expansion of the skull. Combined with a narrowing of the birth canal, resulting from the upright gait as well as physiological reasons, this may have created evolutionary pressure for babies to be born earlier, as well as a need for support during labour (Rosenberg & Trevathan 1995; Trevathan 1996; Dunsworth et al. 2012). The early birth of totally helpless infants forged a durable mother-child relationship and promoted the cohesion of (extended) family groups. Ultimately, it would have become a positive selection advantage for child-raising women not only to be able to bind the father to the family for a longer period, but also to secure help from other family members. The social way of life already typical among ancestral primates would thereby have been strengthened. In the context of long-term close coexistence, coordinated hunting and a lifestyle based increasingly on the division of labour, communication steadily improved. The emergence of a complex language reinforced in turn not only the scope for cooperation and mutual support, but also intellectual ability; social intelligence ultimately made us human (Shultz et al. 2012). The raising of helpless infants seems to have played a key role in this process (Piantadosi & Kidd 2016). The triumph of *Homo sapiens* was made possible in evolution not by egoism, but by ▹ **cooperativity**. It was the human capacity to form groups, which involved, among other things, social regulation and a sense of community.

> **Cooperativity is the ability or tendency to cooperate.**

Thinking, knowledge, symbol systems

With the emergence of the human brain and language, biological systems were elevated to a new level of ▹ **self-referentiality** and self-objectification (Gritschneder 2005) — that is, the ability to relate to and interact with oneself. Not only could individuals now interact better with one another, but they could also reflect in ever more complex ways on themselves, on their own actions and thoughts. It was only the development of language and words that enabled certain issues to be thought about at all. On the basis of information stored in the brain, completely 'dematerialized' thought systems began to develop: interactions between thoughts created new ideas — consciously and unconsciously (!) — and thereby accelerated the growth of knowledge. Since individually acquired knowledge could be passed on and discussed more effectively thanks to language, cultural development was also accelerated. Symbol systems and artistic creativity were engendered (Morriss-Kay 2010), which opened up ever more far-reaching opportunities to grapple with the world. In order to ensure their continued existence, however, these new self-referential ideas and cultural products had to be repeatedly tested against the practical experience of life with and in nature, and selected accordingly by reality.

> **Self-referentiality is the ability to relate to oneself (feedback) and to respond to changes in the system, thereby developing a (stronger) identity that distinguishes the system more clearly from the environment.**

From cooperation to ethics

For all its benefits, long-term social coexistence in groups presumably represented a serious challenge from the very beginning. The closer and more beneficial the social coexistence of a species, the more both communication (exchange of messages, different forms of coordination, etc.) and cooperation (the purposeful coordination of actions between two or more living beings) have to develop and diversify. In the early social systems of the hominoids, aggression and competitive behaviour would have needed to be minimized to enable cooperation. After all, primates are among the mammals which display the strongest forms of aggression (often deadly) towards their peers (Gómez et al. 2016); as mentioned earlier, violent conflicts between (predominantly male) individuals and also between groups are common among our closest relatives, the chimpanzees. Killing is a strategy for eliminating rivals among chimpanzees; it is used when the consequential costs are relatively low (Wilson et al. 2014). The evolution of personal relationships and pacifying rituals (e.g. greetings, dances, ceremonies) served to inhibit aggression. With increasing intelligence, the self-organization of the groups became more sophisticated, although hierarchical domination and leadership by powerful individuals predominated.

Reflective beings with minds

In human groups, antagonistic feelings and behaviours emerged, including both aggression and powerful forms of oppression, as well as unconditional love and cooperation. Finally, human beings, who became capable of ever more abstract thought, also became reflective beings with minds who could inhibit their immediate physical impulses and make rational decisions. Building on this capacity for objectivity, thought was also given to how things ought to be and how people ought to behave — the starting point for the development of ethical propositions. Many philosophers, like Max Scheler (1874–1928, German philosopher, anthropologist and sociologist), assume that as beings with minds, humans must have liberated themselves completely from their biology:

> » So a 'being with a mind' is no longer bound by its drives and environment, but 'free of its environment' and, as we might say, 'open to the world'. […] Such a being, moreover, is capable of elevating the given capacities with which it 'resists' and reacts to its environment, which belong to the animal alone and in which it is ecstatically subsumed, to the status of 'objects', and in principle to grasp the very nature of these objects without the limitations that this world of objects or its givenness suffers through being mediated by its vital instinctual system, sensory functions and sensory organs. The mind, therefore, is objectivity, determinability through the very nature of things. Only a living being capable of complete objectivity can 'have' a mind (Max Scheler 1927 in Scheler 1978, p. 38).

Ethics as a guarantor of survival

It cannot be denied that people only succeed in being (relatively) objective when they put aside their own needs and environmental constraints. However, many findings — for example, from neurobiology, evolutionary research, psychology and economics — demonstrate that people find it difficult to achieve 'perfect' objectivity (▹Chapter 2.1). But this does not disprove the theory that ▹**morality** and ▹**ethics** emerged for functional reasons. Since human culturality is a natural product, so too are the foundations of morality: such a highly social and self-reflective being needed or needs social behavioural regulation for coordinated action. Ethics is vital to our survival.

Morality describes the context-dependent sense in individuals or groups of what is considered appropriate and proper, which often becomes a behavioural standard within a society.

Ethics is the study and teaching of what is considered right and proper (morality and behaviour) as well as the norms and desirable and legitimate maxims of life conduct and human action.

Norms represent a set of values in a society based on moral concepts or formalized ethical principles. They are rules of conduct which are socially relatively constant and often legally binding.

Typically human: fairness

Game theory, a sub-discipline of economics, seeks to establish why and under what circumstances humans choose certain behaviours such as cooperation. This always involves the question of how selflessly or selfishly people act. Game-theoretical experiments with monkeys and children are informative in this regard. They seek, among other things, to prove whether individuals are prepared to set aside their own interests, whether they can act fairly and justly, and whether they may be willing to share scarce resources with others. For example, in some experiments sweets or other titbits must be shared equitably as a condition of any of the parties receiving them. It has been shown that both chimpanzees and bonobos struggle to make and recognize what are (from a human perspective) fair apportionments — they are ultimately selfish maximizers, while human beings recognize and generally respect social norms and rules of fairness (Jensen et al. 2007; Kaiser et al. 2012). Although chimpanzees can recognize unfair behaviour that disproportionately favours one individual, and they can show social disappointment, this does not lead to an ability to enact fair behaviour (Engelmann et al. 2017). And although they seem to recognize behaviour that violates social norms, such as infanticide, they do not intervene to prevent it (von Rohr et al. 2015). Bonobos are peace-loving, yet they certainly show respect for antisocial individuals (Krupenye & Hare 2018). Even young human beings, on the other hand, can be seen to derive real enjoyment from the fact of others being helped (Hepach et al. 2012). Empathic behaviour — the ability, which can also be observed in some animals, to put oneself in another's place and to take their side — is particularly pronounced in humans. Since this can already be observed in very small infants, we may regard it as a species-specific biological behavioural tendency (Haidt 2001; Blohm 2010). However, fair behaviour and the acceptance of unfair treatment are also clearly context-dependent. Local competition increases people's willingness to demand fairness and reject unequal treatment (Barclay & Stoller 2014). It is unlikely that fairness in humans simply came about for its own sake; it is more probable that fairness emerged as a means of encouraging sustainable cooperation with fellow species members (Brosnan & de Waal 2014).

Position in society and status symbols

In the case of allegedly fair or unfair behaviours, it cannot be said that behaviour is good or bad if individuals are only aware of their actions and their consequences to a limited degree. A rule violation occurs only when there are social norms and someone is prepared to consciously break them while fully aware of the consequences (see Spitzer 2009). Interactions within human social groups became increasingly complex over time. Individuals must have realized, on the one hand, that it benefitted their own position in society and access to resources if they protected others or helped them, for example as a chief — this enabled cooperation to be planned and carried out according to plan. At the same time, the necessity and the desire to establish one's own position within the group obviously grew as groups began to significantly outgrow families in size (tribes typically have up to 150 members).

Consumption and possessions as status symbols

It is probably no coincidence that in very different human societies leaders and chiefs, but also other dignitaries and officials, have almost always 'adorned themselves with borrowed plumage' (feathers, striking robes, hats, crowns, etc.; see Fig. 2) or surrounded themselves with status symbols to signal their importance (Ibisch et al. 2010). In the course of the history of clothing, this later took on very sophisticated forms. This furthered not only the development of aesthetics and art appreciation but also the inclination to accumulate conspicuous possessions to convey their owner's importance. This was the beginning of a consumption not driven by basic needs in materialistic societies — and thus also of less sustainable behaviour. In certain societies, membership and social participation were (and are) increasingly defined by symbolic possessions, such as prestigious housing, jewellery — but also sports shoes, cars and electronic devices.

Figure 2: 'Adorned with borrowed plumage': Status symbols borrowed from nature and culturally repurposed endow people with a special significance in society. Indigenous costume from Sarawak. (National Museum of Malaysia, Kuala Lumpur, photo: P. Ibisch)

The discovery of evil and guilt

As abstract thought and symbolic systems grew in importance, which in turn may have boosted intellectual capacity, dissocial individuals soon learned to pretend to hold certain positions, or to create fake facts. For other group members, it then became important to be able to recognize and condemn acts of deception (Hamilton 1975). The competition between those who cheated and those who detected the scams fuelled the further development of mental faculties. At a certain point, human beings' growing intelligence led to the loss of their 'innocence'. Knowledge of oneself and of one's own mortality (i.e. awareness of the future and of death), but also the discovery of ignorance (in the sense of an awareness that there are many connections, but that one does not understand them) led to an increasing sense of uncertainty. This uncertainty and corresponding insecurity may have been compensated to some extent by spirituality and religious behaviour within the framework of emerging moral systems. The recognition of death, and the identification with other living creatures through the recognition of kinship, extended empathy among hunters, who now incurred a sense of guilt for inflicting on animals what they themselves did not wish to suffer — violent death. This too may have been a source of spiritual and ritualistic behaviours, and another of the roots of art. Many of the oldest cave paintings depict hunting scenes and prey. Even in modern hunting, customs are maintained which express respect and honour for the killed animal (for example, the German tradition of 'Strecke legen' ('laying out'), where killed animals are displayed in a ritualised way).

The discovery of the future

Even if people's ideas about the future regularly lead to false predictions (Gilbert & Wilson 2007), it has become commonplace that ideas about an imagined future produce concrete consequences for current decisions and events. Future-oriented purposefulness and a complex, planned way of proceeding are typical human traits that promote both sustainable and unsustainable behaviour.

Sociality and self-reflection lead to self-taming

Increasing sociality and self-reflection brought about the self-taming, indeed the civilization, of what was an often violent primate, man. Early state formation was also the consequence of violence in tribal societies that repeatedly attacked each other (Fukuyama 2012). Up to a certain point at least, it is more bearable for individuals to be suppressed by hierarchical leaders than to live in fear of murderous hordes. The exercise of authority within the group at least promotes efficiency, through the separation of roles, and suppresses conflicts. In fact, evidence shows that the proclivity to peace also increased in the course of cultural development and the formation of ever more complex social systems (Gómez et al. 2016). Never in the history of humanity has it been as unlikely as it is today for individuals to suffer violent death. In general, the ability to plan with confidence, the security of basic supplies, health and life expectancy have all improved rapidly, especially in recent times. However, the way human beings are now enclosed in increasingly self-regulating social, fast-growing systems was bought at considerable cost: namely, at the expense of the proper functioning of the ecosystem (▹ Chapter 3.1).

From biological to cultural evolution, and the growth of human power

Cultural evolution

For a long time, human development depended solely on progress in biological evolution. This changed as social systems, composed of individuals capable of reflection who communicated with each other, intensified and took forward the cultural evolution which enabled survival techniques to be learned and passed on without individuals having to re-invent all the details of those techniques in their own lives. Complex social systems are characterized by emergent properties (▹ Chapter 1.3) — especially by culture, including technological progress, as well as future-oriented and systematic-planning thinking — which for humans simply brushed aside certain ecological-evolutionary restrictions that apply to other animal species.

Humans spread across the world

Particularly important was the capacity of human beings to spread very rapidly across different ecosystem boundaries. Recent *Homo sapiens* fossil finds suggest the early colonization of not only Africa but also Southern Europe. The most notable difference from other animal species, which applied to several human species *(Homo erectus, Homo neanderthalensis, Homo sapiens)*, was their ability to abruptly break out of their inherited ecological niche, that of the opportunistic hunter-gatherer of the savannas. The use of clothing, fire, tools and weapons enabled *Homo sapiens*, above all, to spread in a relatively short time into almost all biomes of the Earth — from hot deserts and oceanic islands to polar areas and high mountain regions. 300,000 years ago, they had already spread throughout much of Africa (Hublin et al. 2017). The appearance of *Homo sapiens* on other continents had far-reaching consequences for the ecosystems they colonized. Superior technology seems to have led to the displacement of other human species, such as the Neanderthals, which had been living in Europe until *Homo sapiens* arrived, probably 80,000 years ago. The immigrants probably also contributed to the eradication of large huntable species (Barnosky et al. 2004; Lorenzen et al. 2011).

Humans become landowners, and develop an 'ethic of honour'

Following the loss of nutritional resources from hunting, modern humans (frequently convergently on different continents) succeeded in developing new sources of plant food and domesticating several animal species during the Neolithic Revolution. The development of agriculture increased the energetic efficiency of social systems. More food energy could be provided to more people with less effort and less manpower. Those who no longer had to devote themselves to agriculture could invest their newly freed intellectual capacities in the construction

The noble savage, or: was it no better in the past?

Ever since the Enlightenment, controversy has raged over whether 'native peoples' still lived in harmony with nature and whether the destruction of our environment and unsustainable behaviour are therefore the products of culture. The German sociobiologist Eckhardt Voland emphatically refutes the hypothesis of a sustainability worldview among traditional ('indigenous') peoples (Voland 2006). Voland cites examples of over-exploitation which he believes demonstrate that even among today's traditional lifestyle peoples, such as the Maya, Yanomami, Piru and others, there is no such thing as a 'sustainable' approach, and that they use everything they can without regard to conservation. He offers a series of examples in which they have also destroyed the basis for their own subsistence. Citing Ridley in support, Voland considers 'serious evidence for the ecologically "noble savage" …' to be very rare. The idea of the 'noble savage' who consciously and deliberately protects nature is modern romanticism, he argues. The legendary stance roughly expressed as 'forgive me, stag, for the fact that I had to kill you' comes, he says, from films. And even if such a ritual did exist, says Voland, 'the stag was dead in any case'. Voland's laconic tone alone should raise the suspicion that the intention here is to prove a point: traditional peoples were not much better than us… ! In contrast, Diamond (2006) and Roszak (1994) report many plausible examples of the sustainable use of natural resources among various traditional peoples, and even corresponding religious codes of conduct.

The sustainability philosopher Meyer-Abich, however, points to something important: the rituals of native peoples at least prevented or reduced overuse, he argues. If, in an animistic culture, a young couple wanted to cut down a tree in order to build a house, they had to justify this to the tree on the grounds of necessity. 'The justification, however, helped all the trees, since to fell them all would have been inexcusable' (Meyer-Abich 1987, p. 73). The green branch that hunters in our culture put in the mouth of a slain deer is also a remnant of a similar ritual of thanks. Such rituals at least demonstrate that people connected with nature know that nature gives them something, and that this brings with it reciprocal inner feelings of obligation, responsibility and gratitude. The Christian ritual of giving fruit or food to needy people at the harvest or thanksgiving festival probably also arose originally from the inner desire to give something back (to someone or somewhere) when given something by Nature (here understood as divine). It is important to bear in mind the following:

1. Native peoples lived in a world in which everything had a soul and a being, and which was given meaning and purpose by an all-embracing world spirit; that is, in a systemic whole — a mindset which, because it was based in the systemic, tended to favour and support sustainability (see also Haskell 2017).

2. The psychological component of human sensibility towards nature influences our actions: emotionally sensitive hunters today still feel the need to somehow exonerate or unburden themselves after killing an animal. European hunting rituals are remnants of this.

3. Voland does not mention that the studies he cites were undertaken at a time when industrial society was already affecting the jungle peoples, and industrial thinking had probably already penetrated their mindset. The examples given by Diamond (2006) and Roszak (1994) contradict Voland's statements. Native mythologies were created out of an ongoing experience of nature, repeated over many thousands of instances, that entailed an understanding of carrying capacity and its cyclical nature. They took only as much as was needed, and nature was mostly content to give it (see Jung 2006).

Similarly not mentioned is the importance of sociality and the development of worldviews (which were almost exclusively sustainable, encompassing circulation, dependency, etc.) and myths in general. The animistic myths that have sustained prehistoric societies for tens of thousands of years are characterized by a spirit of co-responsibility (the individual as part of the whole) and by a summons to dialogue with Nature or its spirits, to giving and taking (sacrificial rituals, restrictive rites, prohibitions). It is therefore questionable whether sustainable action and thinking has anything at all to do with the moral categories of 'good' and 'evil'.

of ever more complex ordered social systems, which were becoming necessary in light of population growth. (Access to) land ownership became a central dimension of human economic activity; land theft and territorial invasion emerged as novel forms of injustice and created new grounds for conflict. An 'ethic of honour' arose, which was often underpinned by religion (Sumser 2016).

Development as emancipation from the chains of local ecosystems

Increasing population density, through accelerating cultural exchange, resulted in ever more diverse occupations and different cultural techniques, which, in turn, have benefitted human and community well-being. Important achievements that further accelerated this development, via positive feedback loops, included not least the development of writing and mathematical systems, and later the systematic sciences. Through the manipulation and transformation of ecosystems, humans have succeeded in purposively altering their living conditions — the history of ecosystem management began; human beings seemed to raise themselves above nature (▻ Chapter 3.1). The pursuit of development was the attempt to cast off the shackles of local ecosystems. Cultural techniques such as religion helped people to deal with emerging contradictions. For a long time, it was possible to ignore negative feedback from over-exploited ecosystems, as people were sufficiently adaptable to find or open up new resources and areas of the world. Nevertheless, it is undoubtedly the case that for human beings the consequences of their own actions on the ecosystem became a selection factor far sooner than was the case with other species (Kivinen & Piiroinen 2018).

From revolution to revolution: ever faster, more, bigger and better

System evolution and innovation

Evolution within the global ecosystem is always open-ended; it is a complex systemic process. By means of particular mechanisms — such as mutations in the case of species, a re-combination of components in ecosystems, or technological discoveries and inventions in social systems — systemic units bring about an innovation which, in interaction with other systems, leads to either more or less efficiency, resilience or adaptability, and can thus determine the sustainable continuation or failure of a system (▻ Chapter 1.3). Innovations often lead to major leaps forward in the progressive development of systems or of their performance, or of their geographic distribution. Such leaps also mark the biological and cultural evolution of human beings — some are often called 'revolutions'. The development of increasingly complex and successful social systems is leading to a significant acceleration in such groundbreaking innovations and revolutions.

Revolutions in the cultural evolution of human beings

The first revolutions on the path towards our becoming what we are today were associated, many millions of years ago, with (among other things) the adoption of an upright gait and with brain development. Then came (even before *Homo sapiens*), a few million years ago, the systematic use of tools and the instrumental use of fire for hunting, cooking, heating and making things. This led to a change in nutritional physiology, among other things. The development of agriculture around 10,000 to 15,000 years ago paved the way for population growth and the emergence of advanced civilizations, but it also set the stage for changes in the Earth's plant cover and for our conceptual decoupling from the ecosystem. Rising population densities fuelled cultural evolution, the ever-increasing use of the human intellect, and the emergence of science. Through feedback, ever greater population densities became possible. Finally came urbanization and, from the mid-18th century, the Industrial Revolution, with corresponding consequences for energy use, manufacturing and trade, transport, mobility and human prosperity. The development of the sciences also included the emergence of theories, ideas and discourses which exerted growing influence on world events and above all on further cultural evolution. Of particular importance in this context are the efforts to construe, from logical descriptive systems (theories), normative principles for human coexistence and the relationship between humans and the world.

Anonymous societies and states lead to alienation, over-exploitation and conflicts

With the emergence of more anonymous and more complex societies and states in which individuals could no longer personally know all other individuals, new sociological and cultural dimensions came into play; group territories and associated tribal knowledge diminished, the suppression and possession of other human beings arose, and self-determination over the use of what was produced and what was won from nature was lost as a result of alienated structures. No longer were the needs of the community alone decisive, but instead the luxury and comfort enjoyed by the rulers. This kind of anonymous social cultural development required an intensified exploitation of nature. Wars proliferated, and with them all kinds of other events, in an ever-growing confusion and complexity, as a result of the size of the population (state formation) and the growth of cities, so that systems of rule also became divorced from practical experience and concerns. Shortages resulting from over-exploitation were balanced out within human population systems (through armed raids and territorial conquests). Both the growth of complexity and the tendency of powerful states to expand their spheres of influence (that is, access to resources) by military means continue unabated to this day. The consequent impossibility of comprehending the full scale of human impacts on nature, due especially to human existential alienation from the natural and social foundations of subsistence through industrialization and capital concentration, meant that the scope for direct feedback — that is, the experience of the consequences of one's own actions — has continually diminished. Environmental impacts have arisen that do not allow sensory feedback because they are not perceptible (CO_2, O_3, dioxin, climate change, etc.).

The cybernetic-informational revolution

However, the increasing difficulty of maintaining an overview of global events is still some way from reaching its peak. Just a few decades ago, a cybernetic-informational revolution took place based on the automated processing of information. This caused the explosion of knowledge we are currently experiencing as well as the emergence of an Internet-based global knowledge or global culture, aided by globalized initiatives and companies such as Google and Wikipedia. This was followed just a few years ago by virtual social networks and ever more global real-time communication possibilities. In the last few years, it has become clear that human-computer interaction is increasingly influencing people's behaviour by automatically filtering and manipulating the way they handle information. In the meantime, the Internet has given us so-called social bots, virtual communication partners that try to influence our decisions. And the visions of artificial intelligence — at least at the level of self-referential algorithms that make decisions without needing people, and of the Internet of Things interacting with each other without human intervention — are almost with us. They pose unimaginable new challenges for our moral and ethical systems, which of course have to grow in line with our achievements and capabilities. Critical scientists point out the dangers of digitalization for mental and physical health — that it is testing human nature to its limits.

Faith in technology, alienation from nature, hybris

Humanity is rapidly creating new kinds of social systems and technologies that are supposed to increase its prosperity and reduce its vulnerability to the hardships of nature. However, all technologies also have unintended and often unforeseeable consequences that need to be assessed and managed. It is clear that it is not the technologies themselves — whether printing, rocket propulsion or the Internet — which are good or evil, but the social systems that use them (without always being aware of the consequences that may follow), or the individuals who manipulate or exploit social systems and technologies, for their own

Figure 3: Living in artificial worlds and relationship disorder: Technical feasibility and artistic emancipation from natural forms can increase the degree to which urban people are disconnected from nature. The emotional connection to a 'disorderly' organic environment not created by man is lost. (Parliamentary buildings in Berlin, photo: P. Ibisch)

ends. The current proliferation of technological possibilities, especially with regard to digitalization and virtualization, is creating new instances of path dependency — that is, existing pathways of human development that cannot easily be abandoned. Above all, the incessant technological advances have led to a blind faith in technology, to ever greater alienation from nature and to an overestimation of humanity's capabilities that ignores the lessons of evolution (referred to as 'hybris' in Bateson 1985 [1972], p. 62 ff.). Art and architecture also reflect our gradual and increasing separation from nature and contribute to the fact that today a large proportion of people have fewer opportunities to experience nature (Fig. 3). In our air-conditioned dwellings, we still surround ourselves with elements of nature such as pets or ornamental plants, testimony to the remnants of a yearning for nature. But they also reaffirm our dominance, since their well-being depends on us (Fig. 4).

Figure 4: Phalaenopsis orchids in a hotel lobby as citations of nature in our urban habitat. People still yearn to have elements of nature all around; but our sense of the complexity and dynamics of ecosystems is being lost. (photo: P. Ibisch)

Is the world we have created now overwhelming us?

We now have to deal with an anthropogenic environment that is global and hyper-complex—and thus far beyond our Pleistocene perceptual horizon. It is a world we have ourselves created but which is increasingly overwhelming us. The many technological possibilities also mean that our social systems and the technological demands made on us by life require ever greater time resources. It is also becoming apparent that growing numbers of people in our modern, technological society are not at peace with themselves. A significant proportion of the population suffers from mental disorders such as anxiety, insomnia, addiction or depression (a total of 38 % of the EU population suffers from at least one of 27 identified disorders in any given year, Wittchen et al. 2011). Perceived and diagnosed levels of stress in so-called 'developed' countries such as the USA are steadily increasing, for various reasons (APA 2016; 2017a; 2017b)—a possible indication that the needs and skills set of *Homo sapiens* are being transgressed. This could also be related to the increase in drug use (not just in the USA), given that stress promotes drug dependence (Wall 2008). Electronic media represent an increasingly important source of stress (Kross et al. 2013).

Is the sustainability revolution on its way?

Humans as ambivalent beings

So what should we think of ourselves? Are we going to be able to effect the 'emergency stop' that is now ecologically indispensable, and the next revolution, the sustainability revolution, without throwing our societies into chaos through anti-humanistic action? It certainly looks as if we should not place our trust blindly in the 'human biological system'. From a genetic perspective, this system is optimized for life in non-anonymous tribal societies, for familiarity and personal relationships. Human behaviour is socially, culturally and biologically contingent — it results 'from a complex, dynamic and highly random interaction between environment, organism and genome' (Kösters 1993, p. 331). The biotic component is the least mutable in this case. Changes in the genome are relatively slow; we are, biologically speaking, still largely the hominid of the Pleistocene, whose path diverged from that of the chimpanzees and bonobos. We see in ourselves a great ambivalence, an inner strife resulting from the dual inheritance that makes us both aggressive-belligerent and empathic-peaceable, both destructive and creative-constructive. But our 'Pleistocene' genetic make-up has given us an amazing capacity for learning and above all for social cooperation and culture. These are central insights for our picture of ourselves, for our **view of human nature**.

Our view of human nature, our values and ethics as important conceptual foundations for sustainable living

Our view of human nature: The question of good and evil is a moral question. Morality is the basic grammar of human social behaviour, that is, the behaviour of people towards each other, or towards the group (Spitzer 2009 et al.). The roots of pro- or anti-social behaviour (e.g. one's predisposition towards giving help, justice, caring; Eibl-Eibesfeldt 1997; Blohm 2010 et al.) lie in our biological inheritance, but these predispositions can be individually suppressed or promoted at the modifying cultural-social level (education, indoctrination, normative pressure) or raised to the status of (formal or informal) norms or laws at the societal level. Morality is thus not in the first instance a category which applies to a love of nature or to environmentally friendly behaviour.

A *view of human nature* encompasses the main drivers of human behaviour and therefore also the scope for changing it. Thus, a conception rooted in learning theory or behaviourism assumes that in principle all behaviour must be learned from birth (there is also a variant based in milieu theory which holds that socialization is the sole cause of all behaviour). This has proved to be reductionist and overly one-sided. From an interdisciplinary point of view (biology, psychology, sociology, philosophy), the view of the human being as a biopsychosocial unit is today considered the most complete (von Hayek 1979; Ciompi 1999 et al.; Wessel 2015). This holds that all behaviour is comprised of a basic biological disposition, a cultural-social modification thereof and individual psychological decisions — both rational and emotional (▹ Chapter 2.2). This reflects systemic interaction at different levels, and also corresponds to the fundamental interdisciplinarity required for sustainable development. In practice, everything depends on whether we view humans as controlled by others, or controlled from the outside (learning and socialization dogma), or alternatively as self-directed or controlled from within (personality), and on the attitude we take towards human beings in accordance with that view (Meyer-Abich 2012). The way we seek to exert political influence on them also depends on that view.

Values are perhaps the most complex psychological dimension because they are always a conglomeration of emotion, reason and behavioural disposition. They arise 'passivistically' through the experience of things, people, situations (Joas 1999; 2006). Values form the core of intrinsic motivation, i.e. of the personality. They are of particular importance for sustainable development, as they reach out beyond the present; they constitute a guideline for action even in new, unpredictable situations. It is possible to distinguish between 'lower' and 'higher' values. Having food every day, having a roof over one's head and security are certainly very practical fundamental values. The basis of such values, but also of social ones, is given

by natural needs, as represented in Maslow's hierarchy of needs (physiological, safety, social relationships, esteem, self-actualization, Maslow 1943, 2016 [1954]; see also Heinrichs 2007, p. 182 ff.). The need for contact with nature has now been added (Gebhard 2009; Eser 2012; Skidelsky & Skidelsky 2013; Nussbaum 2014 [1999]). According to Skidelsky & Skidelsky (2013), basic human needs (for basic goods) must be clearly distinguished from wants, which can be artificially produced through indoctrination (advertising, unconscious learning, etc.) (▷ Chapter 2.2).
Certain values can be elevated at the societal level to the status of social, ethical or legal norms (e.g. the statutory obligation to provide emergency aid, caring for family members, or the prohibition against killing). Legal and social norms, in contrast to values, do not require an inner conviction on the part of individuals, and in fact can if necessary be enforced on them against their convictions. With regard to sustainable development, in the interests of the conceptual transformation now being widely called for, it might be worth asking whether the values currently propagated in politics and the public sphere are actually fit for such a purpose. The values of profit maximization, growth and perhaps also fundamental competition are unlikely to be found conducive to this goal (see Bateson 1970; Meadows & Seiler 2005). An appreciation of nature, however, requires emotional experience (Gebhard 2009; Jung 2012 and others).

Ethics must, if possible, be comprehensible to everyone and able to hold out the prospect of a good life. For example, people striving for sustainable development consider it unethical to ruthlessly decimate natural diversity (including resources) through the maximization of economic interests and excessive consumerism, thereby leaving behind a poorer world to our descendants. Immanuel Kant formulated as a (socio-)ethical maxim (the 'categorical imperative') that one should act as one would like others to do (Kant 2011 [1788]). Since this does not address the future consequences of one's own actions, Hans Jonas saw the need for a new 'ecological imperative': 'Act so that the effects of your action are not destructive for the future possibility of such life' (1984, p. 11). Building on this, the maxims identified in the wake of the 1992 UN Conference in Rio as promoting sustainable development were (distributional) justice, connectedness, moderation/sufficiency, preservation and enhancement of diversity and humility/frugality (Brown & Quiblier 1994). The 'intrinsic value of nature' (Gorke 2010) is therefore part of an ethics of sustainability. In stark contrast to this, the deadly sins identified by Pope Gregory I in the 7th century — greed/avarice, gluttony, envy, pride, lust, wrath (aggression) and sloth (acedia, or 'coldness of feeling') — have become socially acceptable in today's unsustainable society as a result of an unquestioned and unfettered liberality.

Counting on social intelligence

In modern society, our evolved pro-social disposition towards conformity and our openness to indoctrination (Lorenz 1973) has a flip side: we can be seduced by the power of opinion. For that reason, the idea of clearly distinguishing human (basic) needs from desires that can be artificially created through advertising, ready availability, fashion, etc. (Skidelsky & Skidelsky 2013) is a promising one for the sustainability debate. We should rely above all on our social intelligence, which has enabled us to form complex social systems that help us to keep our immediate instincts in check and to tame the chimpanzee-like aggression within us. This applies to groups of a manageable size. However, it is quite understandable that reciprocal altruism (which in tribal societies means, 'if I do something for others, I will probably get something from them in return') has weak biological drivers in an anonymous global society, because the corresponding biological behavioural disposition originally arose within tribal societies where, firstly, everyone knew each other (connectedness through acquaintance) and, secondly, everyone lived in a particular place, so that regular personal encounters were unavoidable. This is why others advocate not merely arguing on idealistic grounds that there is a moral obligation towards altruism (self-sacrifice for others), but also emphasizing the link between altruistic behaviour and beneficial 'rewards', i.e. the promotion of one's own interests (narcissistic gratification) (Mohrs 2002). The task of striking a balance between altruism and egoism with regard to sustainability is certainly a tricky one.

Civilizability and political systems

Social systems and cultural evolution enable us to make the best possible use of our ability to adapt and to learn, and of our 'civilizability'. However, our socio-culturally moderated reason is fickle; it is not firmly attached, and a relapse into barbarism is therefore possible at any time (see Welzer 2008). For that reason, these social systems have to be carefully nursed. They can arise spontaneously and become self-organizing, but some are also deliberately created to serve a purpose. The history of humankind shows the development of ever more complex social systems on ever new levels nested within each other. From the level of simple rival hordes, socio-political evolution led via feudalism and empires to states and organizations marked by high levels of solidarity and cooperation. The more cooperatively the different social systems interact, and the better the balance between vertical and horizontal complexity (▹ Chapter 1.3), the more opportunities there are for rules and policies that promote sustainability. So even though it is possible for complex democratic systems to organize the unsustainable and radically marketized exploitation of resources without regard for the big picture (as can be observed in the current Western states), it is pluralistic systems that remain far more likely to have regard for many more aspects and to develop the appropriate checks and balances[1] (▹ Chapter 4.2). It is no coincidence that movements for democratization and ecologism often go together. It may be concluded from the above that a high degree of self-determination within territorial groups (municipalities) would be a good model for the future of social organization.

States – a curse and a (potential) blessing

In ideal state social systems, on the other hand, there is the potential for arriving at an appropriate balance between the creation of individual freedom and the maintenance of the integrity and functionality of the higher-order social and environmental systems. This balance, however, does not arise automatically, in and of itself; rather, it must be permanently supported by a narrative (i.e. meaningful and orientational story motifs). All experience shows that higher-order systems which organize morality and ethics (such as, traditionally, the churches) can be very powerful. They can function as counterweights to a one-sided focus on selfish profit maximization, organizing approaches based on collective altruism and acting as a 'social glue'. The creation and maintenance of appropriate contemporary systems that can bring out and promote the best in us is therefore indispensable. In contrast, the prevailing radicalized market ideology (Randers & Maxton 2016), which assumes that the individual satisfaction of needs and wants in free markets is the best engine of development, promotes our worst qualities: avarice, profit maximization and the accumulation of status symbols, regardless of the cost (to others).

1 E.g. state separation of powers, bicameral parliaments, decentralization/federal structures, appellate jurisdiction systems.

We humans are creatures with two faces—accordingly, we will only achieve sustainability if we acknowledge our evolutionary heritage and develop a realistic view of human nature. Skills for sustainability—such as foresight and proactive behaviour, i.e. the ability to develop and incorporate scenarios into current decisions, and the ability generally to think in a fair and moral way—are developed when people are able to live in free, balanced and just systems—in other words, when they are able to be truly human. Human beings had developed a deep insight into the structure of nature and the consequences of human intervention by the time of the Neolithic Age, if not before. This created the possibility of living sustainably in and with nature through awareness of natural processes. This provides us with a key thesis:

Our capacity for sustainability

Humans are naturally endowed with behaviours and thought patterns that enable careful and 'sustainable' ways of interacting with their equals and the environment, they are fundamentally 'capable of sustainability', especially if their external living conditions allow it. Their pronounced capacity for cooperation belongs as much to the set of typical human qualities as does their ability to empathize, the capacity to make external reality into an object of thought and to develop theories, and their pronounced 'self-awareness', the 'animation of the mind' (Scheler 1978) or the resulting capacity for 'self-objectification' (Gritschneder 2005, p. 112 f.).

People are able to inhibit their urges and to make objective, informed decisions. They can therefore also develop and apply an ecosystem ethics (▷ Chapter 4.1). At the same time, it is important to bear in mind a well-founded environmental psychological observation, namely that rational insight (knowledge) alone produces neither a change in attitude nor a change in behaviour (see Jung 2015) (▷ Chapter 2.1). Without motivation firmly based on emotion and fundamental beliefs (intrinsic motivation), the corresponding ethics and behaviour will be on a weak footing.

The solution lies in our nature. We just have to recognize it—and cultivate it. Is the sustainability revolution on its way? Only if we can think it into being, if we want it, and if we start it.

What motivates people to sustainability-oriented action?

2.2

Heike Molitor

What makes people behave in an environmentally friendly or sustainable way? Often, we assume that they only have to be informed: once they understand that their own behaviour is not optimal, they will then, as self-determined and rational agents, make changes. Reality shows us that this assumption is wrong. Many people are familiar with our current environmental problems, such as pollution, resource depletion and climate change. Yet not everyone consequently behaves in an environmentally friendly or sustainable way. Nationwide surveys repeatedly confirm that Germans have a well-developed environmental awareness. But this simply does not automatically translate into the desired behaviour. How can the findings of (environmental) psychology be used to promote environmentally friendly or sustainable behaviour?

What motivates people to act?

A glance at advertising

In order to understand what motivates us to act, a digression into the world of advertising will prove helpful. In the advertising of products and ideas, images of nature or natural elements are commonly used, as in the example here of an advertisement for beer with an image of a large lake with an island alongside the slogan 'a pearl of nature'. This advertising campaign, which is supposed to convey qualities such as purity, innocence, clarity, authenticity — all adding up to a 'pearl of nature' — has been running since 1994. The picture appears to be the epitome of nature — an idyll — although it is in fact a dam, and not at all untouched, idyllic nature. Nature is not actually experienced, but the image prompts feelings similar to those we experience on encountering nature directly, and thus positive emotions (Spanier 2008).

Addressing emotions

Many depictions of nature in advertising symbolize a form of idyll or of sublime grandeur or majesty. The stylistic device of grandeur — in contrast to that of the idyll — addresses target groups that 'can be reached by means of emotionalised charges of freedom, adventure, independence, power, etc.' (Spanier 2008, p. 38). Alongside the factual level, which delivers information about the beer, emotions are addressed through symbolic depictions of nature.

Advertising transmits implicit calls for action via positive emotions. The aim is to induce people to buy a product. Useful information or well-founded arguments are rare because 'an idea does not achieve cult status through better arguments, but because it reaches hearts' (Röchert

Figure 1:
The Feldberg Lake District – a possible motif for advertising (photo: P. Ibisch)

2008, p. 11). How can these insights be carried over into the sustainability context? How is it possible to do serious work in nature conservation without relying solely on emotionally charged messages? What answers can be found using (environmental) psychology? Psychological concepts are used to examine **the human-environment relationship** on an individual level. The explanatory power of psychology, and thus of this chapter, is therefore limited to the individual and does not apply to society at large, which is the province of environmental sociology.

The role of psychology in the interaction between humans and the environment

'Psychology deals with the experience and behaviour of human beings in their world' (Bottenberg 1996, p. 2).

'Environmental psychology is a sub-discipline of psychology that deals with the interactions between humans and their physical and socio-cultural environments. Environmental psychology thus takes as its subject matter the influence of the physical and socio-cultural environment on the experience and behaviour of individuals and groups, as well as human behaviour that—intentionally or unintentionally—causes changes in the environment' (Hellbrück & Kals 2012, p. 13).

Theories of environmental psychology in Germany focus on human beings and their feelings, attitudes, values, opinions and the like (Herkner 1993, p. 22). These are phenomena which can be accessed only by the individual person—subjectively—(Bottenberg 1996, p. 2) and which only become recognizable when they are the subject of communications from that individual.

Research into environmental awareness

A central focus of environmental psychology is environmental awareness research. This builds on socio-psychological attitude research into the following questions (Rosch & Frey 1994, p. 296):

1. What are (social) attitudes?
2. How can attitudes be changed?
3. What is the relationship between attitudes and behaviour?

Environmental awareness as an attitudinal construct

Attitudes can be classified on a spectrum from superficially held opinions to life-determining basic convictions. Depending on how firmly they are anchored, they can be harder or easier to change. In the early 1970s, based on the work of Maloney & Ward (1973), a research tradition emerged in the social sciences which regards environmental awareness as an individual disposition and an attitudinal construct (Fuhrer & Wölfing 1997, p. 21; Homburg & Matthies 1998, p. 49). As a rule, the term 'environmental awareness' can be understood in either a broad or a narrow sense, and includes both individual and social aspects.

Range of meaning of 'environmental awareness'

According to Spada, the following characteristics are decisive for environmental awareness: environmental knowledge; environmental perception; environmental value orientation; environmental behavioural intentions; and manifest environmental behaviour. Depending on the context in which it is used, the term 'environmental awareness' is interpreted as having a variable range of meanings. The narrow interpretation includes the dimension of perception (experience and personal involvement). In the medium range, interpretation, environmental knowledge, value ori-

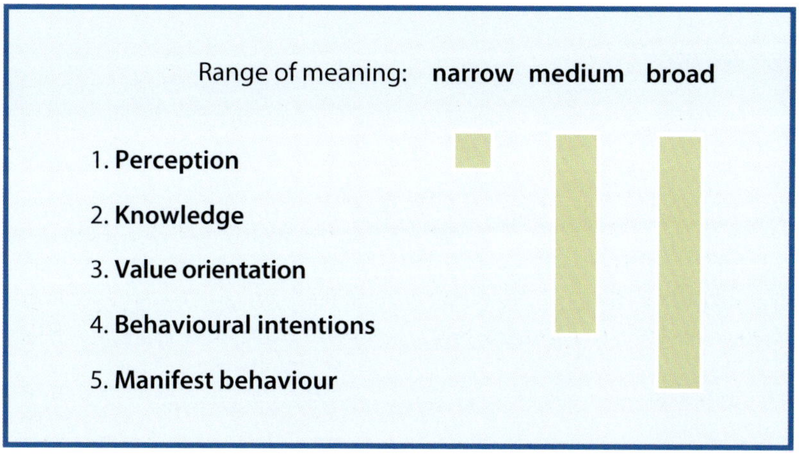

Figure 2: Range of meaning of 'environmental awareness' (see Spada 1990, p. 623)

entation and behavioural intention are also included, and in the broad range, manifest (or actual) behaviour is also taken into account (Spada 1990, p. 623).

The term 'environmental awareness' is thus not clearly defined and needs to be made more concrete. Understanding the links between attitudes, knowledge, value orientation and other factors is of considerable importance for creating effective campaigns or projects that aim to change behaviour (Breit & Eckensberger 1998).

Case study: How aware are Germans of the environment or sustainability?

Representative study on environmental awareness

Since 1996, a representative sample survey carried out every two years by the German Federal Environmental Agency *(Umweltbundesamt)* has provided information on awareness and behaviour in Germany with regard to environmental and sustainability issues. The German-speaking resident population aged 14 and over is questioned about its thinking and behaviour in relation to sustainability, so that 'the future needs and wants of different population groups can inform a community-based, participative, future-oriented and sustainable policy' (Umweltbundesamt 2017c, p. 11).

Protection of the environment and the climate are important issues

Since 2000, when asked to identify the current issues of most concern to them, about 20 % of respondents have named environmental and climate protection. Depending on the immediately topical situation, immigration, migration, crime, peace and security were the most pressing issues in 2016 (Umweltbundesamt 2017c). In 2014, social security and eco-

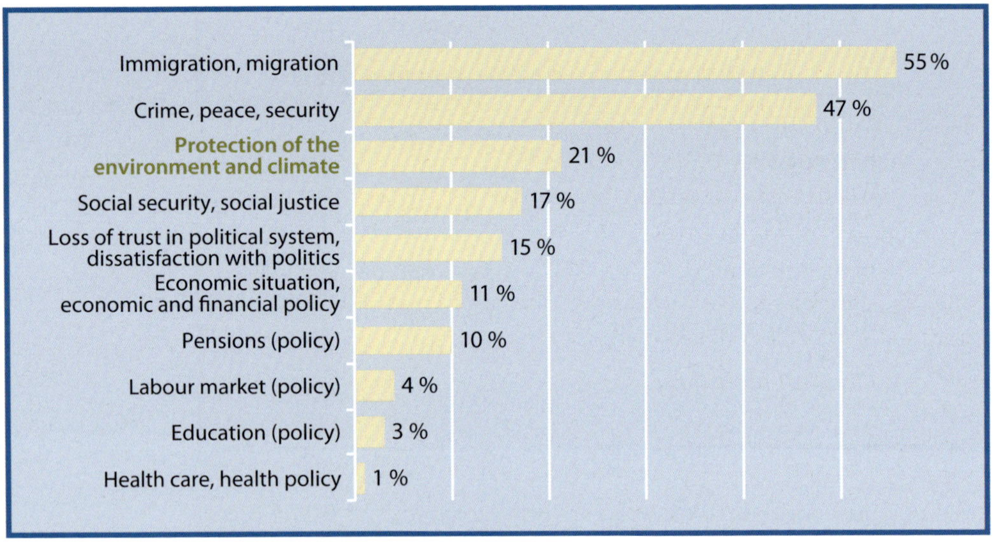

Figure 3: The most important problems of our time, as identified by respondents in the 2016 environmental awareness study conducted by the German Environment Agency (Umweltbundesamt 2017a, p. 15). Wording of the question: What do you think are the main problems that our country faces today? Please enter here the two most important problems from your point of view. Open question, maximum of two answers possible.

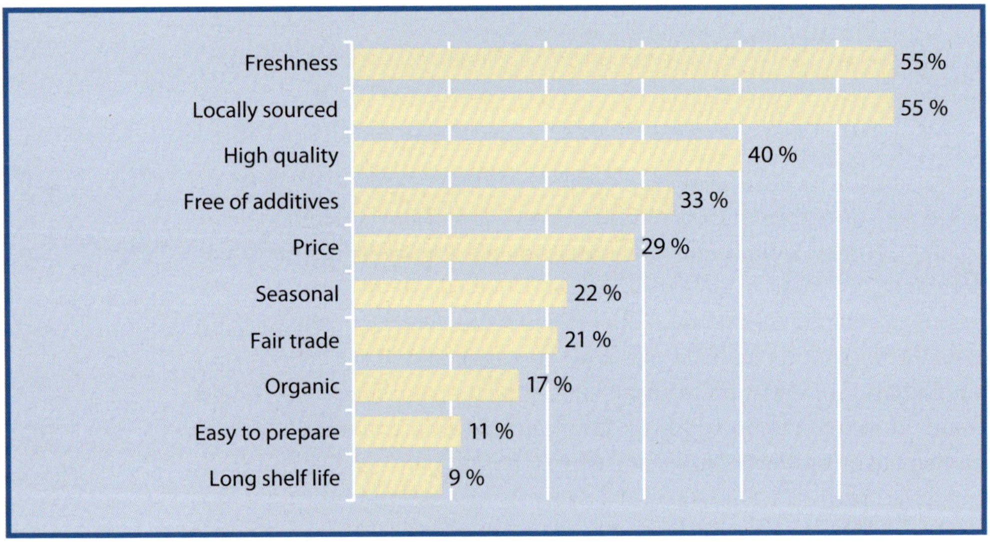

Figure 4: Selection criteria when purchasing food; German Environment Agency environmental awareness study 2016 (Umweltbundesamt 2017a, p. 54). Wording of the question: What is particularly important to you when choosing food? Please select the three most important criteria.

nomic and financial policy remained the leading issues (Umweltbundesamt 2015a).

Sourcing fresh local produce

Looking at behaviour with immediate environmental impact such as people's food purchases, the Environmental Awareness Study 2016 shows that criteria such as freshness and local origin play a decisive role. Organic and fair trade produce are important for one in five when making their purchasing decisions.

Types of purchaser

Three types of purchasers were identified on the basis of the data collected, and for each of these types, specific criteria were particularly important (Umweltbundesamt 2017a):

1. For just over half of the respondents (54 %), very broad and practical issues such as low price, ease of preparation and long shelf life are particularly important. Younger age groups and men (59 %) are over-represented in this group.

2. Roughly one fifth of respondents (18 %) are discerning customers who focus on quality, and especially on freshness and quality in food. This type usually belongs to middle-class, young social milieus with a higher educational background (especially among men).

3. Just under a third (27 %) of the respondents are happy to be described as sustainability-oriented and place particular value on food that is locally sourced, seasonal, organic, fair trade and without additives. This type of shopper is more likely to be female (35 %) and in the age group 50+.

Change for sustainability

The results show that, over time, dietary behaviour is slowly moving towards sustainability. In particular, the local sourcing of products and animal welfare are becoming increasingly important criteria for food purchasing behaviour. Meat consumption is slightly declining overall. Avoiding food waste is a new issue at the attitudinal level. But sustainable behaviour is still strongly related to social, socio-cultural and gender-specific factors. Low-income households are more likely to buy cheap items; young people consume more meat on average; and women tend to behave more sustainably than men (Umweltbundesamt 2017a).

At this point it becomes clear that attitudes and behaviour with regard to sustainability are very diverse and depend on social factors. The inclusion of the social dimension is therefore of particular importance for the transformation of society towards a more sustainable future. It is equally clear that sustainability is becoming increasingly important and that measures that strengthen environmentally friendly or sustainable behaviour enjoy the support of a majority of the public. The behavioural interventions that follow, and which all aim to promote such behaviour, should be evaluated against this background. These measures are intended to bolster existing pro-ecological attitudes.

How can environmentally friendly or sustainable behaviour be promoted? Behavioural interventions

In order to motivate people to behave in an environmentally friendly way, measures (interventions) can be taken that make such behaviour more likely. A distinction is made here between situational and ▷ **norm-centred** approaches or strategies. Situational strategies are concrete measures initiated either before (antecedent) or after (consequent) the desired behaviour (Schahn 2008).

Norms are something like 'a guideline, a rule (or a) guiding, binding principle, a measure of value' (Dorsch et al. 1987, p. 449). 'The aggregation of the norms determining judgement and behaviour' is called morality (ibid., p. 425) (▷ Chapter 2.1).

Figure 5: Situational antecedent and consistent behavioural determinants (with acknowledgements to Schahn 1993)

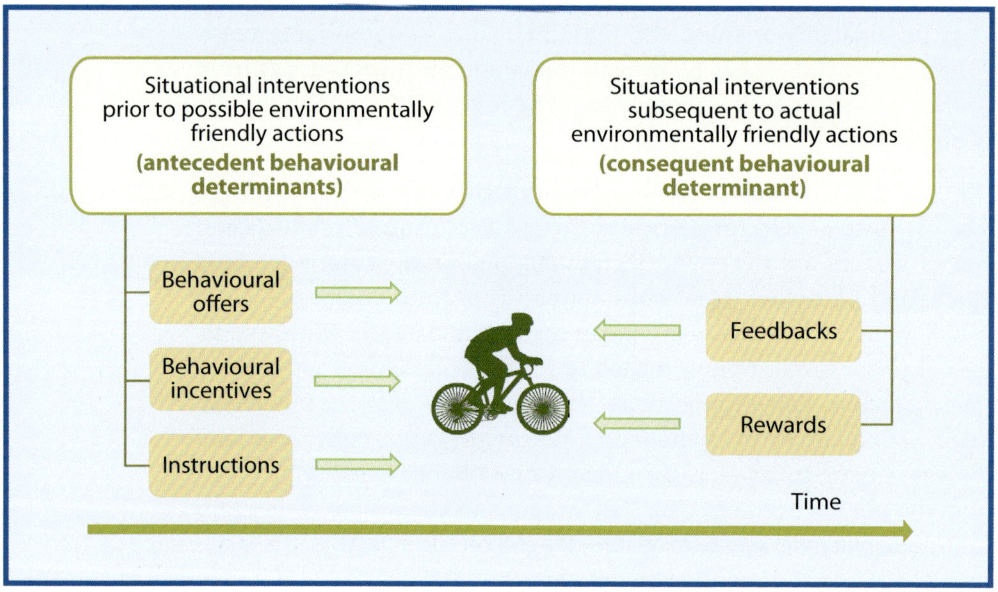

Instructions/ reminders

Instructions and reminders are more effective the more immediate and behaviourally specific they are, provided that the person is motivated to undertake the desired behaviour (Scheuthle et al. 2010). For example, an eye-catching sign on the door of a room saying 'lights off?' can ensure that the light is indeed switched off. The same sign attached to the front door or in the street, however, is less effective, because the message and the 'crime scene' are too far apart (ibid.). In the case of messages or instructions, therefore, the effectiveness is dependent on the placement of the sign and the location where the behaviour is to be carried out.

Behavioural offers

For someone to exercise a behavioural choice, a corresponding behavioural offer must be present. For example, for them to switch from using the car to using the bus, there must be a functioning public transport service. If someone lives in the countryside without a local bus service, it will not be possible for them to make an environmentally friendly mobility choice. Waste separation also requires a behavioural offer. If it is not possible to separate waste for recycling, it will not happen.

Figure 6: Multi-option waste disposal bin – a behavioural offer for waste separation (photo: J. Geyer)

Behavioural incentives

Behavioural incentives are measures that increase the likelihood of a desired action by means of external incentives. Student identity cards with an integrated season ticket for public transport are an example of this approach. These cards enable the use of public transport free of charge. Bus and train travel thus generate no additional costs. Such measures are referred to as situation-specific if they are tailored to the desired behaviour in order to strengthen or initiate it.

Figure 7: Eberswalde GreenCard from the Eberswalde University for Sustainable Development – student ID card with integrated season ticket for public transport (Design: H.-J. Rafalski)

Rewards

Rewards take effect when a desired behaviour has been demonstrated: for example, the bonus system in statutory health insurance schemes in the German health system, whereby fees for preventive health treatment are largely reimbursed.

Feedback

Feedback can take the form, for example, of informing users after an intervention about the money they have saved. If the saving is dependent on them achieving a certain behavioural threshold, it can act as a further incentive to behave in an even more environmentally friendly way.

The effects of external incentives

External incentives such as rewards can have the disadvantage that the desired behaviour only lasts as long as the external incentive applies. If the incentive ceases to apply, the desired behaviour may also cease. However, the opposite can also happen. If the new desired behaviour is associated with an unanticipated positive outcome, then it will probably continue even if the external incentive is discontinued (Scheuthle et al. 2010).

Role models

Norm-centred strategies are based on the activation of already existing attitudes or norms within the target group (Schahn & Matthies 2008). Role models are an example of this strategy. Using people as role models can support or strengthen a specific behaviour in others. Role models demonstrate how that behaviour is actually practised, and are also associated with positive emotions linked to the desired behaviour. The more prominent the person, the stronger the role model effect. Ralph Caspers and Shary Reeves, presenters of the German children's TV programme *'Wissen macht ah!'*, are role models. They are Youth Ambassadors for the

UN Decade on Biodiversity, and use their celebrity status to promote the preservation of biodiversity (UN Decade of Biodiversity 2017).

With all of these measures, it is important that people still feel that they are free to make their own decisions. If the impression arises that a particular behaviour is being forced on them, then reactance may result,[1] meaning that the desired behaviour is deliberately withheld (Scheuthle et al. 2010).

'Behaviour setting' approach

Both situational and norm-centred strategies have in common that they can shape 'spaces' or influence situational behaviour. This so-called 'behaviour setting approach' was developed by Roger Barker (1903–1990) and referred primarily to behaviour in specific spaces, such as observable patterns of behaviour in a church or in a clubhouse. The specific setting influences interaction, communication and attitudes. This approach has proved to be successful in practice. Through concrete interventions such as behavioural options or incentives, spaces can be designed in such a way that environmentally friendly behaviour is more likely (Fuhrer 2010; Hellbrück & Kals 2012). In a university where, for example, green electricity is used, waste separation is carried out, water and energy are saved, and environmentally friendly building materials are used, the ecological behaviour of each individual tends to be better than in conventional spaces. A different space equates to a different setting, and behaviour inevitably changes as well.

How much 'energy' do I have to expend on new behavioural offers? A cost-benefit analysis undertaken by *Homo oeconomicus*

Homo oeconomicus

In this approach, human beings are viewed as *Homo oeconomicus*, who makes decisions based on the balance of costs against benefits, i.e. strictly rationally. Since individuals, according to this notion, act selfishly, they will weigh up the costs against the benefits and their own egotistical interests against the common good. Behaviour is thereby understood as being subject to preferences (the wishes and goals of the decision-maker) and restrictions (the scope for action). Only if sustainable behaviour pays off in economic terms, or results in an advantage for an individual or a group, will it actually be implemented (rational choice theories, Krol 2000, p. 18 ff.).

1 Reactance is a term used to describe the resistance to pressure exerted by someone else to limit one's options for action (Dorsch et al. 1987, p. 553).

Incentive structures seek to encourage people to behave in an environmentally friendly way; that is, the framework conditions are designed so that ecological behaviour brings the greatest individual benefit. Examples can be found in environmental policy, which aims to promote environmentally sound behaviour through market-based incentive structures (taxes and fees, for example). The environmental sociologist Diekmann criticizes this as a purely economic approach. In Diekmann's view, social incentives and intrinsic motivations are just as important, and he describes the view of human nature underlying this approach as *Homo oeconomicus* (Diekmann & Preisendörfer 2001).

Cost-benefit analysis approach

Diekmann & Preisendörfer used the cost-benefit analysis approach to study individual environmental behaviour and found that environmental behaviour is sector-specific. They emphasized that environmental behaviour can be very diverse, finding that it was often not considered across different problem areas (Diekmann & Preisendörfer 2001). People for whom waste separation is important do not automatically also run their households in an energy-efficient way. And people who shop in health food stores don't necessarily spurn long-distance travel.

High-cost/ low-cost areas

In areas associated with low personal costs or effort (low-cost areas), environmentally conscious behaviour is easier than in areas associated with relatively high personal costs or effort (high-cost areas) (Preisendörfer 2000, p. 79). According to Diekmann & Preisendörfer, energy and transport are high-cost areas in which the subjective effort (referred to here as 'costs') is relatively high and the behaviour is less environmentally conscious. *Shopping* and *waste* are associated with lower personal behavioural costs. This means that a sustainable environmental attitude

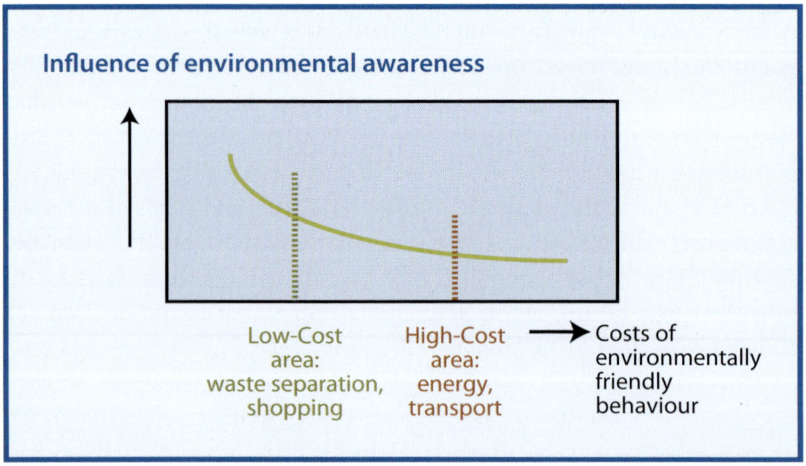

Figure 8: Cost-benefit approach, Diekmann and Preisendörfer (see Diekmann & Preisendörfer 2000, p. 84)

is easier to put into practice, and environmentally sound behaviour is therefore more likely (Diekmann & Preisendörfer 2001, see Fig. 8).

The recognition that knowledge is not the only driver of actions, and that attitudes are not solely responsible for behaviour, leads to the development of more complex models of causal relationships. Theoretical models are employed to describe the complex structure that determines behaviour. Very different models have been developed, emphasizing different factors and offering only approximations of reality.

The psychological explanation for sustainable behaviour

The psychological explanation for sustainable behaviour

In the following section, a more complex behavioural model will be proposed to explain human behaviour in the context of the environment or sustainability. It is based on the empirically tested models of the theory of planned behaviour (Ajzen 1991) and the norm-activation model (Schwartz & Howard 1981). This involves weighing up against each other, in any decision-making phase, various motivators such as the personal ecological norm, influences from social norms and the (behavioural) costs and benefits, leading in a given situation to the intention to behave either in an environmentally friendly or non-friendly way (Hamann et al. 2016).

The personal ecological norm is influenced by (Hamann et al. 2016):

1. awareness of the problem,
2. a sense of responsibility, and
3. experience of self-efficacy.

Awareness of the problem, sense of responsibility, experience of self-efficacy, social norms

Awareness of the problem refers to the perceived threat to the natural environment. This awareness does not directly lead to a change in behaviour with respect to the environment, and needs to be combined with other strategies. If people see themselves as responsible for solving environmental problems, they will be more willing to take action. Anyone who additionally feels responsible for the existence of environmental problems also has a higher willingness to act (sense of responsibility). ▹ **Self-efficacy** in this context refers to confidence in one's own capabilities to exercise influence over events, or in one's own ability to carry out (environmentally) effective actions (Bandura 1991). Anyone who believes that their own actions cannot exert an influence capable of improving a situation is unlikely to behave in an environmentally conscious manner. Anyone who believes that their own actions can contribute to changing

Figure 9: Psychological model for explaining sustainable behaviour (with acknowledgements to Hamann et al. 2016)

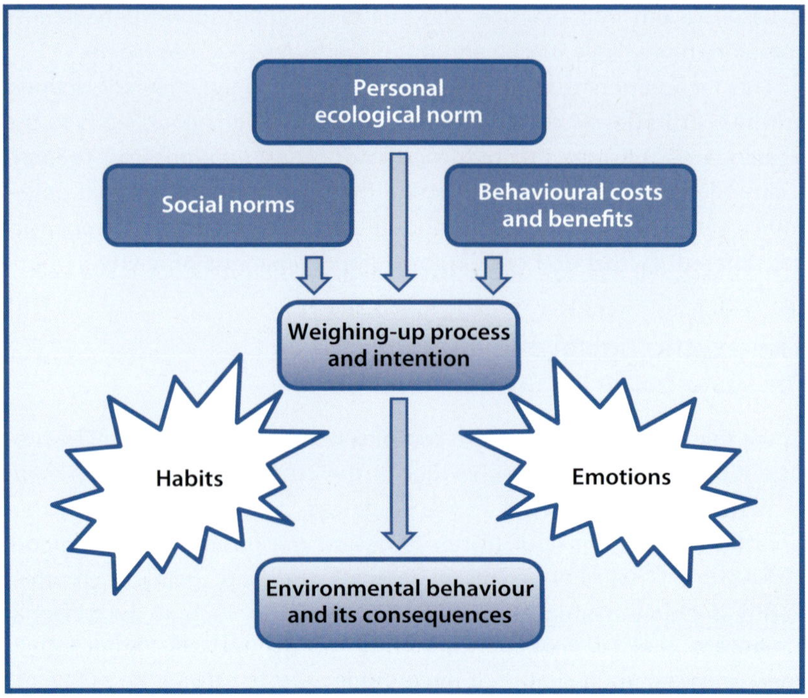

something is more likely to act accordingly. Promoting positive self-efficacy (or 'mastery') experiences is especially relevant in supporting sustainable behaviours (Hamann et al. 2016). Self-efficacy expectations can be increased by opening up genuine new opportunities for action (Hunecke 2013). Social norms include standards shared by many people. These guide the behaviour of individuals. Social norms also include the norms that are shaped by the individual's environment (personal norm). Social norms can exert a significant influence on individual behaviour. Norm-centred interventions are based on this idea (Hamann et al. 2016).

(Behavioural) Costs

(Behavioural) costs consist of both the cost or effort involved in executing a behaviour and direct monetary costs. (Behavioural) costs strongly influence actual behaviour. As pointed out by Diekmann & Preisendörfer (2001), behaviour that is very costly and brings small benefits is less likely to occur. For example, if the time (and thus the behavioural cost) involved in riding a bike to work is substantial, then the use of a car instead (if available) is more likely than if the time involved in cycling is low. If organic food is much more expensive to buy than conventional food, then this becomes a decisive argument for many people for buying conventional food.

Self-efficacy means a person's confident belief that they will be able to master future situations successfully. According to Bandura, there are four basic sources of self-efficacy (Bandura 1977):

- **Personal 'mastery experiences' gained by performing tasks successfully.**
- **Experience gained from observing others, who act as confirmatory proxies for the self and thus serve as role models (as in the self-exhortation 'If others can do it, so can I!').**
- **Verbal feedback from others (encouraging exhortations such as: 'You can do it!').**
- **A person's physical constitution, which can express itself in emotional and physical reactions (sweating, trembling, etc.). The less pronounced these are, the greater their confidence in their own ability to act.**

Rewards and punishments — standard psychological reinforcement mechanisms — (behavioural interventions after the event) can shift the cost-benefit ratio, which means the behaviour can be influenced.

Habits

Intention affects environmental behaviour. The whole process is influenced by ▹ **habits** and emotions. Even though it is often assumed that we humans make rational, well-considered decisions, this model makes it clear that non-rational factors exert a strong influence (Hamann et al. 2016). Habits are acquired behavioural patterns (conscious and unconscious) characterized by the fact that they consistently demonstrate identical behaviour over a long period of time. Relinquishing cherished habits is often associated with high behavioural costs: for example, changing from car to bicycle use. In order to increase one's own environmentally conscious behaviour, the important thing is to seize opportunities to break one's own habitual patterns. Such opportunities can be given, for example, by a period when motorists have to take a notable detour due to extensive road work, or when there is a campaign offering free tickets to get them to switch to the train. If a habitual pattern of behaviour is broken by such an intervention, the likelihood of permanently altering it increases. But there are also acutely sensitive life phases — for example, following the birth of the first child — in which old habitual patterns are broken anyway and new ones develop. This is also a propitious time in which to introduce lasting behavioural change.

Habits are learned sequences of reactions which become embedded and increasingly automatic through repetition (Dorsch et al. 1987).

Emotions Positive emotions (joy, hope, interest, love) and negative emotions (sadness, anger, fear) can influence behaviour that impacts on the environment. Positive emotions associated with environment-related behaviour reinforce such behaviour in the future; negative emotions can lead to reactance, and environmentally friendly behaviour then becomes less likely.

This explanatory model for sustainable behaviour is not to be understood as a phase model; that is, it does not require every phase to be passed through sequentially. Environment-related behaviour should be understood as an interaction between many different factors. Psychological models are not patent solutions that explain behaviour with 100 percent reliability. Nor is it enough simply to come up with a measure based on a behavioural factor. The preference should always be for a combination of measures which seems likely to appeal to the target group.

Multi-dimensional influencing factors What does this mean in practice? The psychological model of sustainable behaviour makes it clear that there are many influencing factors that are important for environmental behaviour. Campaigns that rely solely on appeals, and thus only on altering people's attitudes, have succeeded in bringing about very little change in recent decades. However, as shown above, when people have behavioural offers and incentives, when people are convinced that their actions have an impact, when the effort involved seems proportionate and there is a willingness to abandon existing habits, then sustainable behaviour becomes more likely. The practical implementation of these multiple and diverse influencing factors requires creative and varied ideas and concepts (Hamann et al. 2006).

3
(Un)sustainable systems: analysis and strategic approaches

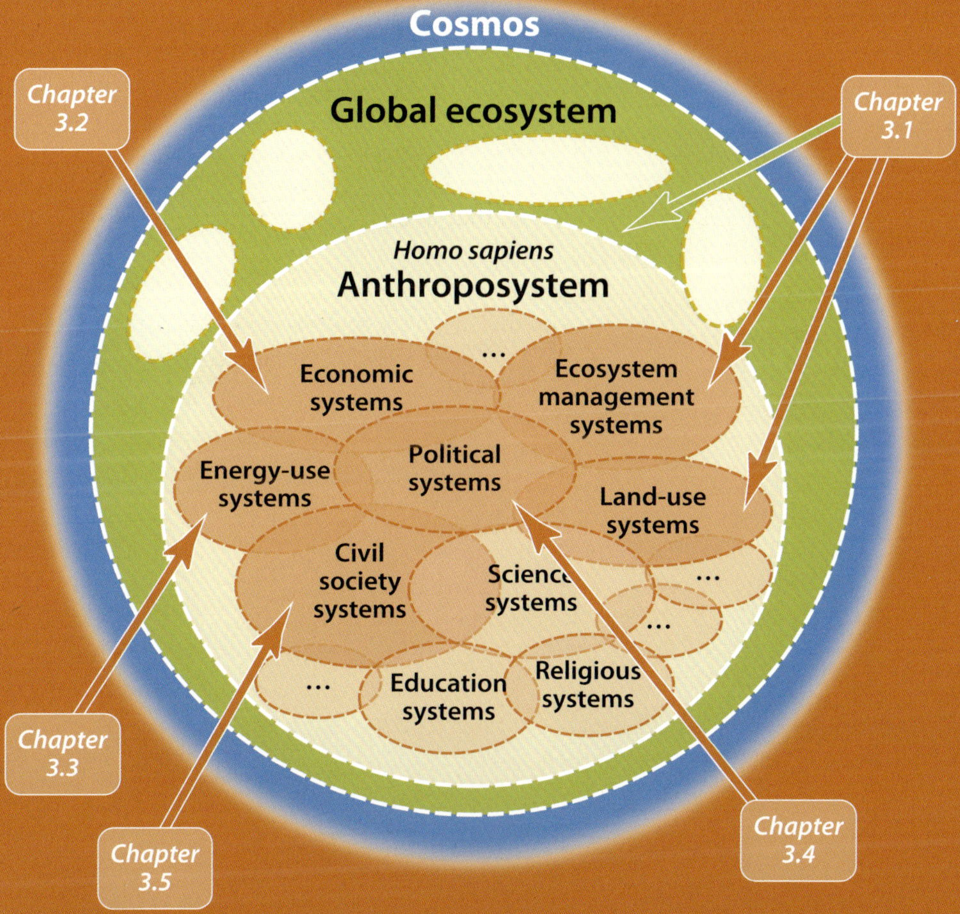

3.1

The foundations: ecosystems and ecosystem management

Pierre L. Ibisch

***Homo sapiens*—modern humans—emerged as an animal species within the ecosystems of (sub-)tropical Africa and spread rapidly over the planet. The species has integrated itself into the energetic structure of the ecosystems and, like all species, has thereby become a force impacting on the greater whole. Through continuous technological innovation and significant energy usage, *Homo sapiens* has succeeded in transforming many of the ecosystems that support it so that the services needed for its life and well-being are provided more effectively and reliably. Its purposive management of ecosystems turns ecosystem functions into ecosystem services. An ecosystem approach to the use of nature raises many technical questions, but also legal and ethical ones. Purely conceptually, nature seems to have become a kind of service provider for the benefit of human development. How is ecosystem work integrated into economic calculations? Does the path to sustainable development involve the further economic instrumentalization of nature? Who actually pays the costs of using the ecosystem?**

The foundations of ecosystem use

The term 'land use' has long been employed for the use and management of the Earth's surface. 'Land' is an old (Proto-)Germanic word that probably referred to a defined piece of the Earth's surface in the possession of an individual.[1] It is related to the medieval word 'landscape' (Old High German *lantscaf*), which was initially used for settlement areas (Müller 1977)—i.e. areas where living space was created for humans. Only later did it become a broader geographical term. So for a long time, the European (or so-called Western) understanding of the landscape and its use was territorial and characterized by the active role taken by humans. Landscape continues to be understood as a space which is subjectively identified by humans and which results from the interplay of natural and human factors ('an area, as perceived by people, whose character is the result of the action and interaction of natural and/or human factors'; definition in the European Landscape Convention, which came into force in 2004; Council of Europe 2017).

To this day, general, professional and legal usage is strongly influenced by these roots. The German Federal Nature Conservation Act is

1 Online Etymology Dictionary (https://www.etymonline.com/word/land).

actually called the Conservation of Nature and Care of Landscape Act ('Gesetz über Naturschutz und Landschaftspflege'). There, 'landscape planning has the task of specifying the goals of nature conservation and landscape management for the respective planning area and identifying the requirements and measures for the realization of these goals, including for planning and administrative procedures where decisions can affect nature and landscape in the planning area' (§9 Para. 1).

'Land use' is actually ecosystem use

This all reflects the traditional relationship of humans to ecosystems. ▹ **Ecosystems** were only recognized and described as such much later (Tansley 1935) (▹ Chapter 1.3). In fact, people rarely use only the land itself (e.g. for construction); rather, they exploit ecosystem functions — they make use of the work of the ecosystem and its components. But to date this has hardly been reflected in everyday language or in laws and conventions. Conceptually, however, there has been significant progress, and human dependence on ecosystem services is receiving increasing recognition. Accordingly, there is also a growing understanding of ▹ **ecosystem management** encompassing (among other things) land use.

> **Ecosystems are complex systems that use energy and accomplish work, in the physical sense. They arise because living beings interact, as system components, with each other and with non-living resources, thereby developing emergent properties which guarantee or even enhance their continued existence (▹ Chapter 1.3). Functionally, ecosystems are bioreactors (self-organizing, through the interactions of their living components) in which energy is captured, transmitted, transformed, stored and, above all, used to accomplish work.**

Ecosystem work

The outcomes of ecosystem work include the production and growth of biomass, temperature regulation, mobility, the transport of matter and information processing. The only organisms that can deliver new energy resources to the earth system from outside are the photoautotrophic bacteria and plants which convert radiant energy from the sun into biochemical energy and provide it to other organisms in the form of food. This biogenic energy, which is stored in the form of biomass and can be used to accomplish work (eco-exergy, ▹ Chapter 1.3), also develops a variety of emergent properties which are crucial for the functioning of the ecosystem.

For the organisms living in an ecosystem, the ecosystem functions are the preconditions for their existence. Particularly fundamental

Ecological niches and the unique status of humans

for each individual organism are the extent, constancy, and reliability within an ecosystem of their access to energy resources. Other factors include: the supply of water and nutrients; the guarantee of a more or less 'safe' habitat; and the possibility of interaction with other organisms, which enables them, for example, to reproduce or to cooperate for the improvement of their own living conditions. As part of the complex adaptive process of evolution, organisms respond and adapt to the perpetually and dynamically changing environmental conditions. The environment includes not only abiotic elements such as soil, water and air, but also all other organisms, which are in an identical mode of adaptation. For some species, this process results in increased efficiency of resource use and the development of more or less exclusive ecological niches which encompass the totality of all dimensions of resource use, essential life services and functions in the larger system. These ecological niches evolve — for the most part slowly, gradually, and on the basis of genetic change. In the case of the evolution of human beings, a primate species of the open spaces of Africa, a number of peculiar features are striking. Human — or rather cultural — development meant — almost certainly unconsciously at first — an ever-greater emancipation from the 'shackles' of local ecosystems and thus from the hardships of changeable nature. Technological innovations and culture — without biological or genetic changes — have repeatedly caused a sudden abandonment or a change and extension of humans' ecological niche (see Ellis 2015) (▷ Chapter 2.1).

Biomes become anthromes

The development of agriculture and livestock farming, the production of food surpluses and the development of food storage, as well as the use of fire and sophisticated food preparation techniques, provided the former hunter-gatherers with much greater nutritional sovereignty. This also enabled their survival in regions with pronounced seasonal variation far from the ancestral ecosystem. Within a few millennia, all the biomes of the earth were opened up for their habitation. Between the years 1700 and 2000, the earth's ecosystems were transformed from being wild and self-organized to being largely human-controlled. Since the early 20th century, humans have dominated and controlled more than 50 % of the earth's terrestrial surface — **biomes became anthromes** (Ellis & Ramankutty 2008).

Biomes and anthromes

Biomes are large ecosystems which have predominantly been formed under the currently prevailing physical conditions of life on earth and exist on both hemispheres: polar deserts, tundras, boreal coniferous forests, temperate deciduous forests, cold and hot steppes, as well as deserts, savannas, subtropical and tropical dry and rainforests. Special, so-called azonal forms arise in mountain regions, among other places. Aquatic ecosystem types can also be classified into biomes, such as watercourses, oceans, coral reefs and estuaries. Biomes can in turn be subdivided into ecologically and (bio)geographically defined ecoregions. The most comprehensive systematic approach identifies roughly 1,000 different terrestrial and aquatic ecoregions (Olson et al. 2001; WWF 2018).

The concept of *anthromes* (or anthropogenic biomes) addresses the fact that human use, especially in recent centuries, has led to a significant change in the nature and functionality of ecosystems (Ellis & Ramankutty 2008). Anthromes are divided into six major groups (urban centres, villages, croplands, rangelands, semi-natural land, and wilderness) and 25 classes. Anthroecology is a new scientific discipline which addresses the ecology of the human-dominated ecosystem of the Earth (Ellis 2015).

Human development through the use of exosomatic energy

Whereas up to now the development of probably all other species has been tied to a local context and a specific resource space, humans have simply overridden the traditional boundaries of species evolution. Their abandonment or alteration of their evolved ancestral ecological niche was based on a cultural 'trick': the human species learned how to use external energy resources. At first, they let other animals work (mechanically) for them. Later, the discovery and exploitation of fossil fuels, the invention of technologies for mobility, and the development of trade enabled ever-new energy and other kinds of resources to be accessed, even from the most remote ecosystems. However, this has given rise to a severe dependence. Humans had to convert ever-larger amounts of energy to serve their needs from outside their physiological system (exosomatic use of energy). Endosomatically, i.e. within their bodily metabolism, humans convert approximately 3.5 to 5 gigajoules (GJ) per year per person. In the course of socio-cultural development and above all industrialization, individual energy consumption (principally exosomatic) increased by a factor of about 100; this was the prerequisite for the growth and increasing local density of the population (see Table 1).

Table 1: Human socio-cultural systems according to their primary subsistence regime, in order of historical emergence (from Ellis 2015, based on various sources)

Socio-cultural system	Median population size	Social complexity[a]	Speciali-zation[b]	Total energy consumption[c] (GJ/Person/Year)	Food energy consumption[d] (GJ/Person/Year)	Population density (People/km^2)
Hunter-gatherer	40	0	0	8	5	2
Simple horticultural	1,500	1	2	no data	5	35
Herding	6,000	no data	9	no data	6	no data
Advanced horticultiral	5,000	7	22	no data	5	110
Simple agrarian	$>10^5$	51	34	18	6	>250
Advanced agrarian	...	...	...	40	9	urban
Industrial	$>10^7$	100	100	118	11	urban
Post-industrial[e]	$>10^9$	100	100	350	15	urban

a Percentage of societies which are highly complex with different status groups (for example, presence of social strata, number of hierarchy levels)
b Average frequency of craft specialization across all societies (various crafts, occupations)
c Total Energy: Home + Agriculture + Trade + Industry + Transportation
d Food energy: food + animal feed
e Post-industrial: societies after industrialization; modern knowledge- and services-based societies

Ecosystem use as a problem of energetics

Just as ecosystems themselves have to be understood primarily as energetic structures — because the exchange of energy between ecosystem components is the fundamental requirement for the existence of life as well as for all ecological interactions and functions (▷ Chapter 1.3) — so ecosystem use, too, is a problem of energetics. The growth and development of humankind in the last few centuries was only possible thanks to the massive use of various energy sources (▷ Chapter 3.3). An important indicator of this is that humans take an ever greater share of **primary production**.

Human appropriation of net primary production

In a first study on this topic, it was estimated that about 24% of net primary production is already appropriated by humans to meet their needs (Haban et al. 2007). From 1910 to 2005, HANPP *(Human Appropriation of Net Primary Production)* is estimated to have doubled (Krausmann et al. 2013). However, over the same period, the human population has quadrupled, and economic output has increased by a factor of 17. This increase in efficiency is partly due to the lower per capita use of biomass for energy purposes and to lower rates of ecosystem conversion; it is assumed, though, that a future increase in the use of bioenergetic resources (including agricultural production of biofuels) could significantly increase HANPP (Krausmann et al. 2013). However, a recent study concludes that as much as 42% to 47% of land biomass has already been lost through changes to ecosystems (Erb et al. 2018).

In potentially productive, biomass-rich ecosystems where humans appropriate a large proportion of primary production, not only is productivity reduced but the system loses many of the emergent properties arising from eco-exergy (▷ Chapter 1.3), which are crucial for functionality (see Fig. 1).

Exosomatic energy turnover in the primary sector

Humankind (temporarily) stepped back from the greater use of biomass because other fuels, principally fossil fuels, became available. And these could also be used to significantly intensify agriculture and increase productivity (Ellis et al. 2013). The so-called *green revolution*[2] is in fact not really 'green' at all, but was actually achieved through the industrialization of ecosystem use (Pfeiffer 2006). On the energy input side of agriculture, there is less and less human (and animal) work and more and more use of artificial fertilizers, fuels for machinery, and pesticides. Many foods require the input of more energy than the crops ultimately biochemically convert (Table 2). In many countries, the exosomatic use of energy in the primary sector of the economy continues to increase significantly due to the progressive mechanization of agriculture (sample growth rates 1995–2007: Turkey 179.3%, South Korea 126.9%, Estonia 148.4%; Andreoni 2017). The energetic footprint of food also wors-

2 The application of new agricultural and breeding methods to increase agricultural production in the global South from the mid-1960s.

Figure 1: The photo shows a 'cleared' landscape in the Caucasus, where a former or potential forest ecosystem (probably including juniper trees among others) has been transformed into intensively managed agroecosystems (on the valley floor) and heavily grazed and now sparse grassland ecosystems (on the slopes). Almost all the biomass that grows back every year is immediately taken by humans and their livestock and energetically converted. Emergent properties of the biomass, such as shading and moistening or habitat formation for animal species, are destroyed. This minimizes self-protection (e.g. from erosion) and self-regulation. (near Dedoplistskaro in Kakheti, Georgia, photo: P. Ibisch)

ens dramatically if the inferior energy yield of the agricultural system relative to the natural ecosystem is taken into account: more complex, multi-layered natural vegetation of diverse functional types converts significantly more light energy into chemical energy than agro-ecosystems consisting of a single crop. The energy input for transportation, processing and preparation is usually many times more than the food energy yield (for example, ten bread rolls made of imported grain, baked in a large commercial bakery and transported from there to the shop, where they are picked up with an SUV vehicle for the family breakfast at home 500 metres away).

Table 2: Energy efficiency of the production of selected agricultural products (taking into account the human energy input, but not the loss of ecosystem energy efficiency incurred by alteration of the natural ecosystem)

Crop	Country	Ratio of energy used in production to biochemical energy stored by plants (input/output)[1]
Manioc[2]	Nigeria	1 : 7.57
Maize[2]	USA	1 : 4.11
Soya[2]	USA	1 : 3.71
Wheat[2]	Kenya	1 : 3.31
Potatoes[2]	USA	1 : 2.76
Wheat[2]	USA	1 : 2.57
Soya[4]	Iran	1 : 2.06
Barley[3]	Iran	1 : 1.94
Milk production (farm)[5]	Estonia	up to 1 : 1.88
Wheat[3]	Iran	1 : 1.49
Rice[2]	USA	1 : 1.42
Maize[2]	India, Indonesia	1 : 1.08
Rice (with draught animals)[2]	India	1 : 0.79
Tomatoes[2]	USA	1 : 0.78
Apples[2]	USA	1 : 0.18

1 Energy not necessarily absorbed/usable by the human body
2 Pimentel (2009) 3 Ziaei et al. (2015) 4 Kordkheili et al. (2013) 5 Frorip et al. (2012).

Ecosystem use, mobility, and the import of ecosystem services

In recent centuries, many local ecosystems have increasingly been managed and transformed to meet single, exclusive needs, and this has generally been associated with a loss of overall functionality and performance. Ever-more powerful civilizations, or population groups in defined areas—which have ultimately defined themselves in national terms—have been able to use other ecosystems outside their own territory for the wellbeing of their own populations, and, where appropriate, to compensate for local losses or unavailability by importing ecosystem services. The development of human civilizations is marked primarily by the evolution of so-called land-use systems, as well as those systems that make natural resources available over greater or lesser distances. The production of renewable raw materials in ecosystems—such as food or wood—became the foundation of increasingly globalizing national economies. The growth, spread and increasing concentration of human societies, as well as greater mobility (initially thanks to horses and ships, then later automobiles and aircraft), led to a completely new dynamic in social systems. Through interaction, hitherto largely isolated social systems formed higher-order systems. This occurred both via conquest and the creation of 'colonial' empires (e.g. the Mongols, the Roman Empire, the Incas, Great Britain) and via a process of growing together in densely populated settlement areas. Besides the fact that the ever larger social systems became entangled in conflicts and even wars over resources and land, the energy and material needs of these more advanced states fostered 'aggression' against the ecosystems that supported them. A well-known example is the destruction of the forests in the Mediterranean region, which fell prey to, among other things, the construction of ships, which were needed in turn to bring in resources from other regions.

Ecosystems as a productive force and a source of human well-being

Ecosystem use as the basis for the primary sector of the economy

The use of the resources provided in ecosystems is the starting point of all human economic activity. This is also reflected in the various classification systems for economic activities. The so-called primary sector (see e.g. Pohl 1970)[3] is traditionally described as comprising the provision of renewable natural resources — such as agricultural products, wood, fish, wild plants and meat from wild animals — as raw materials. Often, the supply of non-renewable natural resources by mining is also counted as a part of the primary economy. But mining can sometimes be included in the secondary sector, the industrial economy, which includes the manufacturing and processing industries (e.g. crafts and trades, energy supply, construction, industrial production). The tertiary sector comprises the many basic services that people provide for each other (trade, transport, administration, care, health, leisure). The services of information and knowledge management are often considered to comprise a separate, quaternary sector (especially information and communication technology). The quinary sector is that responsible for the disposal (removal/storage/processing/recycling) of waste and scrap and for the treatment of water.

3 Based on the work of British economists in the 1930s, who initially distinguished between three sectors.

Ecosystems, then, appear to be relevant only to the primary sector of the economy—there, the exploitation of the energy and material resources generated in ecosystems is quite obvious. However, in the other sectors as well, there are significant interactions with ecosystems, and even hidden dependencies on, and services provided by, nature. Only when ecosystems are damaged or depleted to the point of scarcity does it become clear to people that they are essential for a prospering economy and for human well-being. For a long time, the goods and services provided by nature were regarded by economists at best as 'free productive forces' (see e.g. Marx & Engels 1964, pp. 753) and the economy was considered to include only those services which required human labour. The introduction of the concept of ecosystem services therefore represented an important step forward. The first studies based on this concept were published in the 1980s (the term was popularized by Paul and Anne Ehrlich: Ehrlich & Ehrlich 1981; Ehrlich & Mooney 1983). The United Nations' *Millennium Ecosystem Assessment*[4] not only provided conceptual support for this approach, but also introduced it into politics. Since the mid-2000s, 'ecosystem economics' has established itself as a veritable new scientific discipline; ecosystem services are now discussed everywhere, and for some conservationists the concept now represents the best hope of convincing society of the crucial importance of nature. The Millennium Ecosystem Assessment initially identified four categories of ecosystem services that contribute more or less directly to human well-being: supporting, provisioning, regulating, and cultural. For some years now, work has been conducted on an international classification system which is regularly revised, and which is now restricted to the three categories of provisioning, regulating, and cultural: the Common International Classification of Ecosystem Services (CICES) (Fig. 2).

Ecosystem services

Taking proper account of ecosystems as the dynamically changing and work-performing basis of our existence not only has implications for how we value things in economic terms. It also means, first of all, that economic sectors need to be reconsidered and reclassified so that they can be appropriately understood and evaluated. Among other things, it is important to emphasize that there is such a thing as the 'foundational economy', which comprises all the services that maintain or restore ecosystems so that general ecosystem functionality is sustainably maintained (without attributing greater importance to any individual service).

Maintaining and preserving functioning ecosystems is the 'foundational economy' and the basis of all economic activity

4 Publication of the reports began in 2003; cf. https://www.millenniumassessment.org/en/index.html.

Provisioning ecosystem services

Biomass

- Cultivated terrestrial plants for nutrition, material and energy
- Aquatic plants cultivated in situ for nutrition, material and energy
- Cultivated fungi for nutrition, material and energy
- Terrestrial animals reared for nutritional or energetic purposes or for provision of material
- Aquatic animals reared for nutritional or energetic purposes or for provision of material
- Wild terrestrial and aquatic plants as a source of nutrition, material and energy
- Wild fungi as a source of nutrition, material and energy
- Wild animals as a source of nutrition, material and energy

Genetic material from all biota

- Seeds, spores or other plant material for the purposes of maintaining or establishing a population
- Plant individuals used to breed new strains or varieties
- Individual extracted genes from plants to design or create new biological entities
- Animal material collected for the purposes of maintaining or establishing a population
- Animal individuals used to breed new strains or varieties
- Individual extracted genes from animals to design or create new biological entities

Cultural ecosystem services

Direct, in-situ and outdoor interactions with living systems that depend on presence in the environmental setting

- Physical and experiential interactions with organisms or ecosystems
  - Characteristics of living and ecological systems that enable activities in nature promoting health, recuperation or enjoyment through passive or observational interactions
  - Characteristics of living and ecological systems that enable observation in nature promoting health, recuperation or enjoyment through passive or observational interactions
- Intellectual and representative interactions with organisms and ecosystems
  - Characteristics of living and ecological systems that enable scientific investigation or the creation of traditional ecological knowledge
  - Characteristics of living and ecological systems that enable education and training
  - Characteristics of living and ecological systems that support cultural identity and/or the sense of place and belonging
  - Characteristics of living and ecological systems that enable aesthetic experiences and artistic inspiration

(Indirect or abstract/ remote) interactions with living and ecological systems that do not require presence in the environmental setting

- Spiritual, symbolic and other interactions with nature and/or organisms
  - Elements of living and ecological systems that have symbolic meaning
  - Elements of living and ecological systems that have sacred or religious meaning
  - Elements of living and ecological systems used for entertainment or representation

Regulating ecosystem services

Transformation of biochemical or physical inputs to ecosystems

- Mediation of wastes or toxic substances of anthropogenic origin by living processes
  - Bio-remediation by microorganisms, algae, plants, and animals
  - Filtration/sequestration/storage/accumulation by organisms
- Mediation of nuisances of anthropogenic origin
  - Smell reduction
  - Noise attenuation
  - Visual screening

Regulation of physical, chemical, biological conditions

- Regulation of baseline flows and extreme events
  - Buffering and attenuation of water and wind erosion (including coastal protection)
  - Buffering and attenuation of mass movement
  - Hydrological cycle and water flow regulation (including flood control)
  - Wind protection, reduction of wind speeds
  - Fire protection
- Lifecycle maintenance, habitat and gene pool protection
  - Pollination or gamete dispersal
  - Seed dispersal
  - Maintaining nursery populations and habitats (Including gene pool protection)
- Pest and disease control
  - Pest control and limitation of spreading of invasive species
  - Prevention and control of diseases
- Regulation of soil quality
  - Moderation of weathering processes and their effect on soil quality
  - Decomposition and fixing processes and their effect on soil quality
- Regulation of water quality
  - Regulation of the chemical and physical condition of surface freshwaters (standing, flowing)
  - Regulation of the chemical and physical condition of groundwater
  - Regulation of the chemical and physical condition of salt waters
- Regulation of quality of atmosphere/air and climate regulation
  - Filtering and cleaning of air
  - Reduction of atmospheric concentration of anthropogenic greenhouse gases and global climate regulation
  - Regulation of meso and micro-climates

Figure 2: Classification of ecosystem services (taking into account CICES/Haines-Young & Pochin 2017; author's translation and changes; without so-called abiotic services) (photos: P. Ibisch).

This includes those forms of nature conservation that aim to preserve self-organizing ecosystems which have been anthropogenically altered or disturbed as little as possible — for example, by setting up protected areas and the preservation of wildernesses.

The use of wild ecosystems and the management of cultural ecosystems

In order to analyze how they are used, wild ecosystems — systems which organize themselves and dynamically adapt within a given framework of conditions (including man-made environmental change) — will be distinguished here from anthropogenic ecosystems. After the 'foundational economy', the next level is that of the extractive areas of ▹ **ecosystem management** — the classical primary sector. Here we find, firstly, the depletion and harvesting of ecosystem products, which occurs in a relatively or entirely unspoilt ecosystem and makes use of spontaneous ecological processes. Examples of such wild ecosystem use include the collection and harvesting of wild organisms or their products (e.g. gathering wild fruit or mushrooms, or hunting/catching wild animals).

Figure 3: Reforestation with pine trees to combat desertification on the edge of the Gobi Desert, Saihanba Forest Park, China, as a primarily non-use form of ecosystem management. In fact, the purpose is not so much the restoration of an ecosystem but the establishment of a pine plantation; the original forest vegetation was probably considerably more complex with regard to species diversity and ecosystem functions. (photo: P. Ibisch)

This is distinct, secondly, from the conscious and targeted management of cultural ecosystems, which sets up and preserves individual ecosystem types and conditions so as to optimize the extraction of individual, potentially renewable resources: important examples are agriculture and forestry as well as the management of aquatic ecosystems (fishing, aquaculture). Mining also basically belongs to this category, especially if it is in competition with other possible land uses and causes the destruction or substantial transformation of ecosystems—a notable example being opencast mining for the extraction of coal, ore, sand, etc. But underground mining also usually leads to subterranean and surface ecosystem changes (due to, for example, the pollution of groundwater or the pumping or draining away of water).

Non-consumptive ecosystem use

In addition to these forms of ecosystem use, there is another form which is not based on consumption, which takes place in natural settings, and for which the existence of ecosystems and their components is of direct relevance. In this form, the ecosystems and their parts are not used or consumed in material terms. It includes all nature-based activities which exploit the so-called cultural ecosystem services (e.g. nature-based tourism, wildlife photography, wilderness education and therapy, etc.).

Ecosystem management: The systematic and goal-oriented application of human, financial and other resources for the preservation and restoration of spontaneously changing and self-organizing ecosystems, and for the regulation and transformation of ecosystems and their uses, can be collectively described as ecosystem management. The conscious decision to refrain from such interventions and from direct uses of ecosystems in favour of self-organization and self-regulation of the system is therefore also a form of ecosystem management. Different forms of ecosystem use can compete for space—for that reason, holistic ecosystem management also includes the spatial-temporal balancing out of all uses, and of non-use, in seeking to maximize the functionality of the system as a whole.

The use of ecosystems not only serves to create 'added value' for human development; at the same time, it usually causes **'added damage'**, which often has a direct or indirect negative impact for people, reducing their overall well-being. However, it is regularly the case that the consequences of use—for those who benefit and those who suffer from the damage—are unequal in scale or degree.

'Added damage'

'Added damage', as a contrasting counterpart to the concept of 'added value' in business, is defined as the sum of all negative environmental effects that result from an economic activity. These effects can be triggered both directly and indirectly. The term was first introduced in 1992 and initially applied only to emissions; today, it also includes ecological depletion, any direct ecological damage, and also negative effects on humans (Prammer 2009). Added damage becomes a problem for business when the negative effects weaken or call into question the economic rationale for production. Many effects, however, primarily have an impact at the national economic level, since they exceed the economic boundaries of the individual firm. This raises the problem that the costs associated with added damage are regularly externalized — meaning that costs such as those for the purification of water and air, or indirectly generated costs such as those arising from climate change, have to be borne by everyone.

But the ecological assessment of added damage is a difficult enough undertaking in and of itself, since it first requires the identification of consequential effects and then an assessment of the depth and extent of their impact. And in the case of ecosystem use, these are very complex. The fact is that ecosystem use not only sets up a value chain — at every stage of the use and processing of natural resources, material assets, income, jobs, security of supply, etc. are created — but also a damage chain: every step, from extraction and transport to processing and finally to the production of waste and residues, is (potentially) associated with damage to the economy of nature, or ecosystem functionality, and to human health, as well as to businesses and the economy, which must then be placed in relation to 'added value' in order to make possible a net overall view of ecological and economic impacts.

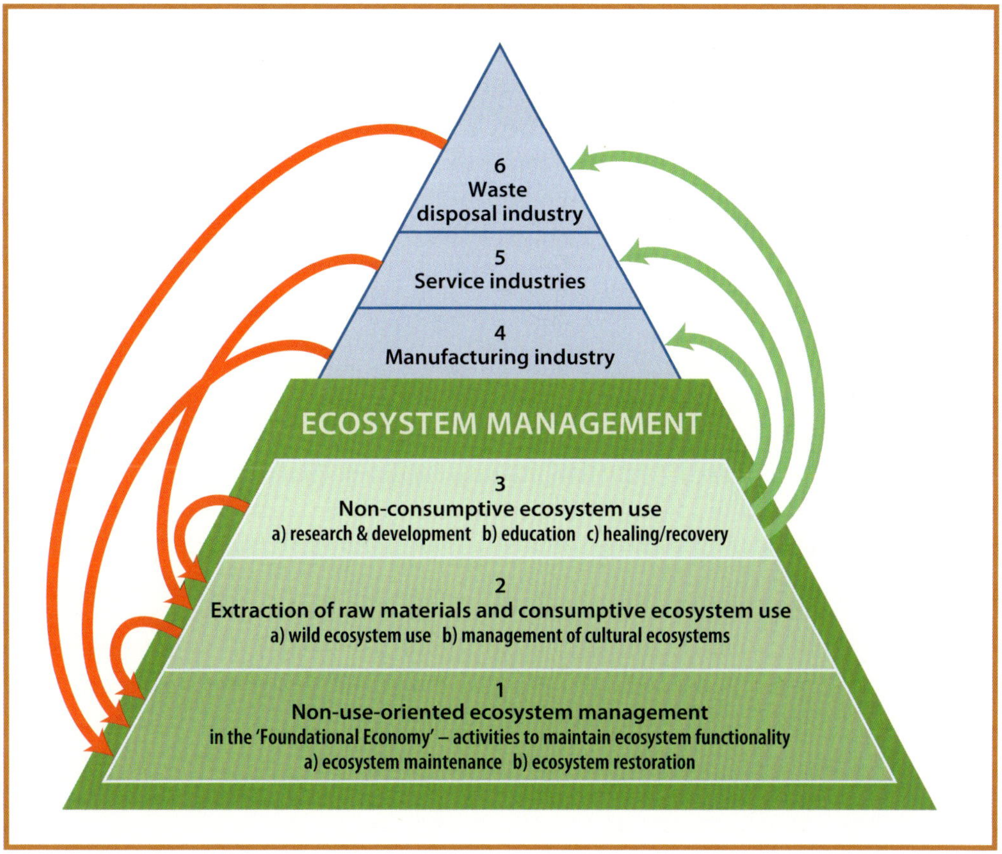

Figure 4:
Pyramid of economic sectors from an ecosystem perspective, showing the position of ecosystem management (green arrows show how it supports dependent sectors, red arrows indicate degradation of ecosystem functionality)

Damage to ecosystems can be broadly divided into the following categories (see also Salafsky et al. 2008):

1. Physical disturbance or destruction of parts of the working ecosystem (inter alia by removal/harvesting of biomass or of individuals or their parts, reduction/destruction of populations, disruption/destruction of vegetation structures)

2. Alteration of the physical, chemical or thermodynamic properties of the earth's surface, oceans or atmosphere (e.g. sealing, excavating or moving soils, draining, irrigation, damming or diversion of rivers; altering albedo or the capacity of ecosystems to absorb or store energy; release of new non-ecosystem substances such as pesticides, plastics or other waste)

3. Interruption or loss of the interactions between living systems (e.g. physical interruption of the interactions between individuals and species by infrastructure; loss of genetic recombination; alteration of well-established interaction networks through the introduction of species new to the system)

4. Loss of biodiversity and ecosystemic knowledge (loss of genetically stored information, resulting in loss of survival and backup solutions long tried and tested)

5. Increase in vulnerability to disturbances and loss of adaptability and reparability (reduction in self-organization and self-regulation capacity; reduction in climatic buffering and changes in water supply and retention through multiple feedback effects such as loss of greenhouse gas storage and sink capacity; climate change-induced disintegration of functioning systems; increased sensitivity to extreme events; see e.g. Pecl et al. 2017)

The various changes causing stress and reducing functionality in the ecosystems do not usually have a singular impact; rather, they are linked together in complex impact chains. They reinforce each other and can escalate systemically. The systemic interactions occur at all levels, including that of introduced substances. For example, some weedkillers (such as glyphosate) and certain insecticides are only toxic (or many times more toxic) when applied in combination with other agents (Vincent & Davidson 2015; Hasenbein et al. 2016). The changes occur at the local level but also contribute to regional and even global damage.

The effects of ecosystem use are dependent on context

The effects of certain types of usage and damage are context-dependent, and also depend in terms of scale above all on the nature of the ecosystems affected. Species-rich, complex and highly self-regulated systems, such as tropical forests, are particularly vulnerable and, after severe disturbances, also suffer increased vulnerability to other stress drivers (such as climate change). Mountain ecosystems on steep slopes are particularly vulnerable: for example, damage to the vegetation cover quickly leads to erosion and loss of soil (Fig. 5).

Figure 5: Long-term alteration and weakening of an ecosystem due to monoculture forestry (Norway spruce), clearcutting and soil erosion in a mountain forest ecosystem (Transcarpathia, Ukraine, photo: P. Ibisch)

Figure 6: Cultivation of oil palms on Borneo, Malaysia, as an extreme form of ecosystem use, where the work of the native ecosystem is completely superseded. It is replaced by the creation and management of a purely cultural ecosystem. The crop comes from a different continent (West Africa), and the associated pollinator has been introduced (the weevil *Elaeidobius kamerunicus*).
The palms are produced in nurseries or in vitro and represent a small number of selected genomes. The necessary plant nutrients are supplied by fertilizers, and the plantation is regularly sprayed with the herbicide glyphosate to make it accessible for maintenance and to exclude possible competition from other plants. Habitats and resources for native species are minimal. In addition to the observable local effects, the substances find their way into the rivers and the ocean as well as into marine ecosystems such as coral reefs. (photo: P. Ibisch)

Figure 7: Simplified representation of the various aspects of ecosystem management that must be analytically and strategically linked in order to achieve a reduction in ecosystem degradation and improve human well-being

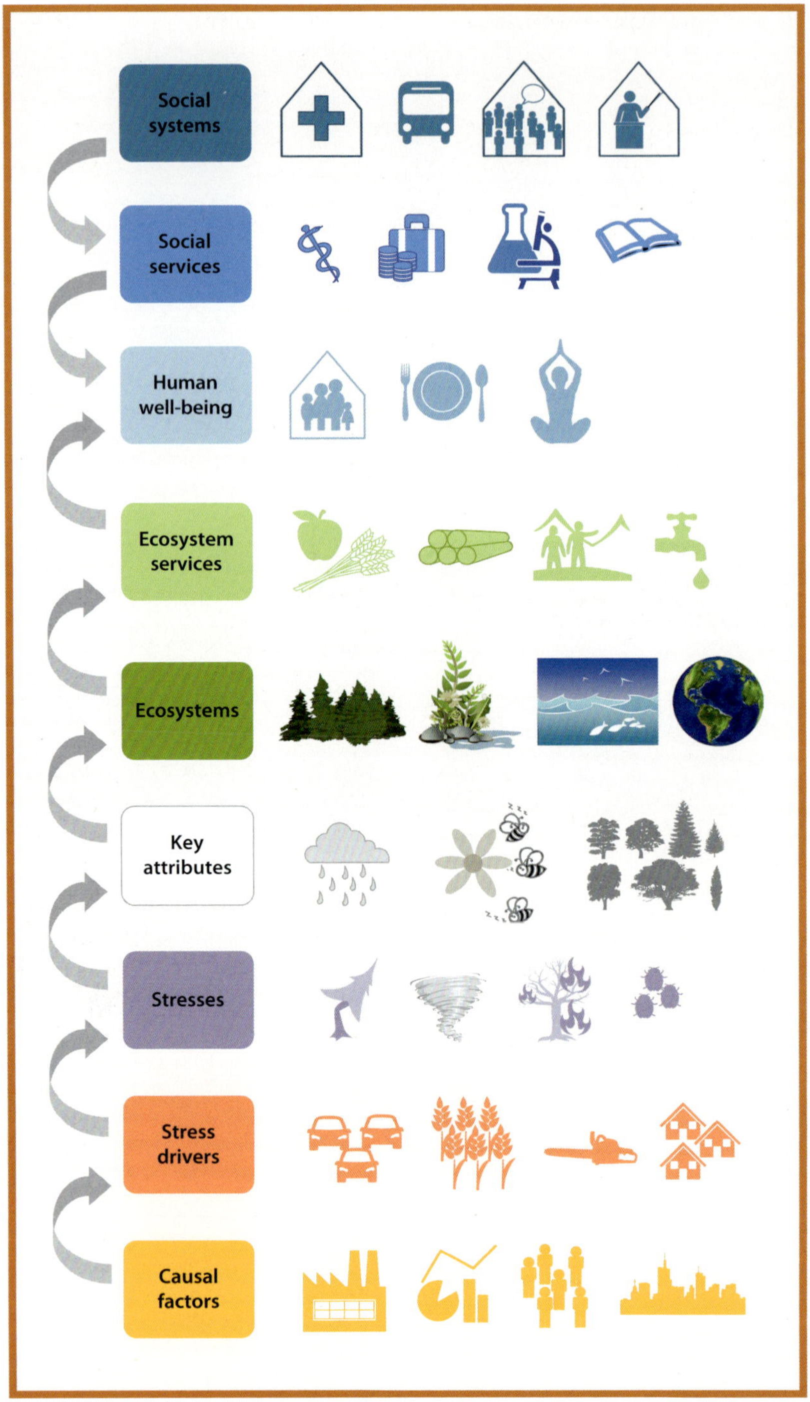

Components of holistic ecosystem management

Holistic ecosystem management not only looks at managing the use of ecosystem services, but also monitors the status of, and changes in, the ecosystems that support human well-being. It also seeks to reduce ecosystemic stresses that affect the key ecological attributes essential to its functioning (e.g. biodiversity, genetic diversity, ecological interactions/processes). This involves not only identifying and recording the symptoms of ecosystem change, but also influencing the drivers of stress and their underlying causal factors, all of which can interact with each other (Fig. 7).

The application of economics to ecosystem management

How much is nature worth?

After the initiation of the discourse on ecosystem services, the logical next step was to consider them from an economic perspective. Walter Westman set this in train by asking: 'How much are nature's services worth?' (Westman 1977)[5]. But it was not until 20 years later that Robert Costanza and colleagues (1997) published a groundbreaking study that attempted to express the value of ecosystem services and natural capital in monetary terms. Suddenly, a figure of $33 trillion was involved. The article, published in the internationally renowned journal *Nature*, was discussed intensively. Although that calculation proved, over the course of time, not only to have a weak basis, but also to be misleading (see, inter alia, the argumentation set out in the following paragraphs), the article had a huge influence. The supposedly revolutionary step here was that of ascribing nature an economic value, even though it was customary not to include such values in economic calculations, but instead to treat nature as a cost-free productive force. In standard economic theories, capital is the basis for economic activity and value-creation. Every form of production is based on three factors (the factors of production): labour (human capital), land (natural capital), and capital (physical capital). In fact, so-called natural capital often only enters into economic calculations as land, which sometimes needs to be acquired and retained through payments. Nature thereby acquires comparable status with the workforce and machinery, which can be replaced as production factors if they become too old or too expensive; the relevant investment here is the purchase of land. However, when one becomes aware of the many

5 A good outline of the history of ecosystem services is provided by Gómez-Baggethun et al. (2010).

provisioning, regulatory and cultural benefits provided by working ecosystems, the question arises as to what these are actually worth—and who presents and pays the bill.

The Stern Report: the costs of climate change

The debate gathered fresh momentum as it became increasingly clear that worsening environmental problems such as climate change are creating significant economic costs. A strategic consideration was that political and business leaders are more inclined to give serious attention to environmental issues when they are faced with real economic costs. In 2007, Nicholas Stern presented a report on the economics of climate change (Stern 2007) which triggered a significant response across society. Worrying figures were once again being presented: climate change might mean that 5 to 20 % or more of future global gross domestic product could be lost.

TEEB: The Economics of Ecosystems and Biodiversity

The Stern Report inspired ecologists and conservationists to call for a similar analysis for ecosystems. This was carried out relatively quickly, and the result was *The Economics of Ecosystems and Biodiversity* (TEEB 2010). Immediately after publication of the so-called Stern Report, fundamental methodological-conceptual questions were raised; above all, how reliable figures could be calculated for future values to be taken into account in the estimation of climate change costs. Economic science has instruments for the calculation of such values, and these were duly applied in this instance. Central to this is what is called ▹ **discounting**.

> **Discounting is an instrument of financial mathematics which determines how much money must be invested at a given time at given interest rates in order to ensure the availability of a certain amount of capital at a later date; alternatively, the value that a given amount of money will have in the future.**

Can ecosystems be discounted?

In discount-based calculations of the future value of any asset, setting the discount rate involves assumptions about the value that future actors will attribute to it. In the case of typical industrial products such as a car or a computer, it can be assumed that they will lose value due to wear and tear and subsequent technological innovation. Accordingly, such investments can be written off, and it can be assumed that their future value will eventually approach zero. In such cases, the future value of goods can be fairly reliably priced in today's terms, applying a positive discount rate. But what about nature and ecosystems? In this case, population growth and ongoing degradation and destruction of ecosystems can be expected to dramatically reduce the per capita supply of ecosystem services. Is it not therefore more likely that future generations will actually value nature more than we do today? This is not taken into account in current approaches to economic evaluation.

Moral-ethical implications of environmental economics

The economic valuation of ecosystems has significant moral and ethical implications. Not only are present generations increasingly and irreversibly damaging ecosystems (extinct species, for example, cannot be brought back to life), leaving behind a much less well-functioning global ecosystem for future generations (see the ecocide debate); in addition, future generations are also being deprived of their voice on ecological-economic issues, since decisions are being made today on how nature will be valued in the future. And indeed, the first interim TEEB study drew attention to this problem: for example, a 4% discount rate applied over 50 years means that a future ecosystem service will only be valued at 1/7 of the value derived from it today (and after 100 years, at only 1/50). However, if we assume that its value will continue to rise in the future, because nature is not a replaceable, self-depleting machine, the discount rate would have to be negative—and today's generations would have to set an infinite value on the future. And from the perspective of theory, this would seem to be correct, given an infinitely long future; but of course, from an economic point of view, it is not acceptable in practice. In extreme cases, the current generation would have to more or less cease its consumption and expect subsequent generations to do the same (Müller & Clasen 2008). So is this about finding the middle ground between the extremes of very low and very high discount rates (Beckerman & Hepburn 2007)? It is in fact one of the biggest ethical questions of our time, one which addresses above all the very substance of the idea of sustainability. When the satisfaction of the needs of the present generation seems to be incompatible with the interests of future generations, who should back down? It would probably represent a significant step forward in this debate if the greater part of the human population were at least aware of the problem. And moreover, it is also a legitimate theoretical-conceptual question to ask whether it makes sense at all for a dependent sub-component to determine the value of the greater whole on which its life or death depends. Is it at all possible in this instance for the value to be less than infinitely large?

The conceptual and methodological problems of the economic and even more so of the monetary valuation of ecosystems are far greater than has traditionally been reflected in environmental economics. Of course, it is also important in such calculations that all values are adequately captured. Direct utility values (for example, consumptive ones

System	Added value, positive impacts/revenue	Added damage, negative impacts/costs
Enterprise	**Income/revenues** [Y] €	**Wages, taxes and dividends, other operating costs** [Y] € **Infrastructurally reduced tree stock and correspondingly lower production** (due to skid trails, forest tracks) [Y–D] € **Investment in machinery and equipment** [Y] €
Society	**Stimulus/incentive to technological progress** (Industry, Research and development) [Y–D] × **Jobs in forestry and the wood industry** [Y–D] € **Provisioning ecosystem services supplied by the forest ecosystem used** (mainly consumers of wood, mushrooms, wild game) [Y] € **Regulating ecosystem services supplied by the forest ecosystem used** (in particular to the local and regional, but also the global populations) [Y–D] × **Cultural ecosystem services supplied by the forest ecosystem used** (in particular to the local population, recreation seekers, tourists, scientists) [Y–D] ×	**Reduced regulating ecosystem services from the forest ecosystem as altered by use** [Y–C] × **Reduced cultural ecosystem services from the forest ecosystem as altered by use** [Y–C] ×
Ecosystem	**Ecosystem functions in the forest ecosystem used** [Y–C] ×	**Reduction of tree biomass and of dead and old wood** [Y–C] × **Loss of biodiversity and genetic information** [D–C] × **Soil degradation and the loss of microclimatic regulation** [Y–C] × **Diminution of the ecological network** [D–C] × **Reduced productivity due to heightened vulnerability to extreme events, diseases, etc., and to changes to soil and microclimate** [D–C] × **Loss of production due to tree loss** (from e.g. disease, windthrow in more vulnerable stands) [Y–D] ×

like food, or non-consumptive ones like recreation or tourism) are actually comparatively easy to capture. For indirect use values, it is far more difficult, if not downright impossible, since it cannot be assumed that all the regulatory functions and services of an ecosystem are fully known (for example, preservation of soil fertility, regulation of the water cycle, pest regulation). In fact, the recent past teaches us how far our knowledge of ecosystem services has advanced in a very short time, thereby broadening the evaluatory basis. It would be presumptuous to suppose that we now know all the relevant utility values. Usage-independent values such as abstract existence value (the value people ascribe to an ecosystem or species simply because it is important to them that it exists) or legacy value (the satisfaction or sense of purpose which people derive from passing on nature intact to future generations) cannot appropriately be described or captured in monetary terms.

Problems of evaluability of indirect utility values and usage-independent values

Conversely, the problems identified here mean that major elements of 'added damage' similarly cannot be calculated. At the very least, what is needed is the possibility of drawing up, for all sectors of the economy, an overall ecological-economic balance sheet—excluding those forms of added value or damage (and their costs) which cannot be calculated—so that we can determine net profitability, at least in qualitative terms. This would also need to take into account the fact that the impacts of different aspects of added value or damage happen not only in different places, but also in different temporal dimensions. Since ecosystems often respond very slowly and with a time lag (to disturbances caused by use), value and damage added occur in different time periods. Above, forestry is taken as an example of ecosystem use to illustrate schematically which benefits and costs would have to be considered in such a balance sheet (Fig. 8).

Overall ecological-economic balance

More than ten years after the publication of the Stern Report, the impact of the increasing economization of nature in popular and political discourse needs to be carefully scrutinized. For example, although the German national version of the global TEEB study showed how the economic output of companies depends at least indirectly on ecosys-

Figure 8: Proposal for a holistic economic accounting matrix for ecosystem use using the example of forestry. Currently, only a few selected values are used to calculate economic profitability and as a basis for the management of the system. But it is factors which are currently either not ascribed a value at all or whose value is difficult to quantify which would be particularly relevant in assessing whether a given use is really sustainable and ultimately 'profitable' (€ = relatively easy to evaluate in monetary terms / × = can only be evaluated in monetary terms with difficulty or partially or not at all); usual timescale for impacts: Years = Y; Decades = D; Centuries = C)

tem services (Naturkapital Deutschland 2015), it remains to be seen whether such knowledge can actually be applied in practice within the current economic policy framework. Even if environmental costs and 'added damage' were to be universally recognized and acknowledged, there remains the question of 'who pays the bill?' At the moment, it looks as if the costs will be borne by the global ecosystem alone, and thus by future generations as well. It seems, though, that an ever-growing number of consumers are no longer prepared simply to accept this gross iniquity. Surveys show that 80 % of respondents do not want to buy products in future from companies that take no account of ecological and social concerns (Naturkapital Deutschland 2015). However, it is well known that such surveys have only limited value, because they do not fully reflect genuine willingness to take action (▻ Chapter 2.2), and the market share of products boasting various sustainability certificates is still small (e.g. FSC – *Forest Stewardship Council*; organic certification; MSC – *Marine Stewardship Council*). While many of these certifications have become more important, it has not been proven that they guarantee real effectiveness or that the products concerned are sold at their 'true' price, one that will prevent or compensate for their harmful effects.

Ultimately, it is doubtful that the systemic defects of the economy can be nullified by the use of market-based and voluntary instruments (▻ Chapter 4.3). Instead, a fundamentally new recognition of the importance of ecosystems is needed. The preservation of ecosystems is not something that we might perhaps be able to turn to later, once the economy is performing well and jobs have been secured. Rather, this is about recognizing the limits to ecosystem resilience and the necessity, for our own survival, of ensuring the continued functioning of working ecosystems. And that requires a conscious focus on ecosystem-based sustainable development (▻ Chapter 4.1).

Close-to-nature forestry in Germany – caught between the increasing demand for wood, the need to ensure sustainable forest ecosystem services, and society's demands and expectations on forests

Digression 1

Martin Guericke and Peter Spathelf

The most widespread silvicultural concept in Central Europe is that of close-to-nature silviculture. This is an integrative concept for the development and management of (structured) mixed forests, with a particular emphasis on natural regeneration and the minimization of damage to soil and remaining stock. The focus of close-to-nature silviculture is on the individual tree and its vitality, health and quality, with the aim of continuously promoting the structural and species richness of the forest. Harvesting is highly selective (but can extend from individual trees up to small stands); tree species belonging to the local native forest community are preferred for regeneration.

The origin of close-to-nature forestry dates back to the 19th century, when it became apparent, following the extensive reforestation of devastated regions, that uniform single species populations do not make for stable forest ecosystems over the long term. One of the first proponents, in the late 19th century, of the concept of mixed (structurally rich) forests was the Munich forest scientist Karl Gayer. He was followed by Alfred Möller, Professor at the Forest Academy Eberswalde, who advocated what is now called 'continuous cover forestry'. This involves the selection and harvesting of individual trees (i.e. the avoidance of clear cutting), natural regeneration and the development of structured mixed stands as a guiding principle of silviculture. However, for a long time these ideas and concepts remained a marginal tendency within German forestry—the high forest of uniform age was the dominant silvicultural system until the end of the 20th century, in both West and East Germany (Spathelf 1997).

In the 1980s, the dominant concept of silviculture changed significantly due to various factors such as the emergence of the 'environmental movement', 'forest dieback' (German *'Waldsterben'*), and the growing evidence that mixed (structured) forests can be more productive and above all more resilient against disturbances than single-species stands of the same age (e.g. Pretzsch 2003; Knoke et al. 2008; Brang et al. 2014).

The current uncertainties regarding the future of the climate call for effective short- and long-term adaptation strategies to enhance resilience, with the aim of increasing the adaptive capacity of forests and thereby ensuring the sustainable provision of all forest ecosystem services.

The silvicultural measures this requires focus on:

1) increasing tree species diversity
2) increasing structural diversity
3) increasing genetic diversity and
4) increasing the resistance of individual trees to disturbances

A detailed account can be found in Brang et al. (2014) and Spathelf et al. (2015). In adaptive management, what is important is to reduce the risks incurred in long-term forest management (Eichhorn et al. 2016) and in the implementation of specific short-term silvicultural decisions on the basis of continually revised silvicultural guidelines and recommendations (e.g. Reif et al. 2010; Landesbetrieb Hessen Forst 2016). The intensity with which the respective adaptation measures are applied will increase in line with the (local) need to ensure the ongoing provision of specific ecosystem services. Within the wider adaptation strategy, the forest regeneration phase is of particular importance (Spathelf et al. 2014). The selection of tree species for regeneration which are suitable to local conditions and resilient to long-term climate change should not explicitly exclude non-native species in addition to the native ones. A number of species have now been identified which—taking full account of risks and opportunities—can contribute to greater 'silvicultural risk diversification' (Vor et al. 2015).

The basic operational maxims of close-to-nature forest management have received general assent as an appropriate way of handling the tensions between several competing demands—ecological and economic aspects of sustainable forest management, the comprehensive, long-term protection of all forest ecosystem services, and changing social awareness of the forest as an object of economic, cultural, and natural significance (BMELV 2011). In contrast, forest management in the sense of 'orderly and sustainable forestry' (see the German Federal Forest Act, BWaldG §11, and, for more detail, the respective forestry regulations of the individual federal German states) is often seen as controversial due to differing social and individual perceptions and priorities, and is the subject of much contentious debate within forestry policy circles and in popular science forums (e.g. BUND 2009; Lehmann 2010; Wohlleben 2015). Such debates contribute to a fundamental questioning of the role of forest management in society (Schraml 2016).

In terms of federal policy, the German Sustainability Strategy (Bundesregierung 2017) and the Biodiversity Strategy (BMUB 2007) have recently set out a series of milestones and goals, including for the management and development of forests based on close-to-nature forestry principles. However, according to the German Advisory Council on the Environment (SRU 2016), these need to be more vigorously pursued. In particular, they recommend specific localized strategies to legally secure the public ownership of 5–10 % of total forest area under natural development, and a high-quality certification system which should be applied to at least 80 % of total forest area (SRU 2012). Meanwhile, nature conservation experts are calling for greater efforts and more widely applied standards for the development and management of semi-natural forests in order to protect biodiversity (NABU 2010).

From the point of view of economic policy, however, the wood industry sees increasing risks for the forestry and timber sector. In particular, forestry is accused of ignoring future demand for raw materials and of producing without due regard for the market. It is feared that the further decline in coniferous woodland — due to the nationwide enforced conversion to broadleaf trees (BMEL 2014), the further increase in stocks, above all of hardwoods, and additional politically motivated 'land set-aside' — will lead to a scarcity of raw timber in the foreseeable future. According to the results of the Federal Forest Inventory (BWI 3) and felling estimates, an average of approximately 76 million cubic metres (m3sub) of raw wood per year was felled in Germany between 2002 and 2015 (Jochem et al. 2015). By contrast, the annual growth amounts to approximately 98 million cubic metres (m3sub), due to the current composition of the tree species range and the age structure, which has meant a continuous increase in stocks to approximately 336 m^3 per hectare in recent years. At 3.7 billion m^3 total stock, more wood is available in German forests than in any other country in the European Union. At the same time, demand for wood in Germany has been rising since 1991, to around 119 million m^3 in 2015 (Rosenkranz et al. 2017). Different scenarios suggest an annual demand for raw wood in Germany of between 123 and 141 million m^3 by 2020. Accordingly, a shortfall of approximately 20 million m^3 of forest wood per year can be predicted.

Since 2009, Germany has been recording net imports, especially of raw pine wood suitable for timber, of up to around 6 million m^3 per year.

As a result, pressure is mounting for greater exploitation of existing forest areas and for weakening or abandoning the principles of sustainable forest management in Germany. This pressure is also influenced by ongoing economic globalization. In Germany, this factor is significant, and will remain so for future forest management, because about 48 % of the total forest area is privately owned. Close-to-nature forest management ensures a balance between the economy, the ecology, and the legitimate expectations of society. It aims at a mixed, structurally-rich forest that is particularly well-suited to meet the multiple demands and requirements placed on it.

Can peatland use be sustainable? The answer: Yes – as Paludiculture

Vera Luthardt

Peatlands are wet habitats in which peat-forming vegetation grows (Colditz 1994). This means that these ecosystems live from water, harbour a very specific community of lifeforms, and accumulate carbon in the form of peat. If water is removed from the ecosystem, it is fundamentally changed in terms of both habitat quality and carbon turnover — a greenhouse gas sink becomes a source (Fig. 1). Typical organisms disappear, and the soil degrades.

Historically, every form of peatland use, in every part of the world — be it peat extraction or agricultural cultivation — has required the land to be drained. The resulting processes run counter to all requirements of national and international legislation on the protection of climate, soil, water and biodiversity. Despite this, peatlands are increasingly being drained across the world to open them up for exploitation, and such exploitation continues unabated in Europe. The German TEEB study (Naturkapital Deutschland 2015), which devotes a whole chapter to carbon-rich soils, states that 30 % of emissions from agriculture and 4.4 % of Germany's total annual emissions come from peatland sites used for agriculture.

The processes triggered by lack of water continue under extensive use, and even intensify after the use ceases. This leads to increasing problems in the agricultural management of these low-lying areas, and as a result there are growing calls for compensation for low yields, both in dry phases and because of heavy rain events. The critical factor is water management, and so the key to answering the question posed at the beginning of this digression is: 'peatlands must be wet' — that is the motto of the Greifswald Mire Centre (www.greifswaldmoor.de). The GMC develops ways of combining resource protection (peat, climate, water) with creating added value: *Paludiculture* = swamp culture. The paradigm shift — which is far from having been completed — consists in not optimizing the site in terms of its technological suitability, so that traditional technology and historically tried and tested mowing and grazing systems can be used, but instead developing systems of use for the site in its present condition and at maximum possible levels of water retention.

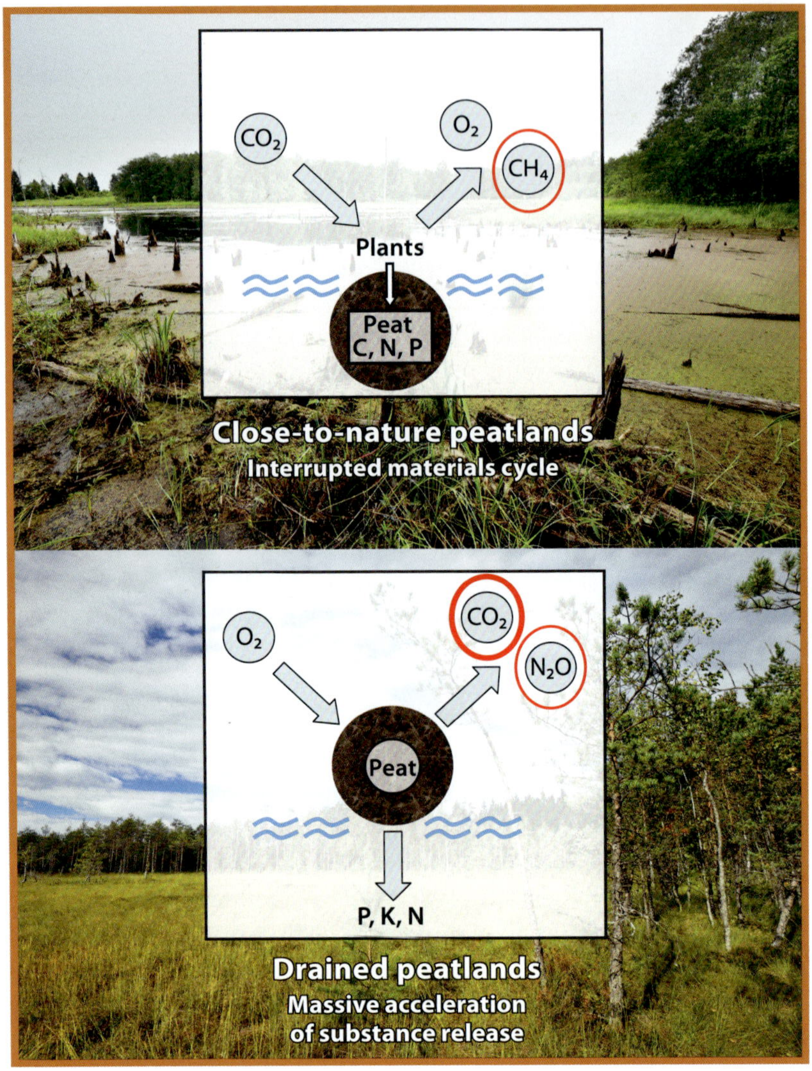

Figure 1: Close-to-nature peatlands accumulate carbon in the form of peat from the incomplete decomposition of dead organic matter under anaerobic conditions. Small amounts of methane are thereby built up in a long-term process and then released into the atmosphere — which means that natural peatlands are not climate-neutral. In the presence of air and absence of water, the peat is very quickly mineralized by microbial decomposition. The final products, in addition to water-soluble mineral components, include large quantities of carbon dioxide and nitrous oxide. Greenhouse gas emissions thus increase by a factor of between 2 and 5, depending on the specific conditions. (photos: Aleksey Stemmer)

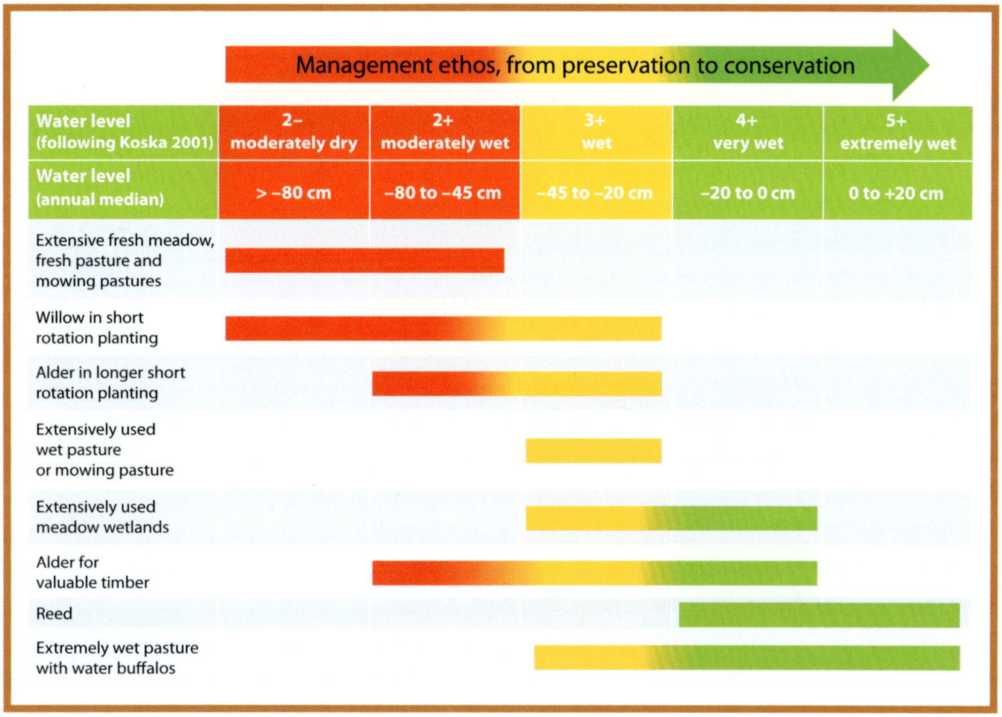

Figure 2: Potential value creation possibilities on a spectrum from preservation to conservation of peatland sites depending on the available water supply (from Schröder et al. 2015)

There are many different approaches to this (Fig. 2) (Schröder et al. 2015; Wichtmann et al. 2016).

In order to be able to balance different demands on the use of peatlands in a transparent way, experts have evaluated the ecosystem services of peatlands at their various stages of development (Table 1). In addition to the essential protection of the last remaining, mostly small, areas of mires that are still close to nature, for the preservation of our biodiversity, swamp cultures represent the way forward towards creating value focused on soil, water, climate and habitat conservation.

Freely available decision support tools for land users (such as Zeitz & Luthardt 2018) can promote the dissemination of knowledge about peatlands and their sustainable use. 'Re-wetting peatlands is one of the most effective and cheapest climate protection measures in the land-use sector and has a reduction potential of up to 35 million t CO_2eq per year in Germany' (Naturkapital Deutschland 2015). However, there are still many hurdles to overcome in securing its implementation, from creating the will and the risk appetite to instituting the formal requirements and support mechanisms and surmounting the legal restrictions.

Sector	Group	Close-to-nature	Drained & used	Drained & abandoned	Re-wetted & unused	Paludi-culture
Provisioning	Food and animal feed	0	+++	0	0	++
	Plant fibres (building materials, straw, substrate)	0	++	0	0	++
	Fuels from biomass	0	++	0	0	++
Regulation & Maintenance	Climate regulation (global & local)	++	−−−	−−−	+++	+++
	Water purification / nutrient retention	++	−−−	−−−	+~	++~
	Regulation of the water cycle	+++	−−−	−−−	++	++
	Habitat of specialized species / gene pool	+++	+	−−	++~	++~
Cultural services	Feeling for nature, recreation	+++	+	+	++	+
	Information & knowledge (processes, archives)	+++	+	+	++	++

Tabelle 1: Comparison of selected peatland ecosystem services in relation to water management and use (+++ strongly positive, ++ fairly positive, + slightly positive; −−− strongly negative, −− fairly negative, − slightly negative; 0 no effect; ~ develops over time) (from Luthardt & Wichmann 2016)

The drivers: economic systems

3.2

Alexander Conrad and Jan König

Economics is not just an academic discipline; it is a core human activity. We are constantly confronted with the question of how best to spend our resources — money, time or labour — in order to realize our goals. We are part of an economic system in which we offer and demand services via markets. These markets represent a coordination mechanism intended to ensure that no resources are wasted, since, according to the theory, they lead to the most efficient use of scarce resources. The theory also says that efficient markets contribute to maximizing prosperity for all, that they ensure that companies reach their optimal operating size and do not grow indefinitely, that resources are only demand-driven, i.e. geared towards customers' needs, and that over the long term an equilibrium is always found, without overproduction or unemployment. But things are different in practice: there is a focus on strong, permanent growth, while simultaneously resources are wasted, nature is depleted, and unemployment is accompanied by increasing inequality of wealth and prosperity. How does all this fit together? Why are we so focused on growth when it also leads to problems that endanger humankind? What brings about economic growth, and what are the fundamentals of economics?

Introduction: What is economics?

Managing scarce resources in a planned and rational way

In the standard literature on ▹ **economics** the subject is usually explained in the following way (see Wöhe & Döhring 2013, p. 33). The needs and wants of human beings are unlimited. At the same time, however, the resources available for the creation of products and services that meet those needs are limited. This creates a tension that can only be resolved through systematic, rational action. It is not only human beings who have to get by with limited resources; all organisms and ecosystems are forced to manage their budgets, albeit in a less planned and considered way (▹ Chapter 1.3).

> **The planned, rational and sensible use of limited resources is referred to as economics.**

Economics as an everyday activity

And yet, economics is an everyday activity that plays a role in many situations. When you go shopping at the supermarket, you have to consider your budget, and what products you need to buy with it. Before taking an exam, you have to decide what result you want and how much preparation is needed to achieve it. Different possibilities and needs or desires are always competing with each other. In the supermarket, you can choose between a range of products, and you have to ask yourself which among them satisfy basic needs (such as hunger and thirst) and which are rather in the category of luxuries. With regard to your time budget, you have to consider how to apportion it between work, learning, leisure and sleep, etc. But we have to do this not only in our private lives and as consumers. Companies also have to practise economics. They have to decide what products to produce, what resources to use in doing so, what the production process will involve, and at what price the product will ultimately be marketed and sold.

Homo oeconomicus

Is it possible to imagine someone who only ever thinks and acts according to the precepts of economics? The answer to this question is given by the theoretical model of *Homo oeconomicus*, which found its way into economics during the period of so-called classical economics (18th century). The model is a way of formulating four assumptions that must always be met if we are to be able to speak of an 'exclusively economically-minded individual':

(1) Every individual strives to maximize their self-interest. (But is this realistic or morally defensible?)

(2) Extrinsic incentives are the triggers for economic activity. (This would require that every action is preceded by a calculation of its potential advantage—a cost-benefit analysis)

(3) Complete information for assessing all alternative options must be available. (How realistic is that? Even in the age of digitalization, not all information is generally available and there are usually gaps, for example between the information available to sellers and buyers)

(4) Decisions are made on the basis of rational choice. (This would include the maximization principle, which states that a maximum return should be achieved using given means. But is it even possible to always see what can be achieved with given means?)

The model thus describes so-called fully rationally-acting individuals, but these are probably very rarely encountered in practice. Rather, we can assume that human behaviour can be assessed as being only rational to a limited degree, or even in fact irrational. Thus, an individual may choose to buy a particular product that does not appear to promise them any direct benefit. It is only if one takes into account the social group to which the individual belongs that the purchase decision can be seen as conferring utility (such as ensuring membership of the group through ownership of a product tolerated or required by the group). Other examples could be given which suggest that our (economic) behaviour depends on multiple factors and often contradicts the assumptions behind *Homo oeconomicus*. *(see Rogall 2013, p. 74)*

Profit and utility

Economic activity thus becomes a means to an end, and its ends can be defined in quite different ways. Classically, they are related to the maximization of utility and profit: consumers are believed to engage in economic activity to maximize their own utility, while firms strive to maximize their profits. But what are ▹ **utility** and ▹ **profit**? While the latter can be easily defined as the difference between income and expenditure, and a comparison of this monetary measure for different companies presents no problems, the definition of utility is more difficult. From a narrow economic perspective, utility is the capacity of a product or service to satisfy a given need (Pindyck & Rubinfeld 2015, p. 120).

Profit is the difference between income and expenditure, or between revenues and costs.

Utility is the capacity of a product or service to satisfy a given need.

Ethics interprets the term more widely, as encompassing a 'good feeling' and even 'social respect' (Gabler Wirtschaftslexikon 2017). But how can utility be measured and compared, since it appears to be open to a high degree of subjectivity? The maximization of utility and profit are the classical ends, and those still most frequently cited in today's economic literature. However, they are not the only ones. Moreover, they are certainly not the most timely of pursuits in light of the environmental and social challenges our society faces (▹ Chapter 4.3).

The economic system

A systems approach to economics

Like the ecosystem (▷ Chapter 1.3, 3.1), the economy, too, is a system: it is based on complex economic processes that result from the manifold interrelations between the actors involved (Sik 1987, p. 17). Here, too, it is true that the interaction of the individual parts can create emergences, so that, for example, the division of labour and specialization can increase the efficiency of the resources invested in the ▷ **economic system**. But it is also true that economic systems can be subject to **negative feedback**, and can tip over.

> **Economic systems are based on complex economic processes resulting from the manifold interrelations between the actors involved.**

Cycles

A simple economic system can be pictured as a kind of ▷ **circular flow model**. The model makes it easy to see how it works and how it reacts to changes. The so-called closed two-sector model (see Fig. 2) depicts a world in which only two actors in the economic system interact (see Core Project 2015, pp. 45 ff.): private households (consumers) and firms (producers). The goal of households is to satisfy their needs as far as they possibly can, and to that end they demand products from firms. They earn the financial resources needed for this by offering firms their labour. The firms employ them in their production processes and produce exactly those products that the households demand. In this closed system, the earnings of households are equal to the revenues of firms from the sale of their products.

But not only is this so-called ▷ **monetary cycle** (monetary sphere) closed and static (no positive or negative growth); so too is the ▷ **commodities cycle** (real sphere). In the economic system, labour is a commodity that can be traded on the labour market. It enters into the pro-

> **Circular flow models show the economic agents participating in an economic system and the transactions between them.**

> **The monetary cycle shows the money flows between economic agents.**

> **The commodities cycle shows the goods traded between economic agents.**

duction process as a factor of production. Consumer goods, which are demanded by households and needed to regenerate the labour employed, are an outcome of this process.

Negative feedback loops and tipping points in the economic system

How might a negative feedback loop operate in the economic system? Let's look at the financial budget of a city. The city collects money in various ways, but mainly through taxation, and then spends it for such purposes as the support of the administration, projects for the benefit of the citizens, and meeting its legal responsibilities. Let's assume that the city spends more money than it receives, and thereby accumulates debts. For these debts, the city has to make interest payments from its revenues. This means that overall the city has reduced resources available to meet its responsibilities, and life in the city may consequently become less attractive for some people. They leave. This diminishes the tax base and lowers municipal revenues. If the debt has still not been repaid, interest must still be paid from the already reduced revenues. The proportion of those revenues available for meeting its responsibilities continues to decline. As this negative spiral continues, it can reach a point where the city is completely incapacitated. The proportion going to service the debt (through interest payments) is now so high that the city no longer has any room for manoeuvre and can no longer fulfil its responsibilities. This development can be accelerated if the city takes on more and more debts in order to preserve its supposed capacity to act. It will also be accelerated if the interest rate on the loans taken out (at variable interest) rises. The negative feedback consists in the fact that the starting value of the closed loop in this case — the revenues — decreases round by round because of the existence of debt and the interest payments made (and decreases exponentially because of compound interest).

Tipping points often appear in economic systems in connection with the formation of so-called speculative bubbles (see Kindleberger & Aliber 2011 for a general detailed presentation of 'Manias, Panics and Crashes', and an explanation of the examples shown here in particular). This is when the prices for a certain traded good (e.g. houses, or company shares, or even tulip bulbs) shoot up to unrealistic heights. The goods in question become objects of speculation. The system is continually fuelled, and usually grows exponentially, as long as there are still buyers willing to pay the inflated prices. However, when there are no more buyers, the system tips over and the price of the goods plummets and stabilizes again at a lower level. The US real estate crisis reached a speculative peak between 2006 and 2007. Compared with 2000, the prices of certain properties had more than doubled. When the bubble burst, meaning the system tipped over, the global economy tumbled into a (financial and sovereign debt) crisis from which it is still recovering to this day.

The dotcom bubble is another example of a tipping point. The collapse of the New Economy happened in the year 2000. Until then, the values of stocks traded on the stock exchanges for companies that offered (sometimes only purported) services only on the Inter-

Figure 1:
The dot-com bubble – daily highs of the NASDAQ Composite index between 1998 and 2002. The NASDAQ index includes all stocks listed on the North American Stock Market for high-growth sectors. With around 5,000 (inter)national corporate securities listed, the index is one of the most important stock market barometers in the world. (data source: finanzen.net 2017)

net had increased more than tenfold, without this increase being even remotely linked to a rise in real value.
One of the earliest (well-documented) bubbles is known as tulip mania. Here, the tipping point was reached in the first half of the 17th century. The value of tulip bulbs, which had become objects of speculation, had increased by a factor of ten in the space of just a few months in 1637, following several already enormous price jumps.

Changes in the system

The system is static when it is not subject to any changes. But how might changes come about? For one thing, the number of households might change, so that more workers become available and there is also greater demand for consumer goods. Similarly, firms might develop a new production process capable of producing a larger number of goods with an equal number of workers. As a result, the system would grow.

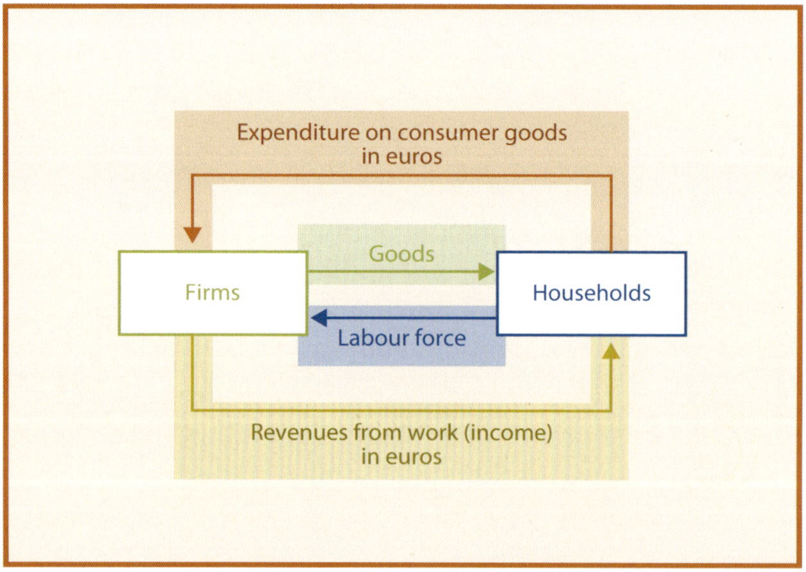

Figure 2: A simple circular flow model of the economy with two sectors (based on Core Project 2015, p. 45)

What role does the ecosystem play in the economic cycle?

The economic cycle depicted above shows a simple, closed system which only considers transactions between firms and households. Labour enters into the production process as a factor of production. But is the system as shown complete? Are all the 'resources' necessary for production included? What is the starting point for the establishment of a firm, and what powers the lights and the machines in the factory halls? What is needed to make bread, sideboards, paper and all the other products that feature in our daily lives? All of this requires things to be extracted from the ecosystem. Nature serves as a means of production (soil, water, coal, plants, animals, etc.) for the provision of services and receives the waste from production and consumption.

In the figure below, therefore, the economic cycle is extended to include the ecosystem and its interrelations with the (economic) actors already identified. It will become clear that the economy is embedded in the ecosystem, and that the limits to the development of the economy are closely linked to the limits to the ecosystem (▹ Chapter 3.1).

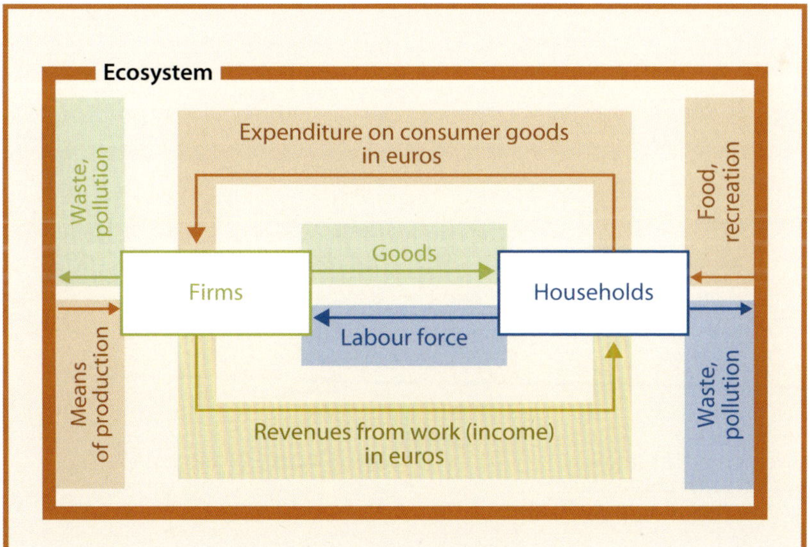

Figure 3: Circular flow model of the economy including the ecosystem (based on Core Project 2015, p. 45)

Decoupling of the monetary sphere

Decoupling of the money and capital markets from the real economy

'The [global] currency speculation of a single day adds up to a larger sum than the annual economic output of Germany or China' (Lietaer et al. 2013, item 1054). The description of the economic cycle gives the impression that the monetary sphere (the circulation of money and capital) is closely connected to the real economy (the circulation of goods). In economic theory, the monetary sphere (the money and capital markets) provides the liquid or investment resources needed to generate the production of goods and services in the real sphere. It might therefore be assumed that the two areas move in tandem. However, this is not the case. For several decades now (since about the 1980s), the monetary sphere has been growing faster and has lost touch with the real economy (see Fig. 4). Compared with the global output of goods and services (GDP), daily cross-border cash flows have multiplied in the years since 1980 and exceed global output many times over. Lietaer et al. (2013) estimate that only about 2 % of cross-border cash flows are related to the real economy. Rather, the remaining 98 % are attributable to speculative financial market transactions (Lietaer et al. 2013, item 1054 ff.).

One of the reasons for this development is the deregulation of international financial markets in the 1980s and 1990s, undertaken in order to give new impetus to stagnating economic development in countries such as Great Britain and the United States (Heiss 2014, p. 107). This

Figure 4: Decoupled financial system – Development of the foreign exchange market (data sources: Data for the World Currency Exchange market for the years 1980 to 1986, based on Lietaer et al. 2013; from 1989 onwards based on BIS 004, p. 76, BIS 2008, p. 102, Statista 2017b; GDP based on UNCTADSTAT 2017)

encouraged and enabled the development of highly speculative financial instruments and expanded the scope for banking activities (including lending opportunities, **money and credit creation**) (BPB 2017). In 1986, the British government lifted admission restrictions for traders in the stock markets. London subsequently became a leading international finance centre. In 1999, the government of Bill Clinton abolished rules dating from the 1930s limiting the scope of economic activities of US banks and separating credit and securities transactions. Both forms of transaction could henceforth be undertaken in one and the same bank. After the turn of the millennium, the way was also cleared for computer-aided high-speed trading in securities. The result was that the volume of speculative trading soared.

Credit as a driver of the uncoupling

The opportunity for banks to increase lending and reduce ▹ **credit rationing**, combined with declining interest rates and low inflation, also led to a steady increase in household debt (BIS 2004, p. 62 f.). More and more private households were able to take out loans (for example, to become home owners) or to pay for consumer spending by credit card—that is, 'on the never-never'. Borrowing now also became possible for households with a relatively poor credit rating and thus a relatively high risk of default (the risk that the loan cannot be repaid). The banks, in turn, protected these so-called ▹ **subprime** loans by means of speculative financial instruments, which also fuelled the growth of the financial system. Figure 5 shows by way of example how private households'

Figure 5: Decoupled financial system – the example of wealth creation in the USA (data sources: OECD 2017a, b)

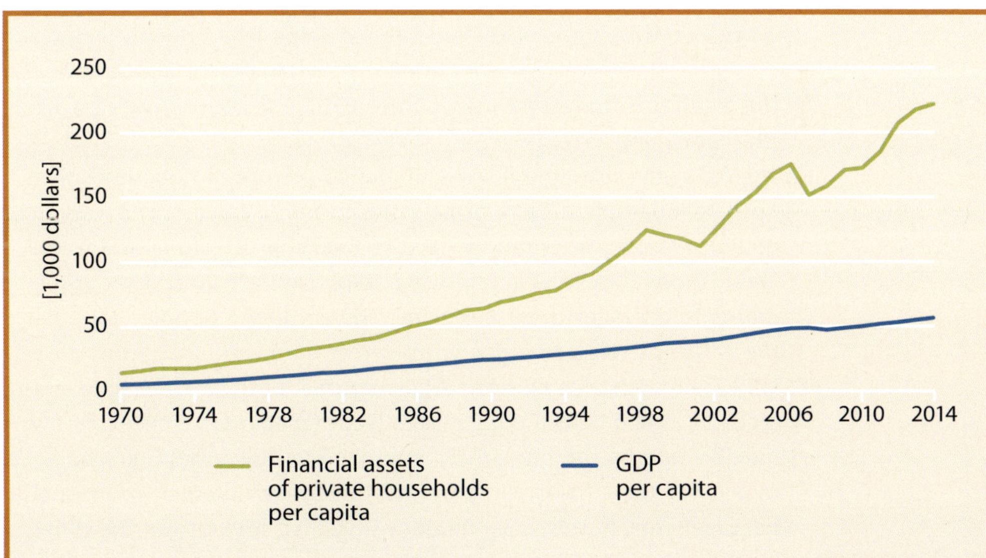

financial wealth (including debt) in the US has grown since 1970 compared to per capita output. This also illustrates the ongoing decoupling of the monetary from the real sphere of the economy.

Credit rationing refers to discrimination against particular classes of borrowers, so that some are unable to get loans because of their age, income, family, social or ethnic background, etc.

Subprime loans are those with a high risk of default.

What are derivatives, and how do they contribute to the decoupling of the financial economy from the real economy?

Derivatives are financial instruments. Their value depends on the future rates or prices of other goods, assets or market-related benchmarks such as interest rates or stock market indices. This means that derivatives trading does not bring about any real economic production. 'Derivatives are pure money-out-of-money transactions' (Felber 2014, item 2723). They are used to hedge against risks (e.g. against fluctuating interest rates or exchange rates, or against loan defaults). Above all, they are used to speculate. They then represent a bet on the development of their reference value (the original financial instrument). Derivatives are traded on exchanges (to about 25% of the total) or simply over the counter (OTC) between private parties (the remaining 75%).

The following example gives a simplified picture of how derivatives operate. Suppose a bank gives loans to home buyers with varying default risks. In order not to have to bear the risk alone and to get 'fresh' money for new lending, the bank can sell these loan agreements to an investment company. The company bundles together the different real estate loans from this bank, and also from other banks, and creates a new financial instrument out of them: for example, a so-called CDO (Collateralized Debt Obligation). This derivative may be sold by the investment company to new investors, who are repaid for their investment by the borrowers' interest and principal payments. In this way, the derivative also spreads out the different default risks. The fact that home loans at varying levels of risk are bundled together makes the likelihood of a total failure of the investment smaller. However, this principle works only so long as bad loans are balanced out by a sufficient number of good loans. In the course of the real estate bubble in the USA, however, more and more loans were granted to ever 'worse' (high default risk) borrowers. Moreover, house prices began to decline, which reduced the value of the collateral. The CDOs became correspondingly more risky.

The solution was to invent derivatives for the CDOs, too, thereby spreading the risk; and when that no longer worked, even more complex 'financial market innovations' were introduced. As a result, the financial instruments moved further and further away from the real values of the real estate, and their risk structures became ever more opaque and incomprehensible. But the financial sector grew.

Economic growth

Improvement in economic performance and its drivers

Gross domestic product

The performance of a country's economy is summarized or aggregated each year in its ▹ **gross domestic product** (GDP) (Brümmerhoff & Lützel 2002, p. 59). If performance improves over time, then the economy is experiencing ▹ **growth** and vice versa. However, it is only possible to aggregate things that have a (market) price, because only with this can a measure of value be extrapolated from the quantitative measures of products and services created. There are two problems associated with this method (Goodwin et al. 2014, pp. 127 f.). Firstly, it does

Gross domestic product (GDP) expresses the total domestic economic output over a reporting period. It only covers goods and services that have a market price. Certain aspects of the economy—whether social (housework, education, income and wealth distribution, etc.), environmental (pollution, natural disasters …) or economic (needs-based production)—are ignored.

Economic growth is when GDP grows (in real terms, i.e. price-adjusted, or 'corrected' for price changes).

not include services that have no market price. These include voluntary work and the care of children or other family members, but also running your own household, or undeclared work in the shadow or informal economy. Moreover, the depletion and damage done to the ecosystem (negative externalities) are not captured unless they are reflected in the price of the service provided (that is, if they have not been internalized). Secondly, the prices of the goods themselves may change, which means that the monetary value of the services they provide rises independent of the real (quantitative) supply of goods to the population.[1] Effects that take place only in the monetary sphere can then overlay and obscure developments in the real sphere. Despite these problems, GDP has established itself as the international benchmark for economic performance and development. But there are alternatives (▹ Chapter 4.3).

GDP as a summary of national accounting

But how exactly is gross domestic product calculated? It is calculated using national accounts, a central statistical programme that systematically presents all economic activity within a national economy (and its sub-regions) over a given period. It records the products and services produced and how they were used. It records the incomes and profits generated through production, and the value added by each industry (or rather, by each stage in the value chain). From these records, the GDP can be determined, using either the production approach (how much value has been added, and where?), the expenditure approach (who consumed the goods and services produced?), or the income approach (who earned what in the process?) (see Brümmerhoff & Lützel 2002, p. 59 f.).

As outlined above, economic growth is equated with an improvement in economic performance. The ▹ **drivers of economic growth** are therefore all those factors that contribute to such an improvement (Blanchard &

1 For example: in an economy, only commodity A is produced. In 2016, 100 items of A were produced. The price of each A in 2016 was 10 EUR. Thus, the nominal value of the output (the nominal GDP) was 100 × 10 = 1,000 EUR. In 2017, 100 pieces of commodity A were produced again. However, the price for A in 2017 was 20 EUR. Nominal GDP (100 × 20 = 2,000 EUR) has thus doubled. However, this growth is not due to an increase in production, but only to the increase in the retail price of commodity A.

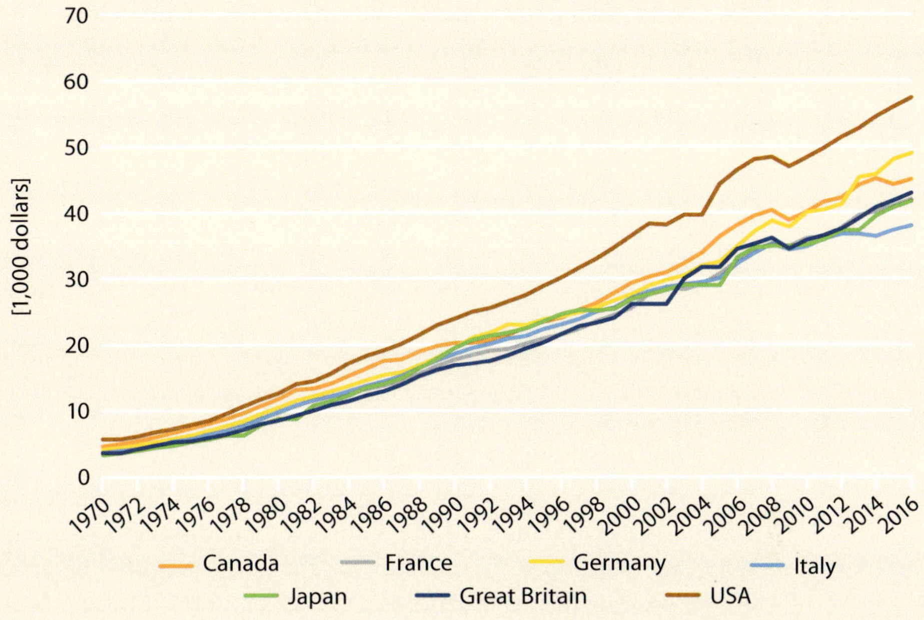

Figure 6: Development of GDP per capita in the G7 countries (data source: OECD 2017a)

Illing 2004, p. 295 ff.; Clement et al. 2013, p. 474 ff.). If performance is equated with output,[2] then the question arises of how this can be raised. In order to investigate this, it can be assumed for the sake of simplicity that the economy consists only of businesses and households (as in the economic cycle depicted above). Then output can be seen as resulting from the production processes in firms where the labour capacity of households is introduced as a production factor. Under otherwise constant conditions (or *ceteris paribus*), output then rises (falls) as the number of workers increases (decreases). Thus, population growth becomes a driver of growth in the economy, as does the improvement or extension of education and training, which enables people to input their labour into the production process.

The drivers of economic growth are all the factors that lead to a quantitative or qualitative improvement in the performance of an economy.

2 Output: the provision of goods and services by a firm, or the outcome of the production process.

Technological progress (TP) as an engine of growth, and creative destruction

The importance of TP as an engine of growth in the economy can be linked to the work of the economist Joseph Schumpeter (1883–1950) (Kurz 2017, p. 84 ff.). He was interested in TP as a consequence of innovation, and he established that innovation is a chief factor in the economic history of capitalist society. Innovation is here understood as the adoption of new combinations of properties of products, or of the production factors used in the production process. Innovation can also be the development of new goods, the improvement of the quality of goods, the opening up of new markets or of new raw materials sources, the reorganization of firms, and new social initiatives and trends. But how does capitalism generate the power for TP and innovation by itself? According to Schumpeter, TP continuously revolutionizes the entire economic system and brings about far-reaching social changes. A central role is played by firms, which are dynamic, energetic, creative actors in this process, overcoming existing 'restrictions'. The entrepreneur enjoys this victory and the social power and status that comes with it.

According to Schumpeter, creative destruction can be understood as a cycle. First, an innovation arises (for example, in the form of a more cost-effective production process) and creates extra profit for a firm. The greater profitability of this so-called pioneer firm enables rapid growth. The competition lags behind. Imitators then enter the market, and the pressure on existing competitors grows. They adapt, and introduce the new technology as well. The new process continues to spread until the extra profit gained by the pioneering firm is lost—because now everyone has been able to reduce their costs. The industry as a whole has achieved technological progress. For extra profits to be enjoyed again, a new innovation is needed. As the rate of innovation is increasing rapidly, a major socio-political debate has arisen in recent years on responsible, i.e. more socially embedded, innovation *(RRI – Responsible Research and Innovation).*

Extensive and intensive growth

Of course, firms do not just employ labour for production. There are also machines and technologies (capital), raw materials from nature, energy, information and much more. An increase in these inputs (the starting materials for production) leads *(ceteris paribus)* to higher output. In such cases, where the increase in output is due to an increase in inputs, growth is described as extensive (or quantitative). Intensive (or qualitative) growth, by contrast, occurs when the number of inputs into the production process does not increase but their efficiency (productivity) improves, because of technological advances, for example. Then the inputs used become more productive, which means that more output (and thus growth) is created on the basis of the same number or volume of inputs.

Growth through positive economies of scale

A combination of both effects (for example, an input of capital as a production factor and an increase in the productivity of certain other factors of production) is also conceivable. Although the labour input remains the same, its productivity can be increased by involving more technology and equipment per worker in the production process (an increase in capitalization). This may make it easier for workers to do their jobs, and they can therefore do them faster or produce more in the same time.[3] Economic growth then results from so-called ▹**positive economies of scale** (scaling up the production process), where an increase in inputs leads to a disproportionate increase in output. As a consequence, a comparatively small input can be associated with a comparatively large output. Assuming that the prices of inputs (e.g. the wages of the workers employed) and outputs (the retail price of the goods) do not change and that the additional outputs can be sold on the market, this results in higher profits for the firm.

> **Positive economies of scale are when a doubling of production factors leads to an increase in output which is greater than double.**

Additional drivers of growth

These profits can now be used to invest in the production process, in accordance with the profit maximization objective outlined above. The following rule applies here. If the company only invests enough into the production process to maintain the existing level of production factors, and especially of capital (by, for example, replacing worn-out machinery), no further growth will occur in the economic system. But if the company expands its production capacity by, for example, procuring more or bet-

3 Whether this is a matter of cause and effect is discussed below.

ter machinery, then (extensive) growth will ensue. This capacity-expanding effect, which is referred to as supply-side growth potential, can then be matched on the demand side by a corresponding utilization potential effect, or an income effect, leading to additional economic growth. The fact that new machinery enters into the production process through investment usually increases the demand for labour (to operate the machines). More workers can be employed, and/or each worker can work longer (more full-time instead of part-time work). Overall, this should (theoretically!) increase household incomes, leading to additional demand for goods and — in the narrow economic definition of the term — to an increase in utility (utility maximization). All the factors that facilitate the achievement of the two goals — profit and utility maximization — are thus also drivers of economic growth.

Money creation by banks as a driver of growth

The account above mainly refers to drivers of growth that are linked to the real economy. However, in the context of the discussion of the economic cycle it was made clear that the real economy does not exist in isolation; rather it is linked to monetary flows (income, consumer expenditure, etc.). We saw that production factors are needed for the production of goods and services. But how are these paid for? In the closed economic cycle outlined, all the goods produced by the firm were consumed by households. The money earned by the households was completely transformed into consumption and flowed back to the companies as income.

So where do companies get the money needed for investment (or expansion of their potential productive capacity)? An answer to this question may be provided by the following scenario (Binswangen 2015, p. 19 ff.). Let us assume that private households do not consume their entire income, but save part of it. These savings could be made available to companies in the form of a loan, which would finance investments. But why should households do that, and on what terms? This would only make sense in economic terms, or in terms of the goal of maximizing their utility, if the consumption and utility they thereby defer or forgo today (and that is what saving amounts to) were compensated by the possibility of greater consumption and utility in the future. This would mean in turn that households had a higher income in the future, greater than their combined income from labour and savings. The company would therefore have to be paying an additional sum to the households, namely the ▹ **interest**.

Interest is the price paid for a transfer of money or capital. Borrowers pay it as a price for the risk taken on by lenders. Savers receive it as remuneration for the transfer of capital and the associated deferral of consumption.

Interest through profits

But how do companies generate this interest? They generate it by increasing their, profits. But where does profit come from? It comes from surplus earnings, that is, from revenue from the sale of products which exceeds expenditure, i.e. the costs of production. However, this will not be possible with a constant ▹ **money supply** (money circulating in the economic system), because that would mean that money was only being shifted around. With money supply constant, profits would only arise if the costs of production were to decline, for example because household incomes were reduced. But then less would be available in the form of savings for investment, and as a result there would be no economic growth. In other words, the money supply has to be increased so that there is always enough money for the payment of interest and for the consumption of the additional goods produced. But how is the money supply expanded? Through, for example, ▹ **the creation of credit** (Issing 1993, p. 53 ff.).

The money supply is the stock of money (coins, bills) in the hands of non-banks (e.g. private households). Several different measures are used to calculate monetary aggregates. The aggregate M1 is basically cash. M2 and other aggregates also include bank deposits, whereby the availability of deposits declines as aggregate money supply increases, and money accordingly circulates ever more slowly through the economic cycle.

Credit creation is creating money by lending.

Banks as agents of transformation

In the following section, the economic cycle as analysed so far will be supplemented by another actor — the banks. Banks are so-called intermediaries. They collect savings from households and transform them into — among other things — investment loans for companies. As part of this transformation process (Neuberger 1997, p. 21 f.), they create a single large loan out of many small savings. But they also transform the terms of the agreement, because whereas savings can usually be withdrawn by savers at any time, loans often take a long time to be repaid.

The creation of credit and money

How is new money actually created? Within the eurozone, euro coins and notes may only be issued by the European Central Bank (ECB). But so-called commercial banks can also create money (Issing 1993, p. 53 ff.). They do this through lending: they generate electronic money (bank money) through electronic payments into the current accounts of private households and companies. Bank money is thus money that the bank itself can create. Initially, it exists only within the bank, in customers' accounts. It is a kind of promise made by the bank to pay out bills or coins to customers on demand. When a customer takes out a loan from the bank, the bank credits that loan amount to the customer's account. The bank thereby has two rules to consider. It must reserve a portion of its own capital (its equity) proportionate to the risk represented by the loan agreement. This sum does not correspond to the full loan, but only to a fraction of it calculated on the basis of the statistical probability of default. In addition, the bank must deposit 1 % of the money created as a so-called minimum reserve in its account with the ECB. This money is supposed to ensure that the bank always has sufficient funds to meet customers' withdrawals. This shows that only a certain part of the credit issued existed in advance as money (or better, bank money). Most of the money came out of nothing. The customer can use the credit, among other things, to pay bills or to transfer money to another bank. The receiving bank then has a new deposit, which in turn can be used as a basis for lending. This means that the money initially created multiplies. This is the so-called multiplier effect (Clement et al. 2013, p. 385). Based on the minimum reserve of 1 % alone, 100 euros added to the system in this way can create 100 times its own value (multiplier: 1/r in the simplified form with r = minimum reserve ratio/100).

Banks also transform risk: risk-free savings deposits are transformed into risky loans. This last transformation is achieved by the intermediaries by means of certain forms of securitization. For example, banks will assess the risk associated with a loan and will then require borrowers to pay a risk-based interest rate or to put down a form of security (mortgage, surety, pledge, etc.). With these revenues, they build reserves which they can then release if and when — in spite of their risk assessments and supervision and monitoring of borrowers — someone defaults on a loan. The reserves thus released are used to repay the savings of the households. Money can now be created by the fact that the banks are carrying out this transformation process particularly efficiently, and, just like the company earlier, are giving out ever greater sums in loans (that is, their output is growing disproportionately) on the basis of the collected savings (here representing the inputs). The credit the company receives from the bank can be used, for example, to take on additional workers, or to have additional or better machinery made for them by another firm. The virtual money on the books of the banks becomes, step by step, real money (it becomes cash), and the existing money supply increases.

Figure 7:
Capital growth at simple interest and at compound interest

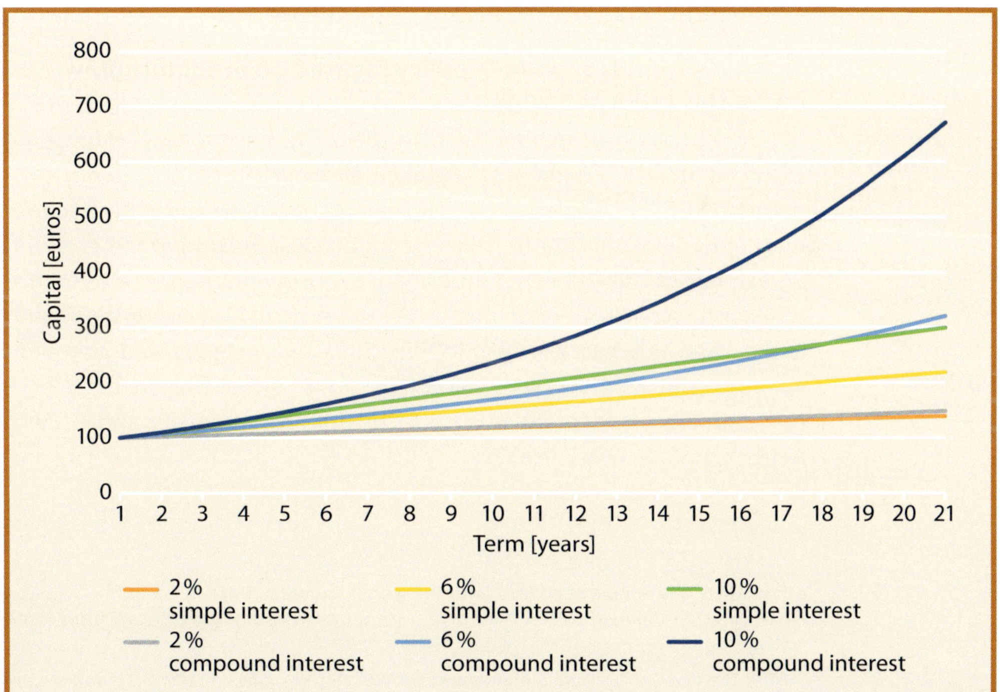

Exponential instead of linear growth

Certainly, the money that was previously 'created' can also be 'destroyed' through the repayment of loans and through paying out to savers, and the growth of the money supply thereby interrupted. However, this is countered by the compound interest effect (Clement et al. 2013, pp. 320 f.). As a rule, interest on savings and loans is not calculated each year solely on the initial savings or loan amount and then added to the amount saved or to be repaid. In other words, the increase in the savings and repayment amounts is not linear. Rather, the interest from one period is included in the sum on which the interest for the following period is calculated; that is, the savings or credit sum to which interest is applied increases by the amount of the interest from the previous period. The result is exponential rather than linear growth. This also means that the profits that have to be made in order to pay the interest from the transfer of savings to borrowers have to increase in line with the duration of the capital transfer. The consequence is that the growth of the money supply is also exponential rather than linear, and that the increase in economic output, which is the basis of profit, also follows an exponential growth curve and continues to drive up the growth spiral.

The focus on growth, and its consequences

To what extent is economic policy focused on economic growth, and is this focus justified?

The political importance of economic growth

'We don't have a desperate need to grow. We have a desperate desire to grow!' (Milton Friedman). A glance at the election and policy programmes of the parties and governments of market-oriented economies underlines this quote from Friedman.[4] Just looking at the coalition agreement of the last Federal German government, there is a reference to (economic) growth on almost every other page (Die Bundesregierung 2013) — and this has a long tradition. Since 1967, steady and adequate economic growth has been one of the four central objectives of economic policy, in accordance with §1 of the FRG's Stability and Growth Act.[5]

4 Friedman was honoured in 1976 with the Nobel Memorial Prize in Economic Sciences for his work in the field of so-called monetarism (notwithstanding protests from some colleagues).

5 §1 of the Federal German Stability and Growth Act identifies appropriate and steady economic growth as one of the so-called four magical goals of economic policy.

But what is the reasoning behind this supposed importance? It can be summarized using the following arguments (Hinterberger et al. 2009, pp. 30 f.) by way of an example.[6]

(1) Growth is equated with a better supply of goods, which is the basis for material prosperity. But does the supply really improve if the growth is based in turn on a shorter use life for products, under the motto of 'planned obsolescence'?[7]

(2) Growth requires positive net investment,[8] and this is, as shown above, a precondition for a high level of employment, which in turn is a key objective of economic policy. But does net investment actually lead to more employment, or rather to the substitution of labour by capital?

(3) Growth defuses distributional conflicts, because wage increases are easier to achieve in a growing economy. But is the growth in real wages in line with economic performance, and do the statistics not show that the gap between poor and rich is continuing to widen despite economic growth?

(4) Public services can be better financed (and expanded) in growing economies. In other words, growth is required to maintain the stability and sustainability of the social and welfare state. However, this would require that the state does not live beyond its means and settles its debts, so that rising tax revenues are not simply 'eaten up' by rising debt service costs — interest payments on loans taken out.

(5) Growth facilitates structural change, as workers from shrinking industries can find employment in growing industries. This connection is questionable in terms of the speed with which our digitalizing economy is changing. Small, less labour-intensive start-ups are 'destroying' entire industries with their innovative and usually digital business ideas. The labour force thereby freed up cannot easily be integrated into the

6 The question or comment following each assertion is intended to indicate how robust the arguments, and thus the cornerstones of the growth paradigm, are.

7 This refers to a planned reduction in the useful life of products. The result is that products need to be replaced sooner and more regularly, which can lead to growth in an otherwise 'saturated' market.

8 Net investment = gross investment less depreciation, which means that positive net investment exists when companies invest more in machinery and technology than was written off over the reporting period due to wear and tear; i.e. when they not only renew the capital stock, but expand it.

growth sectors, which are usually highly specialized, mobile and — as already mentioned — less labour-intensive.

(6) Growth is due to, among other things, technological advances that can help to better manage environmental challenges. According to this argument, economic growth and environmental protection are not in opposition, but rather have a positive association.

But is that really the case? Do the above conclusions necessarily follow, especially with respect to solving the formidable challenges related to our ecosystem?

Can economic growth improve environmental protection?

The environmental Kuznets curve

We can answer both 'yes' and 'no' to this question and it can be linked to an economic model that goes back to Simon S. Kuznets: the so-called environmental Kuznets curve, or EKC (von Hauff & Jörg 2013, p. 35 f.). It illustrates a reverse U-shaped relation between environmental damage and economic output. The graph suggests that environmental damage first increases along with increasing economic output up to a certain maximum point, but that once that point has been reached, environmental damage should decline while economic output continues to increase: this is the point when further increases in output are a consequence of the technological progress that results from solving environmental protection problems. Another possible partial explanation for this relation is that preferences among the general population change along with increasing prosperity (resulting from growth), away from material or economic life aspects (quantitative growth) to non-material or non-economic aspects (qualitative growth) — such as an intact and clean environment. The switch from a dirty industrial society to an environmentally friendly knowledge and services-based society could also help explain this relation.

The paradox of the Kuznets curve

Empirical studies have so far not been able to indisputably prove the relation described by the environmental Kuznets curve. Although some studies show that some individual pollutants do decline as economic output continues to grow, counterexamples can also be found, as well as examples that show the relation between environmental pollution and economic output as N-shaped. According to this interpretation, pollution first decreases with continuing growth of output only to then increase again.[9] Moreover, in some countries which introduced and maintained

9 There are other problems with the EKC. They are discussed in Uchiyama (2016).

The environmental Kuznets curve (EKC) and empirical evidence

Figure 8 shows an idealized EKC. Figure 9 links the evolution of GDP with the evolution of various environmental pressures over the period 2002–2014. It can be seen that while GDP has continued to increase, some forms of pollution have declined significantly at the same time. However, the values for the volume of waste also highlight that there are examples which disprove the universal validity of the Kuznets hypothesis. The illustrations ignore other possible relations. For example, it is possible that with increasing GDP, the pollution of the environment by pollutant X might decline because a production process that generated this pollutant is replaced by a new one. This new production process could increase GDP while reducing the environmental impact of pollutant X, which is no longer generated. However, the production process could also be linked to new, as yet unconsidered pollutants (or to general environmental damage), or to a displacement of environmental damage abroad (this actually happens with animal feed imports, palm oil, and the importation of many industrial products from China). It is also conceivable that exceeding the maximum point (for environmental pollution by a given pollutant) would result in irreversible damage to the ecosystem or to biodiversity. A reduction in levels of that pollutant would then be measurable, but in terms of the ecosystem it would be more or less irrelevant.

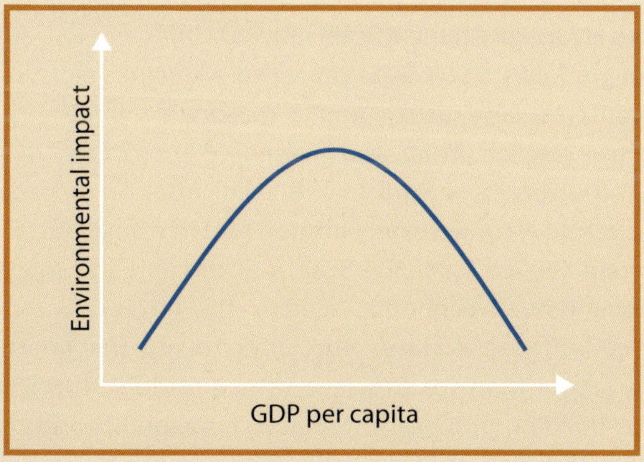

Figure 8: Environmental Kuznets curve (based on von Hauff & Jörg 2013, p. 35)

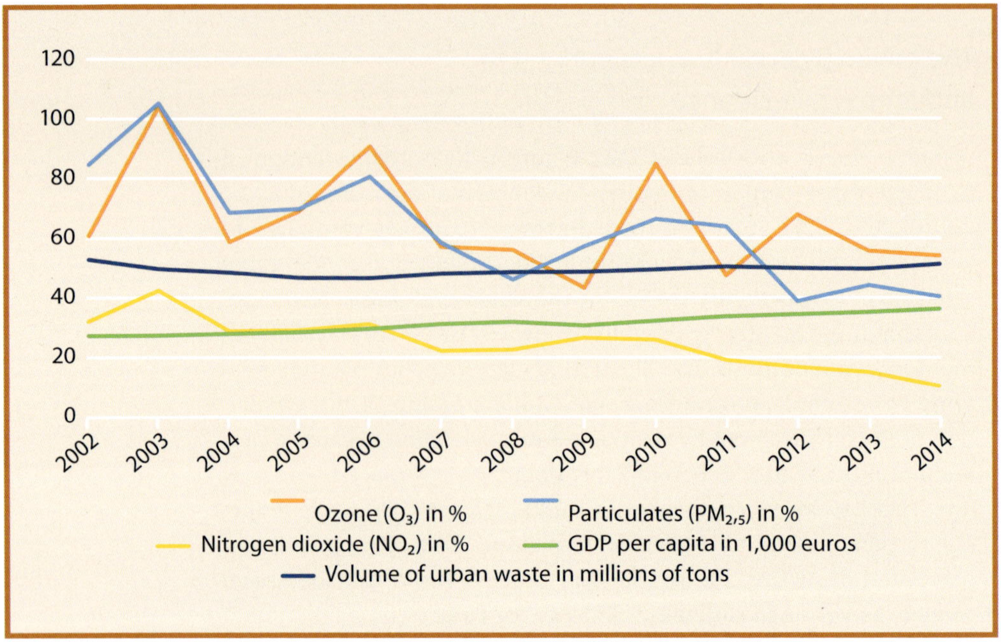

Figure 9: Development of economic output and selected environmental impacts in Germany (data sources: Statista 2017a, Umweltbundesamt 2017a, b)

high environmental standards right at the beginning of their development, it has been observed that the relation has not curved back but continued in a straight line.

Has the economy always been geared towards growth?

Growth orientation in the economy

Looking at the early history of economic thinking (back when developing economic systems were beginning to emerge) can provide answers to the question of whether the economy and an orientation towards growth have always been inextricably linked (Jhingan et al. 2012, p. 12 ff.). The ancient Hebrews (2500 BCE) practised an 'economic system' which prohibited interest and usury, had a reasonable price structure and a limit on the rate of profit, and mandated repayment of debt after 7 years and the return of land to the seller after 50 years in order to ensure that assets were not concentrated within a small number of families. In ancient Greece (800–300 BCE), a distinction was made between natural and unnatural economic activity: the former served to satisfy needs in line with social status, the latter served the sole purpose of enrichment (acquisition for the sake of acquisition). Lending at interest, but also building up wealth (the accumulation of assets, 'hoarding') was therefore regarded as particularly unnatural behaviour diametrically opposed to the functional role the economy was supposed to fulfil.

Under mercantilism (16th to 18th century), however, there was genuine pressure for growth, because economic growth guaranteed the funds needed to finance a continuing hold on power and the expansion of the nation state's sphere of influence. From then on, various schools of economic thought (classical, socialist, neoclassical, Keynesian, monetarist, etc.) addressed the consequences of this economic policy and put forward proposals for an efficient economic system. The focus of the discussion was not so much on economic growth itself as on the question of how people's welfare can be increased. Among the means examined for achieving this goal, however, the notion of growth was (and is) very frequently present.

Alternatives to the focus on growth

This raises the question of whether there is an alternative to the focus on growth, whether an economic and financial system with no or limited growth is conceivable, and whether such a system can be described as sustainable. Various approaches to sustainable forms of economic activity have subsequently evolved that address this issue (e.g. ecological economics and the steady state economy), and some of them are already being applied and developed (▷ Chapter 4.3).

The power source: energy supply systems

3.3

Vanja Mihotovic

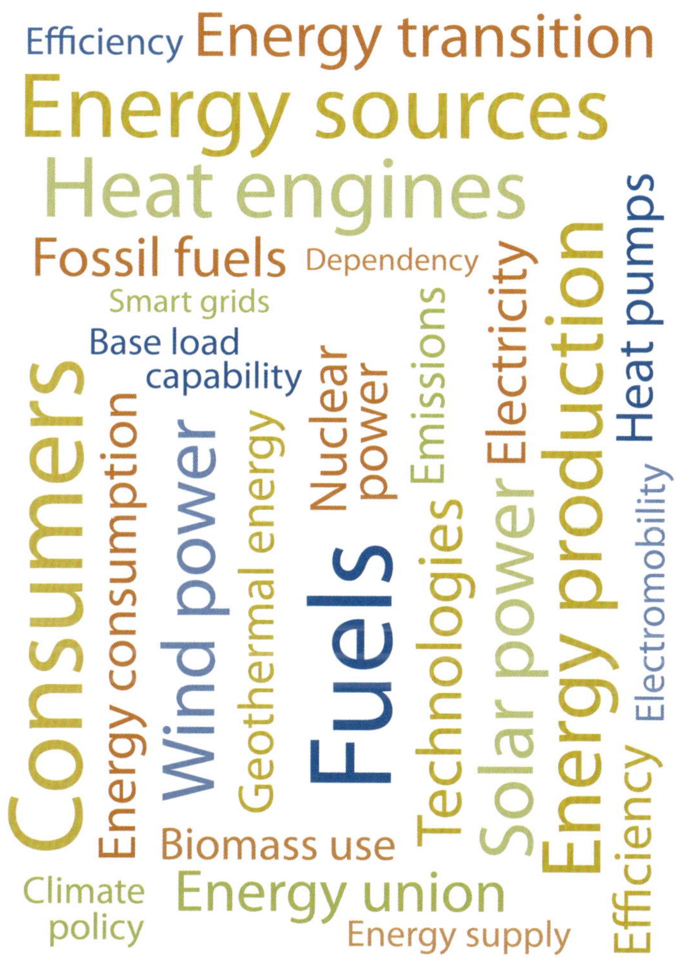

The industrial revolution of the late 18th century was the catalyst for profound changes in the economy and society. It was accompanied by an accelerated development of urban spaces, transport, technological innovations and science. The technological development of energy supply systems such as the steam engine paved the way for this. Energy plays an important strategic role in modern societies: everything depends on the secure supply of low-cost energy. There is still an enormous reliance on fossil fuels and nuclear energy. Ecological, economic and social problems across the globe triggered a debate on the 'energy transition' beginning in the 1970s. Economic considerations and limits on resources ('peak oil') are making a rethink imperative. The transition from conventional fuels to renewables is under way in many countries. But what is the current position? What needs to happen to enable the transition to a society with a sustainable energy supply? How much energy-saving is necessary (sufficiency)?

Energy supply now

A turn away from fossil fuels?

Fossil fuels such as coal, natural gas and oil have shaped national economies, industry, trade, transport, agriculture and people's lifestyles around the world. Fossil energy reserves are still a crucial factor in the geopolitical balance of power. Supplier countries remain in a strong position (Le Monde diplomatique 2012). Given the negative consequences for humans and nature — such as global warming and pollution — of the intensive use of fossil fuels, industrialized nations are being forced to restrict consumption. For a long time now there has been a political consensus in most industrialized countries in favour of reducing dependence on fossil fuels and switching to renewable forms of energy.

Demand for fossil fuels continues to increase worldwide, while crude oil is becoming an increasingly scarce resource. Wide-ranging studies show that as early as 2005, global maximum oil production was reached on conventional crude oil — ▹ **peak oil** (Kerr 2011). Since then, costly and polluting oil extraction from unconventional sources — such as the deep sea, tar sands and oil shale — has steadily increased to compensate for the decrease from disappearing conventional crude oil deposits. In view of the severe dependencies, a decline in production while demand continues to increase could lead to global economic turmoil. According to an OECD study, crude oil is still the most important form of energy, with a global share of 32.9 % of ▹ **primary energy demand**. Forecasts show, however, that its prime position could be lost to natural gas in the coming years (OECD 2016). But why are companies, politics and society so heavily dependent on fossil fuels such as oil, natural gas and coal?

The term peak oil defines the point in time at which global oil production can no longer be expanded to meet growing demand, but begins to decrease due to geological, energetic, technological and economic restrictions. The global oil production maximum for crude oil has been passed (Fisher 2008).

Primary energy is energy that is available in its original form (coal, natural gas, sun, wind, nuclear fuel). It is converted into secondary or end-use energy through processes in which a substantial proportion of the original energy is lost.

Fossil fuels consist of chemically stable carbon or hydrocarbons with a relatively low flash point. Advantages in use are its high energy density, the possibility of storage and transport costs. In addition, deposits can still be developed relatively inexpensively. The industrial revolution was based on the development of technologies for converting large amounts of energy into mechanical work. In the last century, especially in the industrialized countries, prosperity grew in step with energy consumption. Easy access to fossil fuels contributed significantly to prosperity in many areas of the world. The downside to all this is the enormous problems caused by emissions. Sulphur oxides cause acid rain (forest dieback). Carbon monoxide, nitrogen oxides and particulates damage the respiratory tract. While the latter pollutants can be reduced by filters, there are no technical systems for reducing the massive CO_2 emissions produced during the combustion process. In addition, there are substantial emissions from agriculture, for example the methane emissions from livestock. The rising levels in the atmosphere are regarded as another major cause of increasing climate change (Gramelsberger & Feichter 2011).

It is now considered indisputable that global warming as a result of human activity is posing great dangers. Climate science assesses the consequences of this development as ranging from very serious to catastrophic ('climate catastrophe'). Due to the greenhouse effect of carbon dioxide, global mean temperature increases and severe local changes across the world are now expected. Enormous impacts will result (▷ Chapter 1.1), for example:

- periods of extreme heat with severe consequent upheavals in agriculture
- storms and floods
- sea level rise (melting ice, thermal expansion of water)
- thawing of permafrost
- loss of biodiversity (extinction of plant and animal species)
- spread of diseases
- migration movements, armed conflicts

Figure 1: Annual global primary energy consumption in Mtoe (oe = oil equivalent). In 2016, primary energy demand was 13,276 Mtoe (data source: BP 2017)

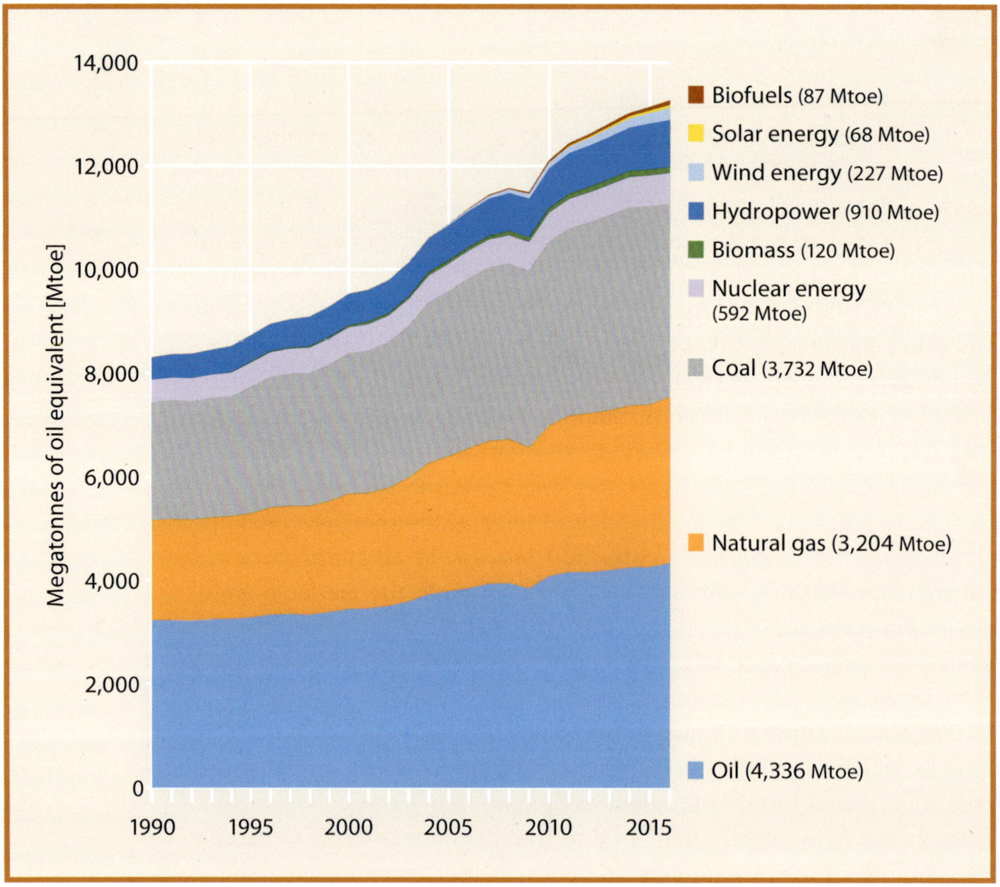

In order to stop this development and reverse it, there is no alternative to phasing out the use of fossil fuels. This is a daunting task for the global community.

Global energy use

The industrial revolution brought the industrialized nations to prosperity and power by way of an enormous consumption of energy. A similar development can now be seen in the so-called 'emerging' countries. In particular, the intense drive for development in China, India, Brazil, Mexico and South Africa has caused a 64 % increase in primary energy demand in those countries between 1990 and 2016 (see Fig. 1). The energy mix in 2016 was dominated by fossil fuels: oil, natural gas and coal accounted for over 85 % of global demand. Renewable energies accounted for only 10.1 %. Between 1990 and 2016, manmade CO_2 emissions also increased from 20.6 billion to 35.9 billion tonnes (BP 2018). Scenarios drawn up by the World Energy Council project that emissions will double or even quadruple from 1990 levels by 2050 (WEC 2013).

If we want to understand the interconnections in energy use, it is important to acknowledge that the use of energy is regionally specific

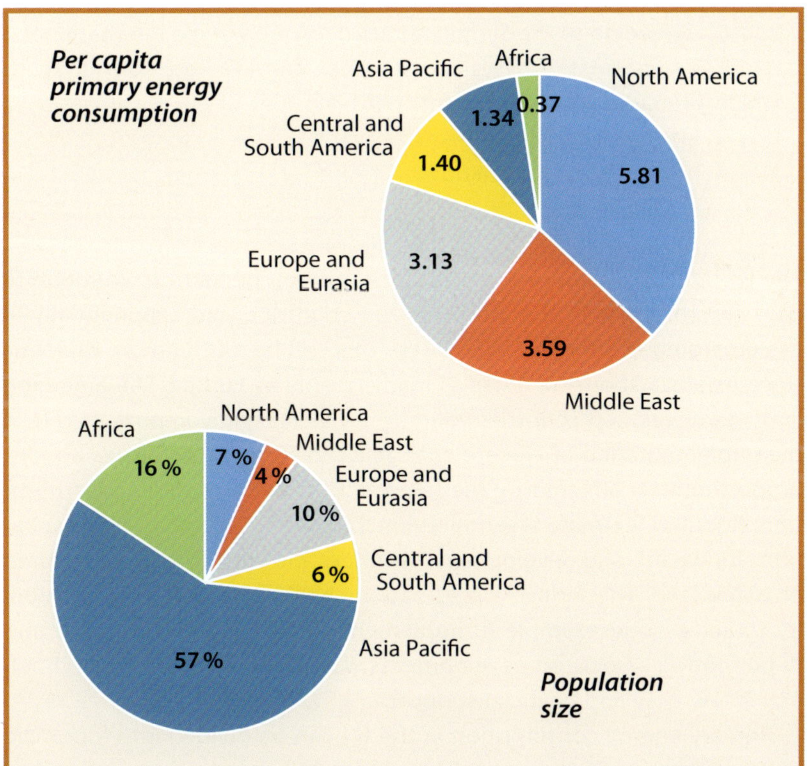

Figure 2: Per capita primary energy consumption (average 1.79 t oil equivalent; top) and population size (7.24 billion; bottom) in 2014 by region (data source: OECD 2016)

Regional differences in energy use

and very unevenly distributed across countries. The average per capita primary energy consumption worldwide in 2014 was 1.79 toe (oe = **oil equivalent**), see Fig. 2 (OECD 2016). In terms of regions, primary energy consumption was highest by far in North America (5.8 t), followed by the Middle East (3.59 t), and Europe and Eurasia (3.13 t). In Africa, consumption was the lowest by far (0.37 t). The USA plays a key role in cross-country comparisons, because primary energy consumption there at 7.04 toe per capita is multiplied by more than 325 million inhabitants. Among the countries with the highest consumption, energy producers like Qatar (22.3 t) are notable. In the most populous states, China and India, consumption was only 2.1 t and 0.5 toe per capita respectively.

Oil equivalent

Due to the dominant position held by oil in the world market, oil equivalent (oe) is still the reference energy unit for all primary energy forms. 1 Mtoe (million tonnes of oil equivalent) corresponds to 41.868 PJ (petajoules). The joule is the basic physical unit of energy according to the SI unit system (French Système international d'unités = international system of units). The kilowatt hour (kWh), which primarily serves to quantify the amount of electrical energy (e.g. in an electricity bill or electricity tariff), is easier to grasp. 1 kWh is equal to 3.6 MJ or 0.086 kgoe.

The *World Energy Council* expects global energy demand to continue to rise — more slowly in the industrialized countries, and exponentially in the emerging economies (WEC 2013). This will be all the more dramatic since Asia has about six times as many people as Europe. The emerging and less developed countries will play an increasingly important role in the implementation of climate change policies. These countries are disproportionately affected by the impacts of climate change (desertification, flooding, extreme weather events), while having the same aspirations for wealth and development as others and no desire to be lectured or patronized. The industrialized nations have a particular responsibility to set a good example in implementing the energy transition and in promoting associated developments worldwide by means of technology, know-how and financial support.

Primary energy consumption in the EU can be divided into four main sectors: transport 33 %; private households 26.5 %; manufacturing industry 24 %; and the service sector 11 %. The service sector is becoming increasingly important because of the digital economy, data centres with online data storage and cloud computing (Le Monde diplomatique 2012). By contrast, the agricultural sector now accounts for only 2.2 % of primary energy consumption. Within the EU there are significant regional differences.

The energy situation in Europe

The Energy Union since 2015

In July 2014, the European Commission adopted guidelines for the creation of a 'new' European Energy Union. Its task is to deepen and expand the internal energy market and cooperation in the light of global climate policy agreements. It aims to create the conditions for a sustainable and secure energy supply at affordable prices. More than 50 strategic measures are to be implemented by 2030 in the areas of renewable energy, energy efficiency, markets, infrastructure, transport, buildings and carbon pricing. The specific actions are spread across several core areas (European Commission 2015a):

1. **Energy security:** In 2017, the EU imported about 53 % of its total energy needs. It is also highly dependent on a small number of supplier countries (Russia, Algeria, OPEC) (David et al. 2017) and relies on certain suppliers and supply routes. One of the measures included in the EU guidelines is the development of the European high-voltage grids into a common electricity interconnection with a trans-border exchange of at least 10 % of generated capacities.

2. **Fossil fuel phase-out:** 94 % of traffic is dependent on petroleum products, 90 % of which are imported (European Commission 2015a). The aim is to reduce dependence on fossil fuels and diversify supplies. There is consensus in the Union that the CO_2 Emission Trading Scheme (ETS) has a key role to play in achieving the reduction target by 2030.

3. **Energy efficiency:** 75 % of the stock of buildings in the EU is not energy efficient (European Commission 2015a). The Union supports national incentive schemes to promote building insulation measures.

So far, there are hardly any funding instruments available from the Energy Union to increase sufficiency (▹ Chapter 1.2). Policies that focus primarily on growth are more likely to neglect sufficiency. Strenuous efforts are being made in the EU to decouple economic growth from resource consumption. Climate policy measures are limited to the goals of substitution and energy efficiency, with only modest success for climate protection.

The energy situation in Germany

But what is the current energy situation in Germany? In the course of reunification, economic upheaval in the new federal states led to a significant reduction in greenhouse gas emissions, due above all to the slump in industrial production. Since the 2000s, primary energy consumption in Germany has stagnated despite sustained economic growth. Greenhouse gas emissions declined by 28.1 % between 1990 and 2015 (Umweltbundesamt 2016a). In 2015, 318 Mtoe of primary energy were used in

Figure 3: Lignite mining in Niederlausitz, Brandenburg. As recently as 2017, 168 million tons of coal (hard coal and lignite) were mined in Germany, but with a strongly downward trend. Coal accounted for 22.2 % of primary energy consumption in Germany (BMWi 2018). (photo: P. Ibisch)

Germany (OECD 2016). Of this, 79.5 % was delivered by fossil fuels. The proportion covered by nuclear energy declined, to be replaced initially largely by fossil fuels. The share of renewable energy in the energy mix has increased fivefold since the 1990s, and in 2015 was 12.5 % of primary energy demand. This trend will continue to strengthen. Even the large energy suppliers are now turning away from conventional energy sources and producers towards renewable alternatives.

Energy production technologies

The main forms of final energy used by private households are thermal (heating), chemical (petrol) and electrical (electricity). Energy is first supplied, transported and processed as primary energy, to be made available as power or heat, or stored for later use, via supply networks. The following section focuses on common technologies and resources for conventional energy production, as well as technologies that have substantial potential for a sustainable energy supply.

Figure 4: Map of power generation technologies

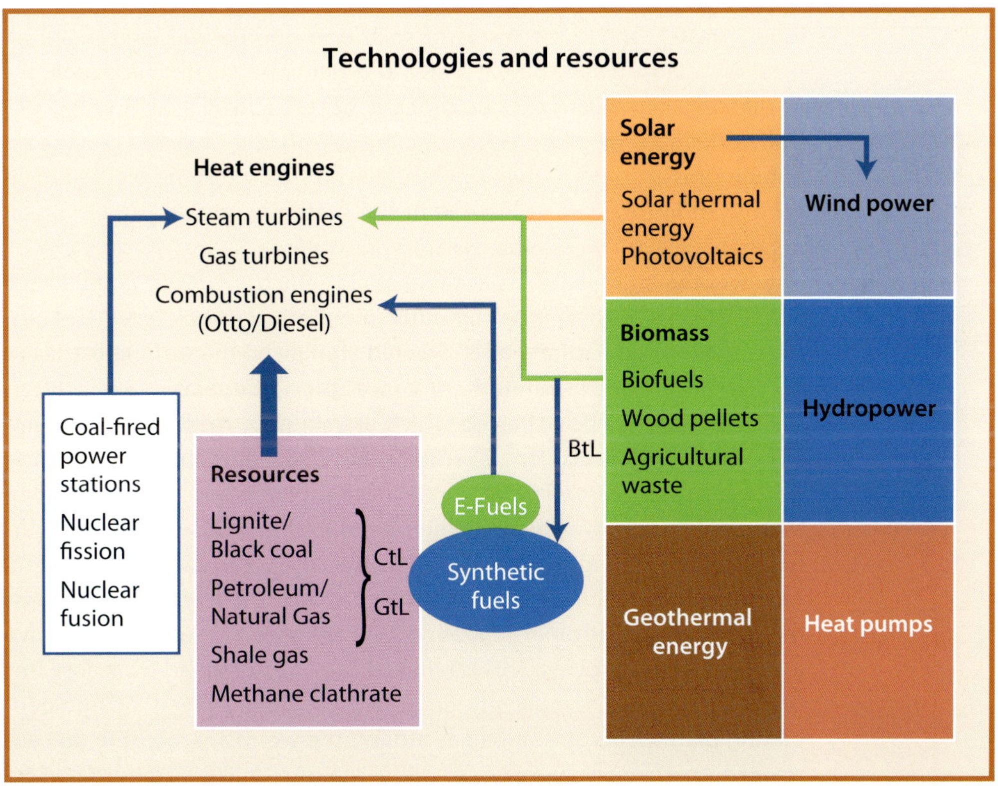

> **CtL stands for the chemical process *coal-to-liquid*, in which liquid fuel is produced from coal. It is also called 'coal liquefaction'.**

> **The GtL process *(gas-to-liquid)* converts natural gas into liquid fuel by means of a chemical process.**

> **After treatment, biogas can be processed into liquid fuel using the same procedure (BtL = *biogas-to-liquid*). The liquid fuel has no sulphur or organic nitrogen. It is also easier to transport and can be used as a petrol substitute, among other things.**

Conventional energy production technologies

Heat engines

Steam turbine process

In 2015, 85 % of primary energy demand worldwide was met by fossil fuels. Coal, mineral oil, nuclear fuel, biogenic fuels (energy crops, residues, waste), and some forms of solar and geothermal energy are all used on a large scale to heat water and generate steam. Subsequently, in the so-called steam turbine process, heat is converted into mechanical work and finally into electricity. In this process, steam is conducted at high pressure and high temperature onto steam turbine blades which set a shaft in rotation. At the end of the shaft a generator rotates which generates electricity via the dynamo principle.

Gas turbines

Since the mid-1980s, coal fired steam-electric power plants have been progressively replaced by much more efficient gas turbine plants. Here, the ignited natural gas is directed straight to the turbine blades to set the drive shaft in rotation. At combustion temperatures of 1,180 °C, ▹ **efficiencies** can be achieved which are superior to those of all other heat engines. In addition to energy efficiency, they have the further advantage of possessing a significantly better rapid start capability than coal-fired power plants. When energy demand increases, gas turbines can be brought into operation quickly to balance loads. Often, heat engines are linked together in a *combined cycle*. Here, the exhaust gases from a gas turbine, at about 650°C, are used subsequently to drive a steam turbine process. This is referred to as a 'Combined Cycle Gas Turbine' (CCGT). Since overall efficiency can only be improved in large generating plants, energy companies around the world are focusing on centralized energy supply. For example, a major city like Hamburg can be supplied with electricity from a single gas turbine plant in the 380 to 420 MW power range.

> **Efficiency: the ratio of work performed to heat energy taken in. The theoretical maximum is defined by the Carnot cycle or principle. Energy that cannot be transformed into mechanical work is released as waste heat into the environment, thereby impacting on the ecosystem.**

Petrol engines and diesel engines

Piston engines are a widespread form of heat engine. Here, fuel is ignited on the power stroke of a petrol or diesel engine. The 'working gas' expands and makes the working shaft revolve by means of an alternating movement via a gearbox (piston engine). This is used to drive machinery, vehicles or power generation. Piston engines provide the basis for combined heat and power plants. The advantages of piston engines are quick start capability, high power density and low maintenance.

Stirling engine

Another example of a heat engine is the Stirling engine, which can only operate at low efficiencies because of its underlying design. In a closed system, a working gas oscillates back and forth between a cold and a warm chamber, setting a working piston in motion. It can be operated using almost any external heat source, including solar thermal energy.

Efficiency enhancement and limits of heat engines

In general, technological innovations over the last century led to heat engines being taken ever closer to their systemic performance limits. The physical limits of a system can be described using the methods of thermodynamics, as in the Carnot cycle. New kinematic concepts, innovative materials, better sealing methods, further pressure increases and concepts for uniform combustion will continue to play an important role in future developments. However, meaningful possibilities for increasing the efficiency of heat engines are already pushing at the limits of technical performance.

Nuclear energy

Nuclear power stations

A nuclear or atomic power plant is a thermal power plant for the production of electrical energy from nuclear energy by means of controlled nuclear fission. The conversion into electrical energy comes about indirectly, just as in conventional thermal power plants. The heat generated by nuclear fission is used to generate steam to run a steam turbine. Since the nuclear disaster in Fukushima in 2011, concerns about nuclear power and resistance among the public have increased (▻ Chapter 3.4). In 2015, nuclear energy accounted for 12 % of the EU energy mix, with big national differences. Whereas in France nuclear energy covers 38% of the primary energy demand and is to be further expanded, the proportion in Germany is 7.5% and decreasing. In Germany, there is a broad social consensus for phasing out nuclear energy. There is a total of 143 nuclear power plants in the EU, 58 of them in France alone. The last German nuclear power plant is due to be shut down in 2022.

Nuclear fusion reactors

Although research has been ongoing in the field of nuclear fusion since the 1960s, there is still no viable reactor in existence. In nuclear fusion, deuterium and tritium fuse under high pressure and temperatures. In terms of comparative efficiency, nuclear fusion can produce 150 times more energy per kilogram of matter than nuclear fission (in line with the equation $E = mc^2$). The technical challenges lie in controlling plasma flows at temperatures of 100 million to 1 billion °C. Research into delivering a controllable nuclear fusion process will continue for decades to come. Nuclear fusion could play a unique role in future energy supply. Undoubtedly, the taming of nuclear fusion will represent a milestone in technological development. Its impressive efficiency is an argument for its implementation. However, the technology involves very high risks because the fusion process sets free 'fast' neutrons that penetrate their environment, leaving radioactivity on the supporting structures. The disposal of radioactive waste — which will remain radioactive for a significantly longer period of time than modern humans have existed — is difficult to reconcile with sustainable development, as it involves a threat to humans and the environment (Umweltbundesamt 2010). Around the world, there are 14 repositories for low-level radioactive waste in operation; so far, there are none in Germany, and the question of final disposal is still unresolved. In the past, waste was temporarily stored in the disused Morsleben and Asse II mines. A new revision of the Site Selection Act for identifying repository sites was adopted by the federal government in March 2017 in order to end the 40-year social conflicts over this issue. The search for suitable sites should be completed by the year 2031. Storage is scheduled to begin around the year 2050 (BMUB 2017).

Shale gas as an energy source

Another way for European countries to reduce their energy dependency is to meet their natural gas demand from their own shale gas deposits. Such deposits are abundant. However, reservations among the general public mean that their exploitation in Europe has been slow. In so-called fracking (hydraulic fracturing), cracks are artificially created in the dense shale from which the trapped gas can escape. Large quantities of a mixture of water and sand containing chemically active substances are pumped under high pressure through piping systems into deep caverns. The technology is ecologically controversial. Soil contamination, uncontrolled gas leaks, the destruction of landscapes and the pollution of groundwater are possible consequences of fracking. Nevertheless, European chemical and pharmaceutical companies are pinning their hopes on the prospect of domestic commercial use of this process. The increased use of fracking has allowed the USA to regain its position as one of the biggest energy producers. In contrast, there are only a few test wells in Europe, used exclusively for research purposes.

Hydraulic fracturing in Germany

A commercial application of hydraulic fracturing has yet to be realized in Germany. Nationwide, about 300 frac jobs have been carried out, most of them in Lower Saxony (GtV-Bundesverband Geothermie 2012). The environmental risks of fracking are an especially controversial topic in Germany. However, there is not yet sufficient data to enable conclusive statements on opportunities and risks. Despite this, in August 2016 the fracking law was adopted, enabling a general ban on fracking of unconventional deposits, although fracking remains permitted in deep layers of dense sandstone. Further insights are anticipated by 2021 from an independent expert commission looking at four test wells (Deutscher Bundestag 2016).

CCS technologies

Carbon dioxide capture and storage

CCS technologies are not part of energy production in a narrower sense. Rather, they complement thermal power production by providing a technical means of capturing and storing CO_2. CCS stands for Carbon Dioxide Capture and Storage, sometimes referred to as sequestering or separating CO_2. In this process, the greenhouse gases released by combustion are separated at the thermal power plants, compressed into pellets and stored in underground caverns. Taking a conventional hard coal-fired power plant as an example, it can be demonstrated that CCS technologies can reduce greenhouse gas emissions by up to 70 % (Kuckshinrichs & Hake 2012). Large-scale energy suppliers operating major coal-fired power plants have a particular interest in this technology. For that reason, big investments are being made into researching the different options. In the so-called oxyfuel process, hard coal is burned with pure oxygen in order to simplify the process of combustion and CO_2 capture. A competing option is the pre-combustion process, which relies on the conversion of coal before combustion by novel gasifier technologies into carbon monoxide and hydrogen. Carbon monoxide is oxidized and separated in the flue gas (Kneer et al. 2006). Hydrogen then serves as the energy source. The goal is to bring these options to market by 2020. CCS technologies are highly questionable from a sustainability perspective. Firstly, CO_2 capture, compression and storage comes at the cost of efficiency. Additional energy consumption of 25 to 40 % is estimated for these processes (Kneer et al. 2006). In addition, the disposal of CO_2 in deep layers of rock represents a transfer of cost and risk to future generations (Quaschning 2015). It is still completely unclear what would happen if there were leaks in these underground caverns. The energy companies repeatedly claim that the technology is safe, but do not want to be held liable for possible errors. According to the International Energy Agency (IEA), there are currently 13 CCS projects, facing daunting technical challenges (Handelsblatt 2015).

Technologies for the energy transition

Energy transition

The term 'energy transition' (*Energiewende* in German) describes the transition from an economy based primarily on fossil fuels (including nuclear energy) to a sustainable supply by means of renewable energies.

A sustainable energy supply

At present, the energy transition is focused on the electricity, heating, building insulation, and mobility sectors, and in the future it will also look towards phasing out fossil raw materials in the production of plastics and fertilizers. A sustainable energy supply includes production as well as technologies for the distribution and use of energy. In this context, the concept of sustainability means an environmentally friendly, demand-oriented use of energy that does not consume more natural resources than are formed over a given period. At the same time, the functionality of ecosystems must not thereby be constrained or destroyed. For example, the energy production costs of the technologies (photovoltaics, wind turbines) must be such that the energy gain is significantly higher than the service life and energy costs of the product itself.

Solar energy

Solar energy is infinitely available in its original form. The use of solar energy in the economy can be divided into at least two major groups: solar thermal and photovoltaic (Krippner 2016).

Solar thermal energy

With solar thermal energy, solar energy is converted directly into heat to generate domestic hot water and to provide process heat for process engineering and the chemical industry. When solar radiation is focused with the aid of parabolic mirrors or optical lenses, the thermal energy can even be converted directly into mechanical work using a Stirling engine, and into electrical energy via kinematics. At central European latitudes, large-scale solar collectors are often used in which heat absorbers deliver energy to a water circulation unit. The circulation is regulated by a pump. The access to the hot water storage tank is opened as soon as the absorber temperature is higher than the water temperature in the tank. Solar thermal energy is particularly efficient in southern Europe and equatorial latitudes (Krippner 2016).

In addition to solar thermal energy there is also photovoltaics, in which sunlight is converted directly into electricity. The basis of the tech-

Figure 5:
Use of open land – Finowfurt solar park, Brandenburg, Germany. It intrudes on an area formerly left over to forest recovery with successional growth of pine trees (an ecological succession area). (photo: P. Ibisch)

Photovoltaics

nology is the photoelectric effect on semiconductor materials. An electron breaks free from its bond in a solid by absorbing a photon of light. The first silicon solar cells, developed for the aerospace programme in 1954, had an efficiency of barely 5 %. Today's semiconductor cells achieve efficiencies of 20 to 40 %. Due to the development of innovative materials and drastic reductions in production costs, photovoltaics has become competitive with fossil fuels in just a few years (Quaschning 2015). Photovoltaics has significant advantages in terms of plant construction. Plant sizes of up to 100 MW are possible today simply by interconnecting modules. Moreover, no additional intervention in nature is necessary. The use of flat surfaces on rooftops and facades could theoretically meet a third of the electricity needs of cities. Furthermore, private households can achieve a degree of energy autonomy through this decentralization of supply.

Energy returned on energy invested (EROEI) for solar

Solar electricity production depends on the location, the weather, the time of day and the season. Solar energy is regarded as not possessing ▹ **baseload capability**. The main cost driver is the amount of energy required to produce the highest purity semiconductor materials. Depending on the location, the ▹ **EROEI** is between 5 and 20 for solar cells, whereas for wind turbines it is between 60 and 90. Studies show that even at very unfavourable locations, the energy production costs for PV installations paid off within 6 years at the latest (Quaschning 2015).

Critics of solar parks point to the loss of outdoor and even forested areas, as well as aesthetic aspects. Significant restrictions have been imposed on solar parks for some years now to limit the loss of commercially usable land. Regulations on the tendering of open spaces prescribe the funding conditions within the Renewable Energy Sources Act (EEG 2017). They apply to roadside verges, sealed areas, and commercial and industrial areas among others.

Baseload capability refers to the capacity of a power supply system to supply electrical energy without interruption. Whereas a solar installation provides electricity only in the daytime, a gas turbine can be operated around the clock. In contradistinction to peak load, the term 'baseload' means the amount of electrical power that must be permanently available for any district supplied.

Energy returned on energy invested (EROEI) is a ratio used to express the efficiency of an energy source over its lifetime. It relates the energy produced by the system to the energy needed to produce and operate the system. The greater the ratio, the more efficient the system. For example, there are silicon-based PV modules which only generate five times as much power over their lifetime as was required to produce them.

Wind power

Wind energy

Wind power is an indirect form of solar energy. It is the result of local pressure equalization between high pressure and low pressure areas. Nowadays, wind power is used on an industrial scale to produce electrical energy via the mechanical work of rotating turbines. With the introduction of the Renewable Energy Sources Act (EEG) in 1991/2000, legal and financial framework conditions were created for technologies promoting the energy transition, so that, for example, energy from wind turbines could be cost-competitive with conventional forms of energy. In 2009, the conditions applying to offshore wind power installations were significantly adjusted once more (EEG 2012).

Comparing wind and solar power

A wind power installation has a modest energy yield compared to solar installations: whereas a wind turbine delivers only 0.1 kW/m^2 of swept area at medium wind speeds in Germany, a radiant energy of 1 kW/m^2 can be measured on sunny days. Only at a wind speed of 11.5 m/s can a level comparable with photovoltaics be reached (Quaschning 2015). This explains why offshore wind farms are increasingly being constructed in the north and solar parks on open spaces predominantly in the south of Germany. Since it is reliant on favourable wind and weather conditions, wind power is considered unsuitable for baseload.

Wind turbines

Wind power is relatively easy to install in technological and energetic terms. Offshore wind speeds are significantly higher than in onshore areas, so that in northern regions of Germany the potential for expansion is much greater than in the south. Larger — and therefore more efficient — wind turbines meet with a lower degree of acceptance from the general public. The expansion of wind energy is leading increasingly to incursions into forest areas now that most available open spaces have been exploited. This results in significant fragmentation effects and forest area losses. In addition, the concerns of nature conservationists regarding the harm done to bird and bat populations by collisions with wind turbines are well known. Another challenge is the expansion of overhead power lines needed for the distribution of electric power to the south of Germany, which leads to further fragmentation of the landscape.

The Fraunhofer Institute for Wind Energy Systems has calculated that it will be possible in the year 2050 for Germany to meet 100 % of primary energy demand from renewable energies. In the scenario developed, 74 % of demand was met by wind energy, both onshore and offshore (Knorr et al. 2017). At the end of 2016, 27,270 wind turbines with a total capacity of 45,910 MW were operating in Germany. Every year, between 1,500 and 2,000 new wind turbines are added. According to a study by the Fraunhofer Institute for Solar Energy Systems, in order to meet climate policy targets, by the year 2050 around 57,000 3 MW wind turbines would need to be operating (Henning & Palzer 2013). According to the study, 1,800 full-load hours per year would then suffice, a figure in the medium range.

Use of biomass

Use

'Biomass' refers to substances of organic origin. Biomass is stored solar energy that can be used as needed. As an energy source suitable for baseload, it represents a sensible supplement to the energy mix of wind, sun and hydropower, as long as it does not conflict with agricultural food production or the preservation of functional ecosystems. In terms of global energy use, biomass is one of the main sources: currently about 1 % of the world's annual biomass production is used to cover about 11 % of primary energy demand. In poorer countries such as Ethiopia, Mozambique or Nepal, biomass represents over 80 % of all energy sources. Often, the major part is firewood, the intensive use of which contributes to global desertification. Globally, biomass is rarely used sustainably. In 2015, about 5 % of electricity and 9.5 % of heat in Germany were provided by biomass. About 5.5 % of the fuel used was derived from biomass (FNR 2016).

Supply and processing of biomass

In the simplest example of its use, biomass is burned in heat engines. In so-called combined heat and power plants (CHP), electrical energy (usually) is generated from biomass, close to the point of consumption, by means of an internal combustion engine. The waste heat is fed directly into a district heating network. Typical examples of biomass are wood pellets (by-products of the wood processing industry), firewood, bio-oils and biogas. These products are used for power and heat production in combined heat and power plants or as fuel additives for transport and traffic. Energy crops such as wheat, maize, oilseed rape, sugar beet

Figure 6: Use of firewood in a central Andean dry valley, Bolivia. (photo: P. Ibisch)

and tree species from short rotation coppice plantations such as poplars or willows are grown in Germany specifically for energy use. In view of population growth and the related question of how to feed the growing world population, a number of ethical issues are emerging — for example, the use of grain for energy (grain burning) is the subject of intense debate.

Biodiesel and biogas

Ethanol is extracted from wheat, corn and sugar beet via biochemical fermentation processes. Rapeseed oil is produced in oil mills by pressing and filtering. It is processed into biodiesel with the aid of ethanol in a controlled transesterification process. Biogas is another source of energy. It is the product of the fermentation of agricultural waste, manure or food waste in large biogas plants. The gas consists essentially of methane, which can be fed directly into the gas supply network. According to the scientific advisory council for agricultural policy of the German Federal Ministry of Food and Agriculture (FNR 2016), the comparative CO_2 savings achieved by biogas from manure are four times higher, and by wood from short rotation coppicing plantations five times higher, than those achieved by ethanol from wheat or diesel from rapeseed.

However, short rotation plantations are in competition with other crops with respect to land use change. The import of food crops also means externalization of environmental costs (▹ Chapter 4.3).

Concerns over the use of biomass for energy production

Other energy crops with significant acreage in tropical regions include soya, oil palms, purging (or physic) nuts and sugar cane. When the point is reached where the use of energy crops as a staple food is competing with their use as a fuel, importing them for the latter use is regarded as extremely questionable. The spread of oil palm plantations is destroying rainforests in Malaysia, Indonesia, Cameroon and South America, causing significant environmental and social problems. Fig. 6 gives an impression of traditional firewood use in Bolivia. This contributes significantly to the destruction of forests. Erosion, landslides, and dry soils with reduced capacity to absorb water are the consequences. In addition, firewood shortages similar to those in Germany at the time of Hans Carl von Carlowitz (▹ Chapter 1.2) are occurring in many places. In recent years, firewood and wood pellets have been subject to criticism in Germany, especially as it is difficult to prove that imported wood has been logged legally or grown according to sustainable forestry principles. The cost of firewood has almost doubled over the past decade. This has led to an increase in imports (including illegal imports) of firewood from the global South and from boreal coniferous forests. More than 50 % of logging in Germany is now for 'energy' purposes, and does not contribute to CO_2 reduction. Added to this are the emissions caused by extraction and transport, which call into question the overall climate neutrality of the process, as well as the emissions that result, in both the immediate and the medium term, from the degradation of the forest.

Hydropower

Development of hydropower

The use of watermills in Europe dates back to the 18th century. The use of mechanical power from water enabled the operation of mills and small cottage industries. Gradually, hydropower was replaced by steam engines. Only with the electrification of the cities at the end of the 19th century did hydropower increase in importance again. At that time, in overshot water wheels (driven by water from above), up to 80 % of the potential energy of the water could be converted into electrical energy.

Hydropower today

Hydropower is today the world's most important form of renewable energy. Regardless of prevailing conditions, flowing waters are available at all times. Norway meets 100 % of its electricity needs from hydropower, and for Brazil, Canada, Austria and Switzerland the figure is as high as 50 %. Countries with fast, large rivers and mountains with steep slopes, are the fortunate ones. In Germany, hydropower has a very low share of 3.3% of the electricity mix. At 4,200 MW of total capacity, the potential for the development of additional usable locations in Germany is already almost exhausted due to the topographical conditions (Umweltbundesamt 2010), though over the next few years some existing facilities will be modernized and expanded.

Pumped-storage power plants

Pumped-storage power plants play an important role in the energy industry. During periods when there is an excess of electricity supply, water is pumped up to a high basin. When electricity demand from private households rises again, the water is fed to turbines via penstocks, using the potential energy to generate electricity. Because they provide this flexibility, pumped-storage power plants are an ideal complement to wind and solar energy. Here, too, suitable topological conditions in Germany are rather limited, and a pumped-storage capacity of only 7 GW is available. The capacity for a continuously operating supply is 4 to 8 hours.

Tidal and wave power plants

Another technological development for discontinuous energy production that is worth mentioning is tidal and wave power plants, which are still in the research stage. Tidal power plants use the powerful tidal flow between a bay and the open sea to generate electric current via turbines. In wave power plants, the relative motion of floating buoyancy elements is used to generate electricity.

Geothermal energy

Using heat flows in the Earth's interior

High pressure and the process of radioactive decay mean that the temperature in the Earth's core is over 6,500 °C. This results in a constant heat flow between the core and the Earth's crust. Geothermal energy systems use the heat flow indirectly by heating fluids at great depth via a closed piping system. The available heat flow is very unevenly distributed geographically, depending on geological conditions. At locations where continental plates collide, geothermal energy is available at shallow depths, for example in Iceland, southern Italy, and Indonesia. Germany has only a few volcanically active areas and is regarded as being safe from earthquakes and thus, from a geological point of view, uneconomic for geothermal energy exploitation. The costs of acquisition, drilling and maintenance are comparatively high in Germany. The most favourable geological conditions are found in the Upper Rhine Plain. Here, temperatures of over 150 °C are found at a depth of 3 km. On average, about 47 °C can be reached at a depth of 1 km in Germany. For geothermal power plants, however, drilling depths of 10 km are required

Energy production	Advantages	Disadvantages
Steam turbine	technologically straightforward	only viable in combination with (steam) heat generation plants
Coal power station	suitable for baseload, low technological sophistication	low comparative efficiency, CO_2 emissions, waste heat, finite resource
Gas turbine	suitable for baseload, highest energy efficiency among heat engines	CO_2 emissions, waste heat, finite resource
Combustion engine	rapid start capability	CO_2 emissions, waste heat, finite resource
Stirling engine	works with any external heat source, compact construction, transportable	low comparative efficiency, waste heat
Nuclear power station	suitable for baseload	disposal of radioactive waste, finite resource
Nuclear fusion	suitable for baseload	research phase only, disposal of radioactive waste
Solar thermal energy	most direct form of heat use, water heating	weather and sunshine dependent, unsuitable for baseload
Photovoltaics	direct conversion into electric current	weather and sunshine dependent, unsuitable for baseload; land use (often at the cost of intact ecosystems)
Wind power	technologically straightforward, robust	weather and wind dependent, unsuitable for baseload; land use (often at the cost of intact ecosystems)
Hydropower	demand-oriented controllable output, no emissions	few available locations in Germany, severe intervention in nature
Biomass	CO_2 neutral, demand-oriented use	(globaly) rarely used sustainably, ecosystem degradation through logging, competition between animal feed and fuel
Geothermal energy	unlimited availability	few suitable locations in Germany, substantial exploration costs

Table 1: Advantages and disadvantages of different energy production systems

(Stober & Bucher 2014). The challenges are huge, because at this depth, at 300 °C and 3,000 bar pressure, rock becomes plastic.

Geothermal power installations

Near-surface geothermal energy makes use of boreholes of up to approx. 400 m in depth and temperatures of up to 25 °C to heat and cool buildings and technical facilities. A heat exchanger transfers the heat from the fluids circulating underground to a closed water circuit. On the surface, a heat pump brings the system up to the required temperature level. In Germany, around 320,000 near-surface geothermal installations are currently in operation. A major advantage they enjoy is the inexhaustible availability of geothermal energy. Against that, the investment costs are high and there is a risk of earth movements caused by faulty drilling, for example if groundwater penetrates into dry mineral layers.

Heat pumps

How heat pumps work

In a heat pump, a fluid at a relatively low temperature, for example ambient air temperature, is supplied with thermal energy through mechanical work and brought to a higher temperature, for example room temperature (Dohmann 2016). The heat pump has a compressor, which is mechanically driven. Electrically driven compression heat pumps are relatively common. A refrigerant is fed into a closed circuit. On compression of the gaseous refrigerant, the heat from condensation is transferred to a second circuit, for example a room air or hot water circuit. The refrigerant is set to absorb heat from the environment as it expands or evaporates. In private households, the ambient air serves as a heat source, or else a separate brine circuit fed by a geothermal source.

Energy production and sustainability

The unchecked consumption of fossil fuels as well as of nuclear energy and the associated over-exploitation show that, globally speaking, we are a long way from being sustainable. Technologies associated with the energy transition have great potential for sustainable use (wind, solar, bioenergy and geothermal energy). However, they are not sustainable in themselves (see Table 1). Biomass in particular (including for biofuel) seldom comes from sustainable sources today. There are many negative examples (the use of fossil fuels for agricultural machinery and fertilizers, the destruction of the rainforests). There are also indirect consequences arising from the use of agricultural land for energy crops. Unfortunately, the principle of sustainability is often violated when open spaces are created at the expense of intact ecosystems to make room for solar parks and wind farms. The quantities of sustainably usable energies are quite limited. In addition to the need to increase energy efficiency, sufficiency must also be increased. Sustainable management of resources (▹ Chapter 1.2, 4.3) does not simply mean substituting some sources of energy with others; it also entails promoting sufficiency in a much more comprehensive way.

Energy supply in Germany in 2050 – possible scenarios

Taking as its starting point the climate policy framework conditions (▷ Chapter 3.4), the following scenario represents the results of analyzing a number of trends. It is based on real data (Le Monde diplomatique 2012; Quaschning 2015; Prognos AG et al. 2017). Future emissions are dependent on economic, social and political developments which are fundamentally unpredictable.

Climate research models start from a wide range of assumptions. All projections are based on the calculation that global warming must be kept below 2 °C compared to pre-industrial levels. This requires drastic reductions in anthropogenic greenhouse gas emissions by 2050. The scope and impacts of the future developments and key technologies needed to meet this challenge are also difficult to predict. Notwithstanding this, certain clear trends for the future of the energy supply at national and European levels can already be discerned. For example, key issues such as 'next generation storage technologies' and the 'development of new compound materials' to replace rare earths (lithium, neodymium, etc.) are already feeding into current European research framework programmes.

Global trends

The primary energy demand predicted for the year 2050 means that the geopolitical balance of power will shift in favour of the regions and countries that produce renewable energy and fissile material (uranium, thorium). India and China will deliver more than half of the world's nuclear power in 2050 (Quaschning 2015; Prognos AG et al. 2017). At the same time, China will be the largest producer of renewable energy (Le Monde diplomatique 2012). In the EU, too, national strategies vary greatly. Whereas Germany is placing its faith in renewable energy, in France the expansion of nuclear energy is being further intensified.

The phasing-out of fossil fuels

In the second half of the 21st century, the last fossil fuel power plants will go offline. Dependence on fossil fuel supplier countries will be succeeded by dependence on renewable and sustainably produced raw materials. Cogeneration plants will supply heat and electricity from renewable fuels. They will continue to emit greenhouse gases and waste heat, but from renewable raw materials. In sum, due to reforestation and biomass production, more CO_2 will be absorbed from the atmosphere than is emitted.

Primary energy demand will fall from 346 Mtoe in 1990 to half that figure in 2050. This will be achieved in large part through energy-efficient building construction. Thermal and damp insulation are now the norm for both building renovations and new buildings. Increasing emphasis will be put on replacing petrochemical materials with sustainable natural materials. All rooms will also be equipped with independent ventilation systems. Waste heat from industrial processes will be used for private households. Heat exchangers will deliver energy transfer in all areas of life. District heating systems will be expanded to cover very large areas.

Energy-efficient building construction

In addition to the predominantly decentralized supply of wind power, solar energy and biomass by farm-based enterprises and private households in rural areas, there will continue to be central energy suppliers with a technological focus on deep geothermal energy, solar parks, offshore wind power, combined heat and power plants and infrastructure for district heating. The large providers will mainly supply industry and transport, with some provision for private households in urban areas. Most vehicle traffic will be electrified, and cargo vehicles and air traffic will be powered by renewable energy products (Prognos AG et al. 2017).

Combining central and decentralized energy supply

The main suppliers of primary energy in 2050 will be wind power (>36 %), photovoltaic (26 %) and biomass (22.5 %) (Le Monde diplomatique 2012). Because of the replacement of older plants, hydropower will continue to have a low share of 5.5 % of primary energy supply. The remaining 10 % will be covered by geothermal and solar thermal energy.

Energy sources

Pumped-storage power plants and new electricity storage systems will be widely used as energy buffers. No single technology will dominate here, but different concepts will be linked together. The combination of ▹ **electromobility** and ▹ **smart grids** could well play a key role in enabling 'mobile' batteries to be used as a local storage or power source.

Energy storage

> **Electromobility is transportation using vehicles exclusively powered by electricity. Such vehicles usually have to carry an electrical energy storage device.**

> **Smart utility networks are a key technology for matching electricity supply and demand. So-called *smart grids* are increasingly being used in Europe to deal with the problem of energy surplus. These intelligent power grids apply information technology across borders to enable optimal generation, distribution and consumption of energy.**

Renewable electricity will be supplied by solar, wind and water power. Surpluses will be used to generate and store methane and hydrogen (Quaschning 2015). This concept is also called power-to-gas. Already

Power-to-gas

today there are electricity suppliers who use the excess electricity from wind power for electrolysis and feed the hydrogen thereby generated into the natural gas grid. This is an indirect form of storage of electrical energy (chemical energy storage).

Methanation Another solution is based on methanation, the chemical bonding of hydrogen and carbon dioxide from industrial processes to produce methane gas. This can be done by both chemical and biological means. Here, too, the existing natural gas infrastructure can be used for feed-in. When the demand for electricity increases, biogas can be used in combined heat and power plants, or hydrogen and methane in fuel cells, for power generation. The waste heat is then fed into the district heating network.

Closing remarks

If the ambitious climate policy goals are to be achieved, and CO_2 emissions drastically reduced, the use of fossil fuels will have to be abandoned entirely (▷ Chapter 3.4). Their substitution by renewable energy sources can only be one of many measures. This is so because by 2050 the same requirements for useful energy in Germany must be met with much less primary energy (Quaschning 2015). This can only be achieved through the implementation of a number of measures:

- reduction of energy consumption in general: avoiding energy use and consuming less
- converting the transport sector to CO_2-neutral approaches and technologies (electromobility, power-to-gas, use of biomass, etc.)
- expanding the cycling infrastructure in urban areas
- the comprehensive insulation of buildings, avoiding the use of fossil raw materials
- the widespread and consistent application of process heat to heat rooms
- the development and implementation of fast and cost-effective energy storage systems
- the construction of smart grids in order to better balance supply and demand
- sustainable construction, while avoiding the use of steel and concrete

Although politicians accept the need for climate protection, action is still largely geared towards economic policy (▷ Chapter 3.4). One central issue is the required pace of change: slow changes are better for the economy and for prosperity, but the possible impacts of climate change threaten to undo any economic benefits. On the other hand, rapid changes are unacceptable to the wider public because the consequences would include steep rises in electricity prices, transport costs and rents. Given the scale of the challenge, the energy transition needs to be the subject of a much broader public discussion that goes far beyond the current focus on increasing costs of energy.

In conclusion, we can say with confidence that the concepts and technologies for a more sustainable energy supply have been thoroughly tried and tested. The changeover requires a significant degree of socio-political and economic acceptance and far-reaching measures to change supply structures. Climate policy goals are still not being pursued with the required rigour and tenacity. For example, Germany has committed itself to reducing CO_2 emissions by 40 % between 1990 and 2020. But by 2015, the savings amounted to only 27 %, meaning that the efforts required over the remaining years will need to be exponentially greater (ZEIT ONLINE 2017). And yet Germany has a particular obligation to act as a pioneer, especially since the countries with the largest populations have only now begun to develop into consumer societies, with the well-known negative consequences that this brings. It is to be hoped that the German *Energiewende* will be replicated in many other countries, thereby contributing indirectly to climate protection.

3.4

The controls: political systems

Benjamin Nölting, Hermann E. Ott and Heike Walk

The development of society is shaped by various social subsystems such as business, civil society and science. Politics and state institutions, on account of their democratic legitimation, lay claim to a steering function. Accordingly, political systems, understood as the interactive collaboration between politics and administration, have a significant impact on sustainable development—both positive and negative. The idea of a politics of sustainability raises fundamental questions about the status, design and implementation of any policy programme for sustainable development. Only in exceptional cases can these questions be answered unambiguously—especially if they concern the extremely complex process of a total societal transformation for sustainability. So politics deals with the negotiations around problem specification, goals, solutions, and their concrete implementation.

The politics of sustainable development – what is it?

Politics: a process of negotiation between competing interests

In a democratically legitimized constitutional state such as the Federal Republic of Germany, the political system (which includes parties, state and political institutions such as the Bundestag, Bundesrat, the Federal Constitutional Court, and the national and regional administrations) directs and regulates many spheres of life. However, both freedom—whether of individuals, of companies, or of civil society organizations—and the legal constraints on the state represent important public goods that should be protected by politics and the law. Societal goals, policy measures and their administrative implementation have to be negotiated within the framework of constitutional checks and balances to ensure the separation of powers. The political system reflects the diversity within society and cannot simply govern by decree. Rather, different interests have to be weighed against each other, majorities have to be sought, acceptance has to be won for the decisions made, minorities and the vulnerable have to be protected, and the legal process and rules have to be followed. The sociologist Max Weber (1864–1920) put it in the following way:

» Politics is a strong, slow drilling of hard boards that requires passion and at the same time a cool perspective (Weber 1919, p. 66).

Sustainability as a policy goal

Against this background, policy is made on very different issues, among which sustainability is relatively new. This chapter uses the example of the Federal Republic of Germany in setting out how policies for sustainable development can be designed. Germany first committed to the goal of sustainable development at the Rio Conference in 1992 (▹ Chapter 1.2). The aim of the conference was to contribute to sustainable development by coming up with recommendations for politically and legally binding proposals for action. Subsequently, the issue of sustainability was taken up at all levels of German politics, and new political institutions were created.

German Council for Sustainable Development

Although there is no Ministry for Sustainable Development in Germany, several political advisory bodies have been established, e.g. the Sustainable Development Council. The Council was appointed in 2001, and consists of 15 prominent figures holding various important positions in public life. In addition to this council of experts, there is also a Parliamentary Advisory Council on Sustainable Development. This was set up in the German Bundestag in 2004 and is reconstituted in each legislative period. It represents, so to speak, the parliamentary accompaniment to the Federal Government's sustainability policy. Its main tasks include the elaboration of measures and instruments for the implementation of the sustainability strategy, and networking with and supervising the other committees of the German Bundestag that deal with sustainable development.

Another important body is the State Secretaries' Committee on Sustainable Development. This is located at the Federal Chancellery and composed of the State Secretaries[1] from all the federal ministries. Together with their respective departments, they contribute to the joint implementation of the German Sustainability Strategy in many different policy areas.

The Sustainability Strategy of the Federal German Government

The Federal Government's first 🔍 **Sustainability Strategy** was formulated in 2002 with the aim of anchoring sustainable development in Germany's government and administration. Numerous scientific, political and civil society organizations contributed to the strategy. The revised version of the Sustainability Strategy adopted in 2017 is based on the UN's Sustainable Development Goals and covers global responsibilities as well.

1 State Secretaries are ranked immediately below Ministers, i.e. each ministry has one or more State Secretaries who are responsible for various tasks and who represent the Minister.

The German Sustainable Development Strategy

'The Sustainable Development Strategy is based on a holistic, integrated approach: long-term, viable solutions can only be achieved if interdependencies [...] are taken into account. The Strategy aims at economically efficient, socially balanced and environmentally sustainable development, whereby the planetary boundaries of our earth together with the objective of a life in dignity for everyone form the ultimate constraints for political decisions. The Strategy bundles the contributions of the various policy areas to sustainability and works towards greater coherence and the resolution of conflicting goals in light of the large number of systemic interdependencies. It thus directs globally responsible, intergenerationally equitable and socially integrative policy' (Bundesregierung 2017, p. 17).

Setting the political course – simply a case of governing by decree?

All political levels are involved

Sustainable development is a global challenge which confronts all political levels and systems — or in other words, the so-called multi-level political system. This includes the international and European levels. Especially noteworthy in this context is the adoption of the 17 Sustainable Development Goals, with their 169 sub-goals, by the United Nations General Assembly in September 2015; 193 states have thereby committed to achieving these goals (United Nations 2015a) (▷ Chapter 1.2). In Germany, sustainable development is dealt with at the national level by the

Table 1: Sustainability strategies at the different political levels

Political level	Actors and examples of sustainability strategies
Global	United Nations: Sustainable Development Goals (2015)
EU	Council of the European Union: The new EU Sustainable Development Strategy (2006)
National	Federal Government: German Sustainable Development Strategy (updated version 2017)
Regional	County: Landesregierung Brandenburg Nachhaltigkeitsstrategie für das Land Brandenburg (MUGV 2014)
Local	City: Freiburg im Breisgau: 2. Freiburger Nachhaltigkeitsbericht 2016 (Stadt Freiburg i. B. 2016)

federal government, regionally by the federal states, and locally by the municipalities. Sustainability strategies have been adopted at all these levels (see Table 1).

In the light of these cumulative commitments, the political realization of sustainable development should follow, as it were, automatically. However, this is not the case, because sustainability cannot simply be decreed by the state. The reasons for this are outlined below and illustrated by means of three key issues linked respectively to the policy objectives, strategies and practical implementation of a sustainable development policy (Brand 2002; Voss et al. 2007):

Sustainability cannot be decreed from above

1. How much priority is given to the *goal* of sustainable development in politics?
2. What would policy for sustainable development look like in terms of *detailed content*?
3. How can sustainability strategies and policy measures be successfully *implemented*?

These three questions will be investigated in turn in the subchapters which follow, and the necessary conditions, and the prospects, for a politics of sustainable development will be set out.

The negotiations leading to the adoption of 'sustainable development' as a policy goal

When sustainable development is considered from a global perspective, literally everything is at stake. Precisely for that reason, problems for policymaking arise as soon as the first question is asked: how much priority should be given to sustainability policy? On the one hand, the scope and importance of sustainable development are so far-reaching that it is difficult to express it with any precision as a policy goal; on the other hand, sustainability is not the only goal of public policy, and is in competition with other issues and interests. In view of this ambivalence, politics requires ▹ **target knowledge**.

Politics requires target knowledge

> **Target knowledge is required for clarifying and differentiating the goals of sustainable development. This involves setting out the differing views, values, norms and interests of the actors involved and/or affected, and weighing up the arguments for and against their respective goals. Here, the detailed analysis of conflicting goals is especially important.**

Politics of sustainability and civil society

The identification and definition of sustainability problems is often undertaken outside the political arena by civil society, environmental or social organizations or through scientific research. Politics only becomes involved when political actors pick up on these issues and declare them to be political tasks. Thus, some environmental issues were ignored for a very long time and made it onto the political agenda only because of massive pressure from the environmental movement.

Since sustainability is a broad goal, a differentiation or translation into specific goals and measures is required for all policy areas. This specification presupposes, first of all, the existence of a substantive benchmark for a politics of sustainability, which can be outlined as follows:

a) the permanent safeguarding of the natural foundations of life for future generations

b) the establishment of satisfactory living conditions for all people, understood as the satisfaction of their basic needs, first and foremost for the world's poorest

c) solutions to global problems which work by integrating ecological, social, economic and political-institutional development perspectives (Grunwald & Kopfmüller 2012)

Tensions and conflicting policy goals

But it is difficult to break down these abstract visions into concrete subsidiary goals. The breadth of the objectives almost inevitably results in competition, tension or conflicting goals. There is a fundamental dilemma between the desire to produce goods to satisfy even modest basic needs and the desire to conserve natural resources and global ecosystems. The principles of market competition and efficient production require that companies optimize resources and exploit and use them at minimal cost, which can quickly lead to the overuse of public goods. The principle of sustainability, on the other hand, implies the regeneration of resources and the fundamental preservation of the material and immaterial basis for production. The maximization of both principles is mutually exclusive (Müller-Christ 2014). That is why political regulation is needed — for example, of the ecological limits within which economic production and consumption can be allowed to take place (▹ Chapter 1.2).

The need for a process of societal negotiation

Sustainable development is therefore a regulatory goal in the sense of an ideal or vision to be aspired to — like freedom, democracy or justice. It prescribes a general direction for political action, while leaving the concrete content to be negotiated continually among the wider

The Bundestag's Enquete Study Commission on 'Growth, Well-being and Quality of Life'

Enquete Commissions of the German Bundestag are a special institution without parallel in any other parliament. In these commissions, MPs and experts contribute equally to the clarification of issues. Unlike committees of inquiry, they are not about throwing light on scandals; rather, they investigate an issue both scientifically and politically in order to create a basis for legislative action.

The 'Growth Enquete' was set up at the beginning of 2011 by the 17th German Bundestag, and submitted its report in 2013 (Deutscher Bundestag 2013). Five working groups addressed the issue as a whole by focusing on the fundamental aspects of growth; the development of a new indicator for prosperity; opportunities for decoupling growth from economic activity; and the sectoral dimensions of the financial system, work and lifestyles.

Despite this work, politics and academia generally continue to assume not only that a steady increase in gross domestic product is required to create material prosperity, but also that human well-being in general depends on economic growth measured in this way. The awareness of the limits to growth, which follow from the limits to our planet, increased in the short term, but this has not yet had an impact on politics (see Ott 2013).

The outcomes of the Growth Enquete were limited by its non-consensual way of working. The prosperity indicator it proposed was not adopted by the 18th German Bundestag. The biggest impact is likely to be of a more long-term and subliminal nature: the Study Commission has sensitized many politicians to the issue, in addition to prompting and encouraging a number of scientific publications and research projects.

society. Politics has the task of handling the contradictions and brokering the compromises.

An intentional plurality of interests

Meanwhile, sustainability continues to be in competition with other goals and interests pursued within our society. In an open democratic society, a plurality of interests is something desirable and an end in itself. Different interests are regarded, initially at least, as being equally justified — as long as they remain within the framework of the German Basic Law.

Innovations in social niches

A cause such as 'sustainable development' grows out of social initiatives, often paired with technical and social innovations such as renewable energy, organic farming or local initiatives for social cohesion. These develop continually in social niches (▹Chapter 3.5), but sometimes they also emerge out of the centre ground of society (Schneidewind & Scheck 2012). Politics can support and strengthen such initiatives, and sometimes even initiates them. This is demonstrated by discrete political movements such as Local Agenda 21, the agricultural transition, the *Energiewende*, and climate policy.

Conflicts of interest and the distribution of power

If politics supports local initiatives, organic farming or a major transformation, then this involves a redistribution of opportunities, power, status and income. This inevitably creates interest groups of those who benefit from the current situation and who tend to be well-connected, rich and powerful, such as energy companies or the automotive industry. In terms of sustainability policy, they can be described as 'dinosaur industries'. These actors often vigorously resist initiatives towards sustainability within the political system. For example, for many years the automotive industry refused to realign its fleet towards low-emission cars. This demonstrates that sustainability is a socially and politically contested goal. It is based on the value judgement of justice, and has to be legitimized in the socio-political contest over conflicting interests and values. Science cannot establish such political goals 'objectively', but it can analyze and clarify the conflicts, provide scientific supporting evidence and point out different possible solutions.

Politics as a trailblazer

Politics is the central arena in which such conflicts of interest are fought out. The drivers of sustainable development and their opponents are in competition for political attention and power (Partzsch 2015). The task of the politics of sustainability is both to support sustainability innovations and at the same time to support the processes whereby established, non-sustainable patterns of behaviour or production systems dissolve, change and transform in such a way that they no longer counteract sustainable development. Such a process of abolition and cessation — like Germany's phasing-out of nuclear power — is part of this change, and is contested because it involves losers ('exnovation'; see Arnold et al. 2015).

In summary, the answer to the first question — about the political priority given to sustainable development — is that it is difficult to frame sustainability in terms of concrete, specific goals; and this itself gives rise to conflicts of interest within society. An initial task for politicians is therefore to provide orientation to the political debate.

The formulation of political sustainability strategies

Content design

Once the importance of sustainability goals for politics has been clarified, the second question must be addressed: how should sustainability strategies and programmes be designed in terms of content in order to be successful? This question relates to the technical requirements involved in rigorous and detailed policy programmes for a transport transition, for example, or for climate policy. Politics requires ▹ **systems knowledge** so that appropriate technical decisions can be made.

> **Systems knowledge is knowledge about empirical facts, processes and interactions within and between natural and social systems which helps us to understand complex sustainability problems. Systems knowledge analyses causes, influential factors, developmental dynamics and how they are embedded in social and natural systems, and helps us to predict the outcomes of different options for action.**

The policy cycle

The emergence or development of policy programmes can be described in textbook terms based on the policy cycle (see Fig. 1). Sustainability policy, too, goes through this cycle, though usually in practice it is accompanied by certain deviations and jumps:

a) A problem is identified and defined, often in social niches.

b) Politics takes up the problem as defined and places it on the political agenda.

c) In the policy process, goals are formulated and policy measures designed to solve the problem are agreed on through negotiation. Political majorities are formed in the course of this process.

d) The next step consists of implementing the decisions, which is usually the responsibility of the administrations.

Figure 1:
The policy cycle

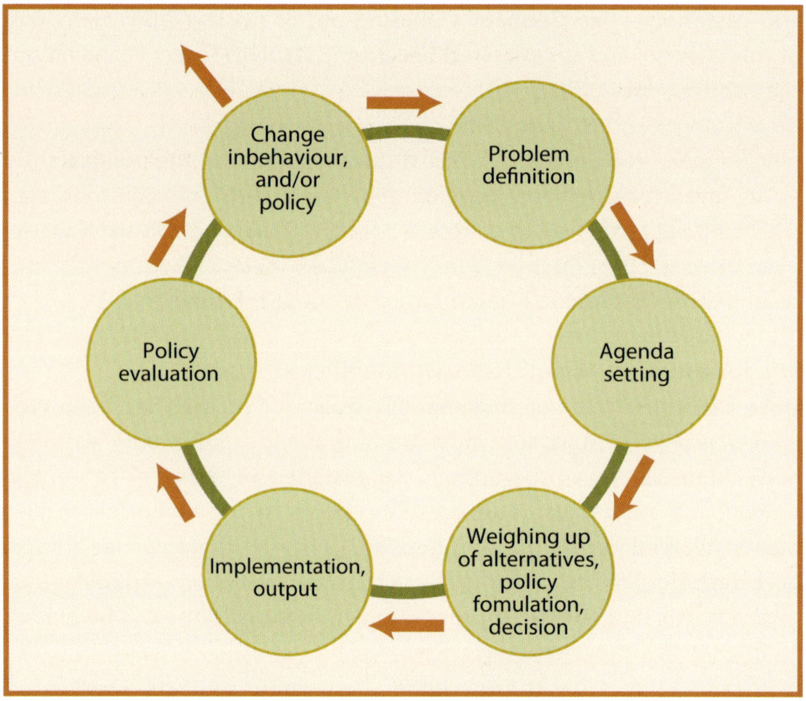

e) Finally, the impact of the policies is evaluated,

f) so that it can be decided whether they should be continued, changed or terminated.

Prevention of side-effects

One goal of policy is to design measures and programmes that solve concrete and specific sustainability problems. When developing evidence-based policy, civil servants usually contribute their specialist expertise and advise their ministers. Frequently, experts are consulted, and sometimes scientific advisory boards are appointed (as was the case e.g. with the **Brandenburg sustainability strategy**). Despite this, politics must always decide under conditions of uncertainty; for example, with respect to genetic engineering, where the effects of genetic manipulation on the structure of the manipulated organism are unknown, since not all of the organism's processes can be attributed to specific genes. The environmental response to any change is also characterized by uncertainty about causal relations. Nevertheless, politics is called upon to intervene as early as possible, and to avoid both deferring the problem and any unintended side-effects, particularly those that have an impact beyond the national borders.

The sustainability strategy for Brandenburg

In 2010, the parliament of the German federal state of Brandenburg decided to develop a sustainability strategy. For this purpose, an inter-ministerial working group was set up under the direction of the Ministry for the Environment, Health and Consumer Protection. This working group steered the complex process. Its tasks included coordinating the different interests and requirements of the participating ministries as well as involving the citizens. In 2010, the Sustainability Advisory Board was appointed, with 13 scientists giving important input into the elaboration of the strategy. In April 2014, the Brandenburg state government adopted the strategy, together with its priority action areas (MUGV 2014):

1) Business and work in the capital region
2) Liveable villages and towns
3) Brandenburg as a model region in its management of energy and climate change
4) A sustainable fiscal policy
5) Education for Sustainable Development (ESD)

Anyone who wants to promote sustainable development in Brandenburg can invoke the state's strategy. However, state policy can hardly be said to be furthering the implementation of the strategy, and the sustainability advisory board was not reinstated in the legislative period from 2014. This demonstrates that the state's sustainability policy in Brandenburg has not yet been consolidated.

Building majorities

Policy measures and decisions always reflect values and power relationships that have to be taken into account, as scarce state resources can only ever be used in pursuit of a small number of goals, and sometimes only one. That is why majorities have to be built in support even of sustainability policies based on persuasive evidence.

The need for integrated policy

Something that is already complicated for individual policy fields becomes even more difficult when complex cross-cutting issues such as climate policy or a transport transition are addressed. Here, policy integration across several sometimes very differently structured, policy fields is required. This complexity is all the greater because the interlinked political levels of the EU, the federal national government, the federal states and the municipalities have to be coordinated. This can lead to a 'joint decision trap'. This is when the political responsibility for particular decisions can no longer be assigned transparently. In response to the second question, that of how to develop coherent policies, the following conclusion can be drawn:

The challenge for policymaking is to formulate informed sectoral and cross-sectoral policies to deal with complex sustainability problems, to handle internal contradictions, and to avoid unwanted side-effects, in spite of uncertainty.

The implementation of the politics of sustainability

Once the importance of sustainability goals has been established in the political system, and rigorous and consistent sustainability policy programmes have been developed, the third question concerns how these programmes and specific measures are to be implemented. Which instruments and administrative structures are appropriate for the implementation of sustainability policy? In order to answer this third question, politics requires ▹ **transformative knowledge** on how to successfully implement sustainability strategies in practice, especially when political goals are not always clear and sustainability strategies are not consistent. This includes process knowledge about what the political requirements are for the achievement of a project.

Transformative knowledge deals with how desired states can be achieved and to what extent the existing conditions are capable of being changed to align with the target and system knowledge. It involves analysis of the conditions for action and the prerequisites for the practical implementation of proposed solutions and sustainability projects.

Table 2: Policy toolbox

In this, politics can employ a set of different types of policy instruments (see Table 2 Policy toolbox, Rogall 2012).

Area/Category	Specific examples
Administrative law	Laws, regulations, prohibitions
Planning	Regional planning, urban and rural planning
Market or financial incentives	Taxes, duties, subsidies
Information and communication measures	Public relations, guidelines, public information campaigns
Participatory instruments	Public involvement, roundtables, petitions, citizens' juries

These instruments all have their own strengths and weaknesses. For example, bans or prohibitions are effective, but their introduction is often difficult and costly, and compliance must be carefully monitored. By contrast, financial instruments are often easier to enforce, but less well targeted, because the target groups can respond in different ways.

It is therefore important to test in every case whether the means are proportionate to the ends. The likely impact can be estimated using the following criteria (Rogall 2012):

a) *effectiveness* –
Will the hoped-for sustainability effects be achieved?

b) *innovation impacts* –
Is there an incentive to develop the proposed solution further?

c) *efficiency* –
At what overall economic cost would the goals be achieved?

d) *fairness* –
Do the tools contribute to social justice?

e) *administrative practicability* –
How great is the administrative burden?

f) *political acceptability* –
Will the proposed tools meet with resistance?

Administrations as key actors

The administrations are key actors in the implementation, and they need sufficient technical and personnel capacity as well as political backing. Otherwise it may happen that resistance to political decisions is rekindled during the implementation phase and leads to political blockades.

An appropriate mix of instruments

Success is less a matter of finding the 'perfect' policy instrument than of choosing the appropriate policy or instrument mix, one in which the weaknesses of individual instruments can be compensated for, and which fits in with the given administrative structures and political style. If these points are taken into account, implementation deficits can be avoided.

Overall, despite good programmes, policy instruments can often be too complicated, and their implementation or the monitoring of regulations too difficult. Thus a suitable mix of instruments and, in any event, an effective administration are crucial for success.

Case study: the politics of the *Energiewende* in Germany – a key to sustainable development

Climate policy and the energy transition are keys to the achievement of sustainable development, because climate change threatens to change life conditions on earth in such a way that it will no longer be possible to guarantee minimum civilizational standards for humankind. The example of the *Energiewende* will be used here to show how politics can deal with the three challenges outlined above (the prioritization, consistency and implementation of the politics of sustainability). Successes and failures are listed in order to highlight the opportunities, as well as the difficulties, involved in energy transition policy.

Is there political support, and a political opportunity, for an energy transition?

Since about three-quarters of total emissions arise from electricity generation, heating and transport, a radical shift in the energy supply is central to climate protection. In Germany, this shift to renewable energies and to a more efficient use of energy is known as the *'Energiewende'* (▹ Chapter 3.3).

The history of the *Energiewende* in Germany begins with the anti-nuclear movement of the 1970s. The resistance to nuclear energy was initially anchored in a relatively small and predominantly younger segment of the population. But this changed after the protests against the construction of a nuclear power plant in Wyhl in Baden-Württemberg, where the more radical anti-nuclear movement formed an alliance with local winemakers and in this way gained entry into the wider population.

The nuclear phase-out and renewable energy

It took another 20 years for the anti-nuclear movement, linked to the environmental movement, to find its way into government (see Fig. 2). With the accession of the red-green coalition under Chancellor Schröder at the end of 1998, the nuclear phase-out became official German policy. The first Green Environment Minister, Jürgen Trittin, negotiated a nuclear phase-out with the energy companies, and a law on renewable energies (EEG) came into force. Due to concerns about the dangers of nuclear energy and a general will to tackle climate change, this strategy enjoyed broad public support. In the 2000s, the fight against climate change increasingly shaped the *Energiewende*, and was, at least rhetorically, supported by all parties in the Bundestag. With the adoption of the Paris Climate Agreement in December 2015, the global dimension of climate protection grew — since then, an economy that takes into account the planet's natural boundaries has become the goal of the international community.

Opposition to the Energiewende

In the traditional energy sector, however, resistance to the nuclear phase-out and the switch to renewables persisted. This had some effect on energy policy. For example, the nuclear phase-out negotiated with the industry by the red-green government was reversed, or more precisely extended by ten years, by the CDU/CSU/FDP coalition that took office in 2009. The Renewable Energy Sources Act (EEG) was significantly weakened and hampered by superfluous complications, which slowed the expansion of wind and solar power systems. The 'phase-out from the phase-out' in nuclear energy, however, had to be reversed in turn following the nuclear disaster in Fukushima in 2011; the Federal Government now decreed a renewed nuclear phase-out by 2022.

Figure 2: Timeline for the *Energiewende* in Germany (with acknowledgements to Carbon Brief 2015)

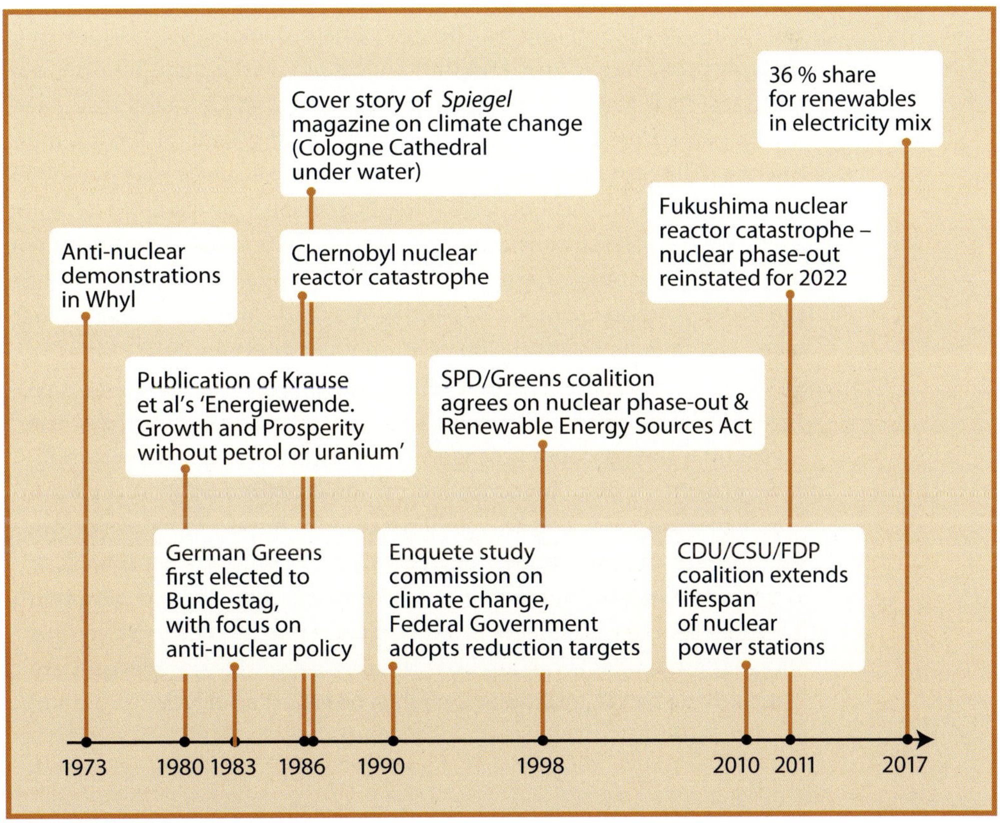

Is there a consistent policy strategy for the energy transition?

A changeover with side-effects

German energy policy is characterized by ambitious goals, but it lacks a consistent strategy. In the area of renewable energies, the EEG has led to a steadily growing share for non-fossil and nuclear energy sources. By 2017, the share of renewables in the German electricity mix had grown to 36 %. However, the zigzagging course set by different governments has led to policy side-effects being tackled either not soon enough or not at all: for example, the problems caused for humans and nature by wind turbines, or what in Germany has been called the *'Vermaisung'* of the landscape (covering and disfiguring it with maize fields) as a result of widespread biogas production.

Rebound effects

Progress has been made on the more efficient use of energy, but it has been cancelled out by growing energy needs (▹ **rebound effect**; ▹ Chapter 4.3). This is largely due to the lack of energy efficiency (sufficiency) strategies — belief in the causal link between prosperity, economic growth and energy consumption remains strong. In the transport sector, the energy transition had no effect — climate-damaging emissions from transport continue to rise, and they cancel out all the gains in other policy areas. As a result, greenhouse gas emissions in Germany rose again in the years from 2014 to 2017. According to the Federal Environment Agency (2018), total CO_2 emissions in the transport sector rose by 5.4 million tonnes, an increase of 3.4 %. Freight transport by road increased by 2.6 %. Higher diesel consumption alone is responsible for 4.8 million tonnes of greenhouse gases, according to a study by Arepo Consult for the Greens (Arepo Consult 2017). Today, it already appears likely that Germany will miss its political climate protection goals for 2020.

The rebound effect: this term describes the process whereby the potential savings from efficiency gains, for example from more efficient technologies, are cancelled out or exceeded by increased use and consumption. This is well illustrated by the example of the car: if fuel consumption in cars decreases, this often leads consumers to choose to purchase larger, less fuel-efficient models. And although a fuel-efficient car leads to lower fuel costs per kilometre driven, this saving often affects driving behaviour: more and longer journeys are then made by car, instead of by public transport or bicycle.

Are appropriate instruments and regulatory measures available for implementation?

Germany was among the pioneers in researching and exploring alternative energy sources. Notably in the field of wind energy, technical developments were initiated by lone pioneers and then financially supported by the ecologically-minded middle classes. Research institutes such as the Ökoinstitut, founded by committed ordinary citizens, developed the initial concepts for an 'energy transition' based largely on the use of renewable energies and on increased energy efficiency. The Wuppertal Institute under Ernst-Ulrich von Weizsäcker developed the idea of ecological tax reform, in which energy prices rise slowly but steadily and thus become integrated into economic calculations.

Hermann Scheer – an important driver of the *Energiewende* in Germany

Figure 3: Hermann Scheer (photo: Verlag Antje Kunstmann)

Hermann Scheer (1944–2010) was a German politician and from 1980 to 2010 a Member of the German Bundestag for the SPD.
He helped to initiate a number of laws promoting renewable energies, including the Electricity Feed-in Act for renewable energy (1991) and the Renewable Energy Act (2000). In 1999, he was awarded the Right Livelihood Award.

'My starting point is not renewable energy, but rather society itself. I have not moved from renewable energy supply into politics in order to implement it. Rather, it is my view of the fundamental problem and my understanding of political responsibility that has led me to renewable energy. [...] It is not renewable energy that we lack, it is time.' (Scheer 2010, p. 270).

100% renewable energy supply is possible

There are now a number of scenarios that illustrate how to organize an economy in which 100 % of the energy supply comes from renewable sources (Umweltbundesamt 2010). This must be seen as a major step forward, because around the year 2000, most experts still considered a power supply with more than 4 % renewables on the grid to be impossible. The instrument mix of regulatory prescriptions and proscriptions, financial incentives, information and communication measures and participatory instruments as well as soft influence (or 'nudging') is now very well developed. Implementation remains difficult, however, because the traditional economic actors have great influence and seek to prevent a consistent political implementation of the energy transition — for example, the phase-out of coal. Among the general public, support for an energy transition is still very high. Yet, at the same time, there is a lack of widespread support for a sufficiency strategy that aims to reduce the use of energy and resources (⊳ Chapter 1.2).

Conclusions on the Energiewende

As a conclusion to the case study of the politics of the *Energiewende* it can be stated that much has been achieved and many interested parties have been brought on board. But at the same time, the energy sector remains largely outside democratic oversight and control. Instead, according to a study by the BUND (2011), it is often characterized by a dense web of parties, politicians and industry that entrenches centralized, inefficient and environmentally hazardous energy use, while environmental protection and sustainability fall by the wayside. In addition, in recent years, there has been growing opposition to specific technologies such as wind energy. Energy consumption is once again on the increase, meaning that political goals cannot be met. The case study of the *Energiewende* shows just how difficult the politics of sustainable development can be.

The scope of, and limits to, a politics of sustainability – the governance approach

Although the guiding principle of sustainable development was formulated at the global level more than 25 years ago, there is a lack of outstanding examples or breakthroughs in sustainability at the national level. Despite some successes, the *Energiewende* is still a long way from achieving its goals; the agricultural transition has stagnated since the 2000s; and the mobility transition has not even started. Sustainable development cannot be identified as the overarching guiding principle of German politics. Isolated successes notwithstanding, it is still the case that too often politics merely pays lip service to sustainable development.

In view of the demands made on policymakers, this (at best) mixed balance sheet should not come as a surprise. The formulation of policy goals, the development of integrated sustainability strategies, and the practical implementation of policy measures are all required to meet the highest standards. In many respects, the political process is thereby tested to its limits. A degree of scepticism about the limits to the politics of sustainable development is appropriate, and any plans for comprehensive ideal solutions should be treated with caution. Nonetheless, politics should not and must not put its head in the sand and give up its aspirations to lead and shape the process (Voß et al. 2007). Business and civil society alone are not up to the task. That is why, in view of global environmental problems, the *German Advisory Council on Global Change*, an advisory body to the German Federal Government, advocates fundamental economic and political-institutional change, or a 'Great Transformation' (WBGU 2011) (▹ Chapter 4.2). In such a transformation, a **'proactive state'** should take responsibility for sustainability issues, take the lead, and at the same time reach out to and involve the general public, business and civil society.

The need for a proactive state

The proactive state

'A central element is a proactive and enabling state which actively sets priorities whilst also offering its citizens increased opportunities for participation, and affords the economy choices with regard to acting with sustainability in mind. New problematic situations call for a different kind of policymaking which takes on the challenge of the transformation through greater political self-commitment in the form of strict rules. It must create a new regulation framework, which society and the economy could and should adhere to. This set of rules encompasses a national climate protection objective, a stringent climate protection law […]; it underlines the credibility of political agendas, establishes the scope for action (or for interpretation), reduces the influence of particular interests, and creates long-term planning horizons, thereby accelerating the transformation process. This increased action capacity on the part of the state must be inextricably linked to active citizenship. Civil society must help to shape the transformation process, as it sets the transformation in motion and gives it the requisite legitimacy.' (WBGU 2011, p. 239).

Governance as an extension of political capacity

However, this requires extended forms of political control and new control philosophies in order to overcome the limits discussed. ▹ **Governance** — understood as an extension of government (hierarchical rule) — represents a promising approach. Hierarchical state rule is here supplemented and extended by social self-organization (Kooiman 2003). Governance opens and extends the management of sustainability problems to non-governmental actors and areas. Especially for complex tasks like sustainable development, this opens up additional scope for action and expands the constellations of actors involved.

> **Governance encompasses all institutional regulatory systems that substantively determine the political decision-making process and its implementation in any given policy field. Governance aims to manage the interaction between different political levels and groups of actors. State institutions thereby cooperate with non-state actors without having direct control of the latter, and the organizational boundaries between state and society are transcended (Benz & Dose 2010, p. 25 f.).**

Participatory processes of reflection

A central element of such holistic approaches is reflection in order to find a way of dealing with complex processes and uncertainties about future developments. Societal learning, transparent, participatory processes of reflection on goals and procedures, and a conscious adaptation of policy to dynamically changing conditions (fault tolerance, experiments and reversibility of decisions) are constitutive elements of a reflexive governance for sustainability (Voß et al. 2006).

Transition management

In the Netherlands, the concept of *transition management* was developed to enable the design of a transformation to sustainability (Loorbach 2010) (▹ Chapter 4.2). In a first step, a small group of pioneers draws up a sustainability vision in a *transition arena* (see Fig. 4). The second step involves stakeholders and experts formulating a *transition agenda*, an implementation plan. In the third step, their ideas are supplemented and extended through numerous experiments and projects. In the fourth step, the ideas and experiences are evaluated and the planned further procedures are subjected to self-critical examination. Transition management takes place on several levels. Strategically, it aims at long-term cultural change in society; tactically, over the medium term, it seeks a reorganization of political-institutional structures; and in the short term, new social practices are developed and tested operationally.

Governance has considerable potential for sustainable development, but it also brings with it new difficulties and problems. The results are

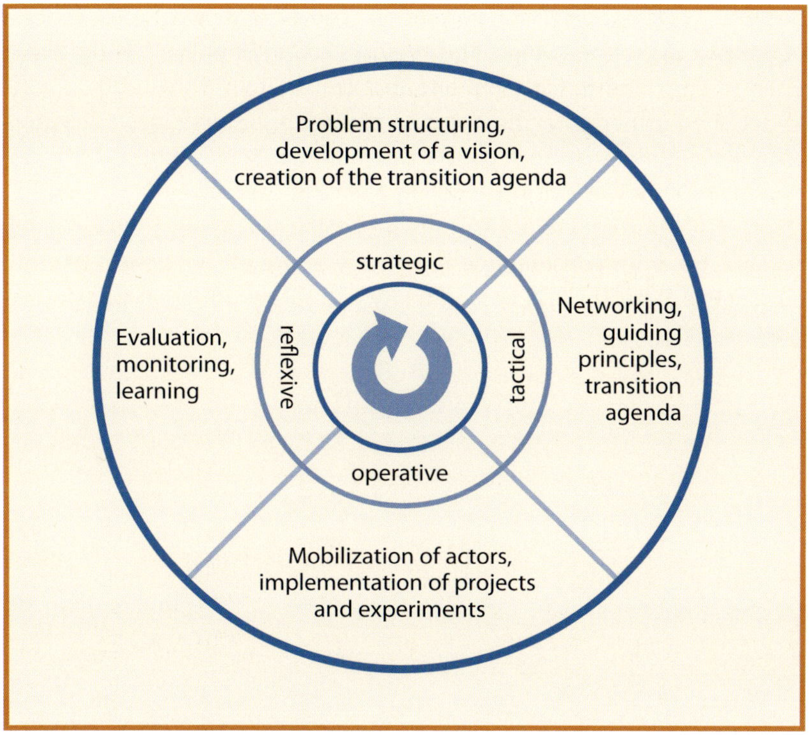

Figure 4: The Transition Management cycle (with acknowledgements to Loorbach 2010, p. 173)

Problems of governance

more open and unpredictable than with hierarchical rule (government), and the processes are dynamic and not always transparent, though this can also apply to traditional forms of political rule. Above all, however, there is the question of the legitimacy of such an extension of political control, because decisions and implementation processes in governance are no longer solely the responsibility of democratically legitimized institutions. Instead, collaborations for governance often involve state, civil society, and private actors (such as interest groups, associations, corporations) and could be termed arbitrary, ad hoc constructs.

Transparent and participatory communication

Nevertheless, governance can make a significant contribution to 'robust' policy decisions and sustainability strategies, which are still achievable in a socially contested field in spite of the unavoidable contradictions and conflicts of interest. The prerequisite for such 'robust' decisions is transparent and participatory communication structures.

Conclusion

To sum up: even with the best of intentions, politics cannot govern by decree and prescribe sustainable development by means of a series of well-formulated laws. Any claims for the possibility of a comprehensive and effective policy programme should be examined very closely, because a sustainability transformation is an extremely complex process of (conscious) social change. This requires a kind of policymaking that is open to error and to learning, one that constantly reviews and corrects its actions. Furthermore, the question of what priority is given to sustainable development by society and thus ultimately by politics is a question of social power. But state policy, even if it can make only a cautious claim (after due reflection) to genuine political control, remains a key social actor, with considerable legitimacy and great influence. Without a proactive state, there is no possibility of a successful 'great transformation' to sustainable development.

Everyone affected and involved: civil society systems

3.5

Heike Walk, Hermann E. Ott and Martin Welp

In addition to the state (political) and market economy systems, the civil society system is another important sphere of the anthroposystem. Hardly any social or political problems are discussed today without the involvement of civil society groups and organizations. They have a key role to play in the implementation of the sustainability agenda in the 21st century. It is particularly in the context of the establishment of international conferences that the role of civil society has begun to attract more attention. What are the core elements that shape the civil society system? What forms of participation are there, and how does democratic participation help to secure public involvement and influence in different fields?

Civil society – a term with a long tradition

Historical associations

In the philosophers' schools of antiquity already, the *Societas Civilis* of free citizens was considered a guarantor of a properly functioning political community (although at that time only men had civil rights). During the Enlightenment in the 18th and 19th centuries, and especially in the context of the French and American Revolutions, the concept of civil society was taken up by numerous intellectuals (e.g. Alexis de Tocqueville, John Locke, Adam Ferguson, Charles de Montesquieu and Immanuel Kant). They all highlighted the positive effects of a functioning civil society. These effects related above all to the social and solidaristic coexistence of a community. An active civil society promoted citizens' trust in the political system (here meaning political institutions such as ministries, parties and administrations).

Civil society is a term with a long tradition, and refers to a sphere between the individual or private sphere and the state. Its spectrum covers citizens' initiatives and civil rights groups, associations and interest groups, cultural and educational institutions, religious associations, economic development organizations and self-help groups of all kinds.

The renaissance of the term

At the end of the 1980s and the beginning of the 1990s, the term 'civil society' experienced a renaissance in Germany due to the peaceful revolution in the GDR and the ▷ **post-socialist transformations** (▷ Chapter 4.2) in Central and Eastern Europe. Prompted by the success of civil rights movements advocating national independence, free elections and freedom of assembly, speech and opinion, the number of academic and journalistic articles on civil society began to skyrocket. At the international level, too, discussion of the growing scope for the creation of an 'international civil society' intensified — not least due to the change in global political power constellations and the rapid development of communication technologies. There was particular emphasis on so-called 'non-governmental organizations' (NGOs) as civil society actors (Brunnengräber et al. 2005). These NGOs took on three major roles in international politics: controlling political processes; mobilizing a critical public sphere; and providing factual information.

> **Post-socialist transformation refers to a special form of transformation involving a transition from a former communist or socialist system to a free-market liberal democracy.**

Definition

Numerous definitions of civil society can be found in the literature, and Kocka (2003) points out that the term is interpreted quite differently in different academic disciplines. Since not all of them can be shown here, what follows is restricted to the definition set out by Jürgen Habermas:

> Civil society is composed of 'those more or less spontaneously formed associations, organizations and movements which absorb, condense and amplify the resonance of social problems in the private spheres of life, and transmit them to the public political realm' (Habermas 1992, p. 443).

Jürgen Habermas

Jürgen Habermas (born in Düsseldorf in 1929) is one of the world's best-known contemporary philosophers and sociologists. At the centre of his extensive and award-winning work is the attempt to renew critical social theory, as developed by representatives of the so-called 'Frankfurt School'. His most important work, *The Theory of Communicative Action*, deals with the significance of communication in social life in (post-)modern society.

Actors and goals The term civil society therefore describes a diverse spectrum of sometimes very different actors. These include associations, NGOs, foundations, citizens' initiatives, business lobby groups and churches. Like the actors themselves, their goals are very different. They range from dealing with general social problems, for example 'increasing educational opportunities' or 'local climate protection', to the concerns and needs of specific groups, such as citizens' initiatives for noise protection or aid to refugees. In general, they are concerned with the provision of goods and services contributing to public welfare and/or with influencing public opinion through debates, protests and other public events. In the early years of the 21st century particularly, a gradual shift in perception could be observed in public discourse — from the notion of civil society to the idea of civic participation. This signifies both an opening up of the political system and at the same time a stronger focus on the involvement of individual citizens.

The significance of civil society on different political levels

The role of civil society has attracted renewed interest particularly in the context of the establishment of international conferences. On the one hand, civil society actors have helped to set the agenda, for example by organizing so-called 'counter summits' (as in Rio in 1992), where critical and sensitive issues are discussed. At the same time, they became indicators of a movement towards the transparent and democratic organization of global, European, national and local institutions. For a proper understanding of current discussions on civic engagement and participation in sustainability, it will be helpful to provide a retrospective review of the significance of civil society and civic participation at the different political levels.

The global level Both the Brundtland Report of the World Commission on Environment and Development (1987) and **Agenda 21** from the UN Conference on Environment and Development in Rio 1992 (United Nations 1992a) (▷ Chapter 1.2) emphasize the importance of civil society and citizen participation for the development of sustainability strategies. Agenda 21, for example, states:

» Through consultation and consensus-building, local authorities would learn from citizens and from local, civic, community, business and industrial organizations and acquire the information needed for formulating the best strategies (Agenda 21, para. 28.3).

The example of international climate policy clearly illustrates the importance of non-state actors. Even before the signing of the Climate Convention in 1992, national environmental associations and local environmental groups had started to organize to achieve greater political influence. For a number of years now, the Climate Action Network (CAN) and many other networked initiatives have put pressure on national policymakers, informing the public and translating scientific knowledge into policy goals. In recent years, the role and status of non-state actors has expanded at the international level. At the UN Climate Change Conference in Paris in 2015, the 'subnational' actors were explicitly recognized as important allies in the fight against climate change (Obergassel et al. 2015, 2016). A year later, at the next UN Climate Change Conference, the *Marrakech Partnership for Global Climate Action* was set up to pool efforts by state and non-state actors. The UN Climate Change Secretariat in Bonn registers and promotes 'subnational' climate measures taken by cities, regions, companies, investors and civil society organizations.

Agenda 21

Agenda 21 is a development and environmental action programme adopted by the UN member states which sets guidelines for the 21st century to bring about a change in economic, environmental and development policies. According to Agenda 21, it is first and foremost the national governments that need to plan the implementation of sustainable development at the national level through strategies, national environmental plans and national environmental action plans. Non-governmental organizations and other institutions should also be involved. Many new forms of participation, model projects and forms of dialogue have been initiated due to the action programme and associated public discussions. In other words, Agenda 21 works on two levels: it improves cooperation between local government and civil society organizations, and it establishes links between global and local sustainability projects.

The European level

Civil society also plays a not insignificant role at the European level. It was strengthened by the Aarhus Convention, signed in 1998 by all European countries and the EU (▹ Chapter 3.1). The Convention covers access to information (an important precondition for civic engagement), public participation in decision-making and access to justice. In Germany, these three pillars are anchored in the Environmental Information Act of the Federal Government (2005), the Public Participation Act (2006) and the Environmental Appeals Act. Another example of the EU's role is the Water Framework Directive adopted in 2000. Implementation in the individual EU member states requires management plans to be drawn up at river basin level informing and consulting ▹ **stakeholders** and the wider public (Article 14). At the European level, the Directive has provided an important impetus for the adoption of participatory approaches in the field of water and land management. Member states have had to adapt their national legislation accordingly and develop procedures for implementation at the regional level.

Figure 1: Meeting of the UNESCO World Heritage Commission in Krakow in the summer of 2017. United Nations multilateral agreements and meetings are also important opportunities for representatives of civil society to engage in consultative, observational or lobbying activities in international decision-making processes. (photo: P. Ibisch)

Stakeholders are all the people, groups and institutions directly or indirectly affected by the activities associated with a project or who have any interest in these activities. The term 'stakeholder' was first used in an internal memorandum of the Stanford Research Institute (SRI) in 1963 and arose from criticism of the one-sidedness of the shareholder approach.

The European Citizens' Initiative (ECI) was launched in 2012. This mechanism allows citizens of different EU countries to jointly call on the European Commission to address an issue or take a legislative initiative. The ECI is thus a non-binding participatory instrument for making proposals for EU legislation.

Germany (the national level)

At the national level, too, civil society groups are the main drivers of the development of global sustainability goals. One of the key studies on how to implement the calls from international UN conferences for action at the national level was *'Zukunftsfähiges Deutschland'* ('Sustainable Germany') (Loske & Bleischwitz 1997). This study was commissioned by the Wuppertal Institute for Climate, Environment and Energy by Friends of the Earth Germany (BUND) and Misereor, and became an important driver of German discourse on sustainability (▷ Section 1.2). Civil society is here accorded a key role in terms of instigating political reform, new alliances, a long-term focus and the political courage for new ideas and visions.

With the establishment in 1999 of the ▷ **Enquete Study Commission** of the German Bundestag on 'The Future of Civic Engagement', a comprehensive inventory of the state of research on 'civil society' was set up. As a result, under the patronage of the Federal Ministry for Family Affairs, Senior Citizens, Women and Youth, a targeted promotion of various programmes, projects and institutions ensued. For example, surveys on volunteering in Germany (German Survey on Volunteering) have been conducted regularly, also since 1999, to help understand the importance of the sector and to increase the visibility of civil society. In addition, since 2009, the Federal Government has produced two engagement reports summarizing the development of civic engagement or ▷ **participation** in Germany and making recommendations for action.

Figure 2: Civil society activities are an integral part of UN climate conferences. The picture shows an action by Greenpeace at the climate summit in Paris 2015: the sun as a symbol of the technological revolution. (photo: Greenpeace)

Enquete Study Commissions are set up by the German Bundestag or by a federal state parliament with the task of gathering material in preparation for decisions on large and significant complex issues. This is done by commissioning reports from acknowledged experts.

Participation and civic engagement are often used synonymously. Participation in very general terms refers to the participation of a person or group in decision-making processes or courses of action in the political field or within organizations. There are many and varied types and forms of participation, including citizen control, delegated power, partnership, consultation and information, and co-determination. These forms of participation can also vary greatly in terms of their specific function, scope and rationale.

The local level

In parallel, and as a result of the UN conferences, the so-called Agenda 21 processes were initiated in many municipalities in Germany in the 1990s, based on cooperation and networking between local government and civil society. These projects served to further collaboration on the objectives and measures set out in the UN Action Programme (Brand & Warsewa 2003). However, the local agenda projects differed widely in their approach. In some, in a relatively tightly organized process, the administration took an active and leading role (top-down), while in others, the process was left largely to develop according to the initiative of interested citizens (bottom-up) (Rösler et al. 1999; Kern et al. 2002). Current bottom-up examples include the **Transition Town initiatives** (▷ Chapter 4.2, 4.3), which implement activities and measures for climate protection at a local level but are also networked internationally and view all local activities as part of a larger transformation movement.

Transition Town initiatives

The term *Transition Town* encompasses a wide range of sustainability projects, such as city gardens, swap shops and repair cafés. Transition initiatives have been established in many countries around the world, from the southern English town of Totnes (the first Transition Town), to Brazilian favelas, to energy self-sufficient communities in Japan. It comprises a loose network of groups wanting to promote sustainable and resource-saving projects against the backdrop of oil shortages (Peak Oil) and climate change.

New forms of dialogue

Different phases can be observed with regard to the forms of dialogue used. While the Agenda 21 projects were carried out in close cooperation with local government in the 1990s, in the mid-2000s sustainability initiatives moved to other networks, forums and platforms — not least by extending their cooperation to economic actors. The debate on civil society and civic participation became more and more public (Olk et al. 2010). Dialogue forums on the Internet, the role of social media (Facebook, Twitter, etc.) in connection with international protests, and finally the experience gained from the protest against the transport and urban development project 'Stuttgart 21'[1] led to a renewal and expansion of citizen participation in politics and administration (Sommer 2015b). As a result, there was also a broadening of participation procedures.

1 The protest against the expansion of Stuttgart railway station escalated in 2010, after demonstrators had serious clashes with the police. Following the demonstrations, a referendum was held on the infrastructure project and very far-reaching participation procedures were initiated by the state government.

From civil society to volunteers: new inequalities

Engagement of volunteers

The movement towards greater citizen participation has been accompanied by a gradual shift in terminology, from 'civil society' to 'volunteers'. In Germany, the terms are usually used synonymously ('Zivilgesellschaft' and 'Bürgergesellschaft') and a clear definitional distinction does not exist. Nevertheless, it is clear that the sharp increase in participation opportunities in recent years is closely linked to the adoption of the notion of volunteer engagement. For example, the Federal Ministry for Family Affairs, Senior Citizens, Women and Youth has set up a sub-committee on the engagement of volunteers. This deals with current legislation that affects volunteering. In addition, a number of organizations and institutions have committed to increasing and developing political participation. The National Network (*Bundesnetzwerk Bürgerschaftliches Engagement*, BBE) which was established by the Social Democratic Party (SPD) in 2002, is a trisectoral network (that is, including the three sectors of civil society, market and state) delivering information, measures and projects for the promotion of civic engagement and volunteering.

Short-term and project-based participation

Larger organizations and political parties, in particular, seem to be increasingly unable to provide a platform for long-term civic engagement. In contrast, short-term, issue-based and project-related forms of engagement are on the rise. Civic engagement is also increasingly linked to professional qualifications and further education. For example, the ▹ **German Federal Volunteers Service** offers practice-oriented continuing education programmes. In addition, it is becoming clear that social inequalities are developing in the area of political participation, too, meaning that citizen participation has become a playground for the so-called 'modernization winners'. For this population group, awareness of the individual (personal and professional) benefits associated with participation is a greater incentive than is the case for other population groups.

The German Federal Volunteers Service was created by the Federal Government following the suspension of conscription (national military or civil service). It was established as an initiative for voluntary, charitable and unpaid work and is intended to promote civic engagement.

Civic engagement: prevalence across the population and by sector

Civic engagement, according to the current report on the topic by the Federal Government, is found to varying degrees among different age cohorts, income groups and sectors (BMFSFJ 2017). The largest proportion of volunteers today is found in the 60–75-year-old age group. In recent years there has also been an increase in the engagement of active young people, especially 14 to 19-year-olds. Women are slightly less likely to be engaged than men, and in different sectors. While men are mainly found in sports, emergency services and politics, women are more involved in church and religious matters, health and social affairs, and schools and nurseries (BMFSFJ 2017).

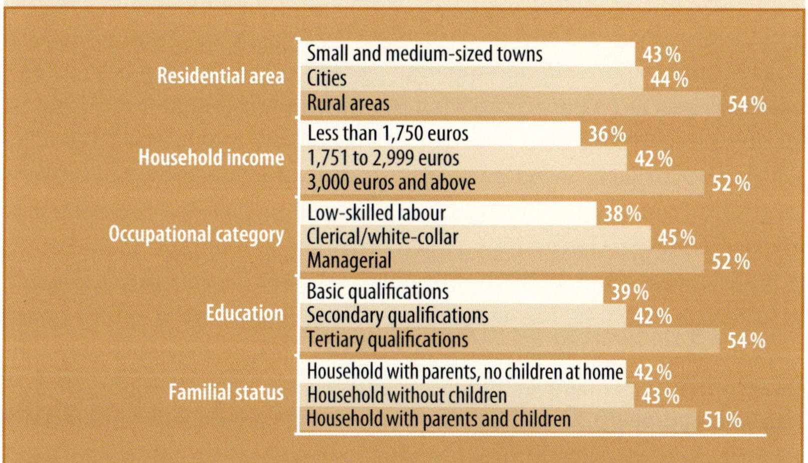

Figure 3: Proportion of different population groups practising some degree of civic engagement (widely defined, including helpers) (data source: IfD Allensbach 2013)

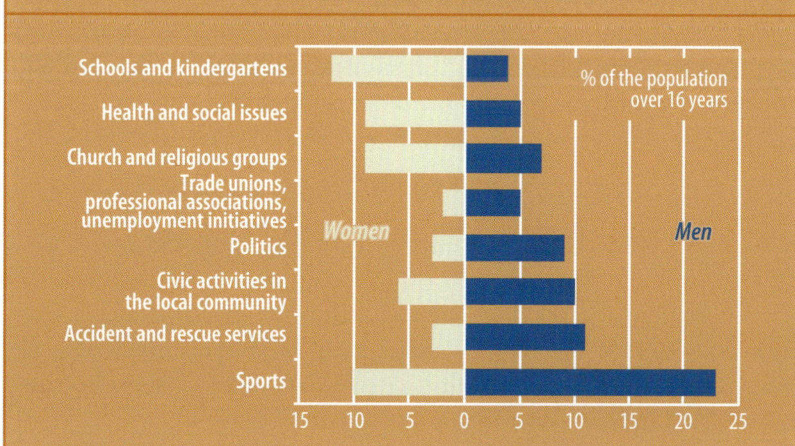

Figure 4: Civic engagement of men and women in different social sectors (data source: IfD Allensbach 2013)

Social inequalities

Empirical findings show a qualitative difference with respect to education and income (Embacher 2009). The lower the level of education and income, the lower electoral and other forms of participation. Citizens who feel themselves not properly understood or respected by the political elites are demonstrably withdrawing from democratic participation. Movements have emerged in the USA as well as in many European countries which actively oppose the political elite and the established media, especially through the use of social network media. This has led to a growing fragmentation of society. This in turn is associated with another significant imbalance, because not all citizens have the necessary resources to enable them to participate — time, expertise, rhetorical skills, self-confidence and assertiveness. These resources are often very unevenly distributed across social groups, so it is not surprising that the new forms of participation are dominated by the well-educated middle classes.

Online participation

The use of the Internet has greatly changed the engagement of civil society groups and the way participation takes place (Voss 2014). This applies both to participation directed toward a specific issue or in a campaign, as well as to participation in the medium of the Internet itself. So far, its use as a source of information has clearly outweighed activism, but its use for web-based participation procedures is growing. Citizens take part, for example, in online petitions to the German Bundestag or in campaigning organizations, which obtain large numbers of signatures in a short space of time for current political issues (in Germany, Campact, to take one example, has received very strong support in recent years, see Campact e. V. 2016). Internet participation is relatively easy and convenient for users; however, the Internet offers such a wide variety of content and platforms that there is a risk that participation will be perceived as short-lived or even superficial and ineffective. The full potential of online participation has certainly not been exhausted. In urban and spatial planning especially, citizens' preferences, local knowledge and individual creativity can be fed into the process through online participation.

Different forms of civic participation

Formal and informal participation

In the relevant literature, a distinction is usually made between formal (written) and informal (non-written) forms or procedures of public participation. Formal procedures are regulated by law: for example, petitions and referendums at the federal state level and municipal levels (Nanz & Fritsche 2012). In addition, citizen participation is also formally recognized in urban land-use planning and planning approval processes. But there are also informal procedures of public participation which are not legally regulated. They are voluntary (and often supplementary), and merely advisory in character. More and more municipalities and state institutions, such as government ministries, make use of voluntary participation processes to bring in ideas and opinions from the public. However, the application of these procedures varies greatly, because there are still concerns among politicians over whether such forms of participation restrict their decision-making powers. Accordingly, the way the results of such procedures are used, as well as the participation procedures themselves, varies widely (Sommer 2015a, p. 13).

Different forms and spheres of participation

Participation may be limited to obtaining information, or it may involve a greater direct share of power, as in sole or co-determination, and takes place in many areas of society. Participation in associations, at universities, in political decisions and in the running of autonomous organizations (e.g. cooperatives, parent-run nurseries, housing associations and group purchasing organizations) are all well-known examples. Citizen participation is thus practised in very different social areas and forms.

Diverse participation formats

The question of which procedure is appropriate for a given participation process depends on, among other things, the backgrounds of the participants, the objectives of the participation exercise (information, consultation or co-determination), the number of participants, the

Figure 5: Citizens' workshop on landscape planning in the district of Barnim (photo: Centre for Econics and Ecosystem Management)

The planning approval process

Planning approval procedures are used for major infrastructure projects (e.g. roads, railways and light rail systems) and certain other large-scale projects. They are carried out in order to deal comprehensively with any problems that might arise for public or private interests affected by these projects. The planning approval process is carried out by the respective authority, which also draws up a planning decision.

An application from the project (usually the federal state or the national federal government) starts the procedure, and the public gets involved only after lengthy informal liaison between the authorities responsible or affected and the sponsor. Following a public announcement (usually in the newspapers or official journals of the municipalities), the plans are displayed for a month in the affected communities. Objections can be submitted up to two weeks after the expiry of the public display period. Comments and objections are forwarded to the project sponsor, who responds in an appropriately objective and factual manner. In a discussion meeting involving the project sponsor, the public authorities, any persons affected and anyone who has raised an objection, open questions and objections are clarified and if possible resolved. The planning approval authority then draws up a planning decision based on the outcome of the consultation procedure. In this, the authority decides on all relevant issues, including on objections on which no agreement was reached.

Critics of the planning approval process complain that participation often takes place too late in the day, that the time allowed for public inspection of the plans is too short, and that the place and manner of publication are not conducive to actually generating broad public participation.

time available and which procedures the organizers are familiar with, and which formats are currently in favour (Walk 2008; Welp et al. 2009). In the 1970s, future workshops, planning cells, town hall meetings and advocacy planning were popular, while in the 1990s round tables, mediation and agenda processes were the norm. Today, some of the leading participation formats are online citizen forums, living labs, and — in the energy sector — financial participation options such as citizens' solar parks and energy cooperatives.

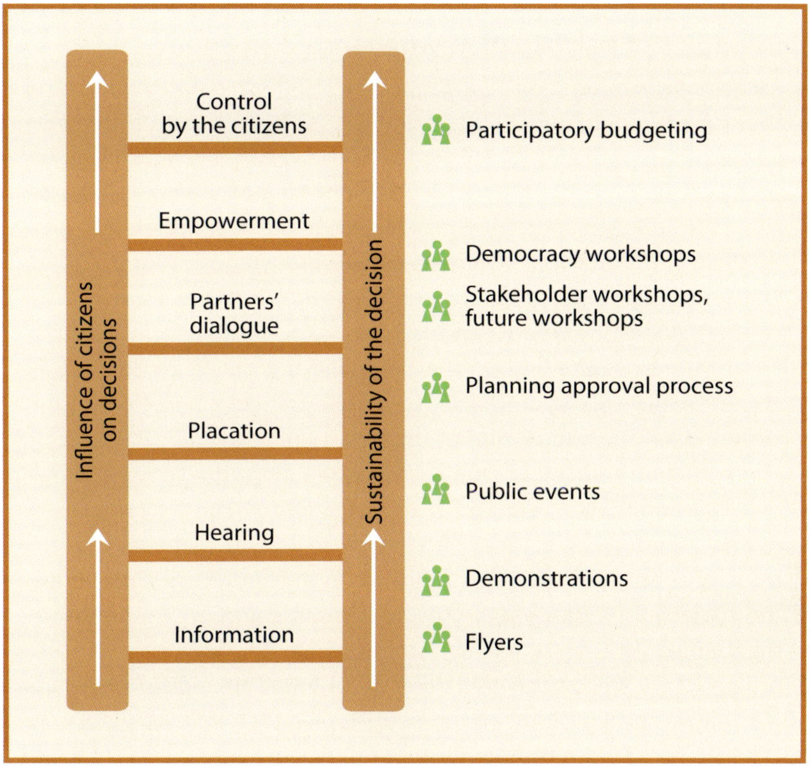

Figure 6: Participation ladder with graded levels of participation (with acknowledgements to Arnstein 1969)

Levels of participation

Sherry Arnstein's so-called 'participation ladder' (1969) is often employed in academic and public debate in order to classify the various procedures with regard to their democratizing effects. The individual steps represent different levels and forms of participation. The higher the step on the ladder, the greater the influence of participation on decisions and the greater the ▹**empowerment** of the citizens involved—also, the greater the power given up by the political system. In any participation process, activities are necessary on several steps (informing, obtaining views, dialogue). Methods that have been used successfully in many practical examples include World Cafés and Open Spaces (see *Toolbox für Zivilgesellschaft und Politik* [Toolbox for Civil Society and Politics], Institut für ökologische Wirtschaftsforschung, undated).

Empowerment is a term meaning a process, or a set of linked strategies and measures, which increases the level of autonomy and self-determination in the lives of individual people or communities.

Examples of participation with long-term impacts

Participatory budgeting is a procedure used for issues concerning the use of public funds (Sintomer et al. 2013). This procedure gives citizens the opportunity to make suggestions either on specific areas of the local authority budget or on the budget as a whole. The idea comes from Porto Alegre in Brazil, where it was first initiated in 1989 as a form of municipal civic participation (Rodrigues Mororó 2014). The format has since been developed further in other country- and culture-specific iterations, and is now used worldwide in very different forms. Depending on the particular budget focus, it is based either on the collection of ideas and alternatives for fiscal consolidation or on the submission of proposals for potential new projects and expenditures. The administration takes the suggestions and recommendations into account in the preparation of its budget. The municipal council decides on the proposals and gives its reasons for why proposals can and cannot be implemented. Subsequently, the citizens are informed about the recommendations to be implemented.

A democracy workshop has a different orientation and target group. It is aimed particularly at young people, who plan and carry out — in self-organized learning seminars, but under the supervision of a lecturer or a political educator — an educational activity (Alcántara et al. 2014, p. 32 ff.). These learning projects can be designed as part of the given framework of teaching or seminar units, or else as supplements to (high) school, involving external cooperation partners and educational formats. The activities carried out range from exhibitions and study trips, to seminars and project days in the usual place of learning, to school and university fairs. In democracy workshops, the focus is not just on the implementation of given educational activities; young people also develop their own didactic methods and projects, such as material on elections or other events or topics. The common idea behind all democracy workshops is that young people learn how politics and participation work and how conflicts can be managed and resolved. In addition, it is to be hoped that they will take away the message that the democratic system should not be taken for granted and is not shaped solely by elections, but requires the active participation of its citizens.

In practice, it has repeatedly been shown that those participatory processes which attach great importance to citizen empowerment have more lasting effects. Lasting effects here mean above all long-term improvements in conditions. Good examples of successful participation processes (in the sense of citizens' empowerment and the sustainability of political decisions) are participatory budgeting and democracy workshops. Numerous guidelines and manuals (Nanz & Fritsche 2012; Sommer 2015b), as well as academic classificatory and comparative studies, provide further discussion of the various forms of participation and a more in-depth analysis of the actual effects (Alcántara et al. 2016).

The greater the empowerment, the more sustainable the outcomes

The link between sustainability and civic participation

> Involving citizens in policymaking — at all levels from the local to the national and the global — is essential for sustainable development.

The pooling of knowledge

Firstly, the basis for decision-making can be strengthened by bringing together different types of knowledge (including common knowledge, historical data, local knowledge). Relevant knowledge is not the sole preserve of public administration and science; many other actors can contribute to robust knowledge and potentially help us to make better choices (Welp et al. 2009).

Capacity building

In addition, the level of acceptance for new sustainable development models is increased, and people feel involved in the transition processes. Finally, it helps to build the skills and competences required to implement processes of change among those involved. For example, participants in democracy workshops are taught didactic and communication skills.

In other words, the ability to apply knowledge about sustainable development in practice, and to recognize problems of non-sustainable development and to make decisions accordingly which enable properly sustainable development processes, contributes to capacity building. In this case, the acquisition of competences is relevant to the question of how everyday and future-oriented knowledge can be built up and structured in such a way that it is possible to transfer the insights gained to new fields of application. And, of course, the focus on capacity building can alter people's values with respect to the importance of civic engagement and other forms of participation.

Communication and cooperation

Citizens' participation can also inspire better internal communication and cooperation between different administration and policy departments. Entrenched power relations and decision-making hierarchies, common phenomena within administrations, can be broken up by participatory processes, and new perspectives and ideas can be brought into established areas and procedures.

Civic participation can thus promote intersectoral cooperation — an important prerequisite for systemic thinking towards sustainable development.

The need to overcome sectoral and administrative boundaries is especially acute for rural development and the associated management of natural resources. The EU Water Framework Directive alluded to earlier — in combination with participation procedures — promotes an ecosystem-based approach which views river basins as holistic planning units (Hartje & Klaphake 2006; Stoll-Kleemann & Welp 2006).

The limits to participation

Of course, participation also has its limitations and problems. Ineffective participation can lead to considerable frustration. Participatory processes can also lead to the implementation of the smallest common sustainability denominator only, or to unsustainable decisions being made (e.g. the cancellation of speed limits). The organization of participation processes places great demands on process planning and communication management. Often — due to differing attitudes, perspectives, competences and opportunities — difficulties and conflicts arise. This is an everyday occurrence within civic participation. The integration of individual interests in a democratic developmental process, one that is also coupled with sustainable models, should, however, be effective over the longer term.

Widening democracy

Civil society associations, organizations and networks will be instrumental in implementing the sustainability agenda in the 21st century. Civic participation has thus also become a way of expanding democracy and a criterion for transparency in political systems. The forms of civic engagement are changing rapidly with the emergence of new forms of work, new technical possibilities and a new understanding of the role of citizens. Politics, legislation, planning and research are being challenged to find ways of maintaining the dialogue with civil society. This can mean the creation of new spaces and opportunities for citizens' initiatives, or offering processes, procedures and methods through which actors can contribute in purposeful, structured and transparent ways (Welp et al. 2009).

However, the extent to which these spaces are actually used depends on many people being willing 'to take the plunge and get more involved' (Beck & Ziekow 2011). Even though we have observed a substantial growth in forms of participation over several years now, and although the buzzwords civil society, engagement, volunteering and civic participation are frequently heard in public and academic discourse, there are also many critical voices, especially within public administration. Such voices see in the growth of public participation only complications and additional burdens for the structures of decision-making and control. In the coming years, it will therefore be necessary to broaden the 'culture of participation' that some have called for (Alcántara et al. 2016) and to combine it with the 'culture of sustainability'.

Combining the cultures of participation and sustainability

4
The transformation to sustainability

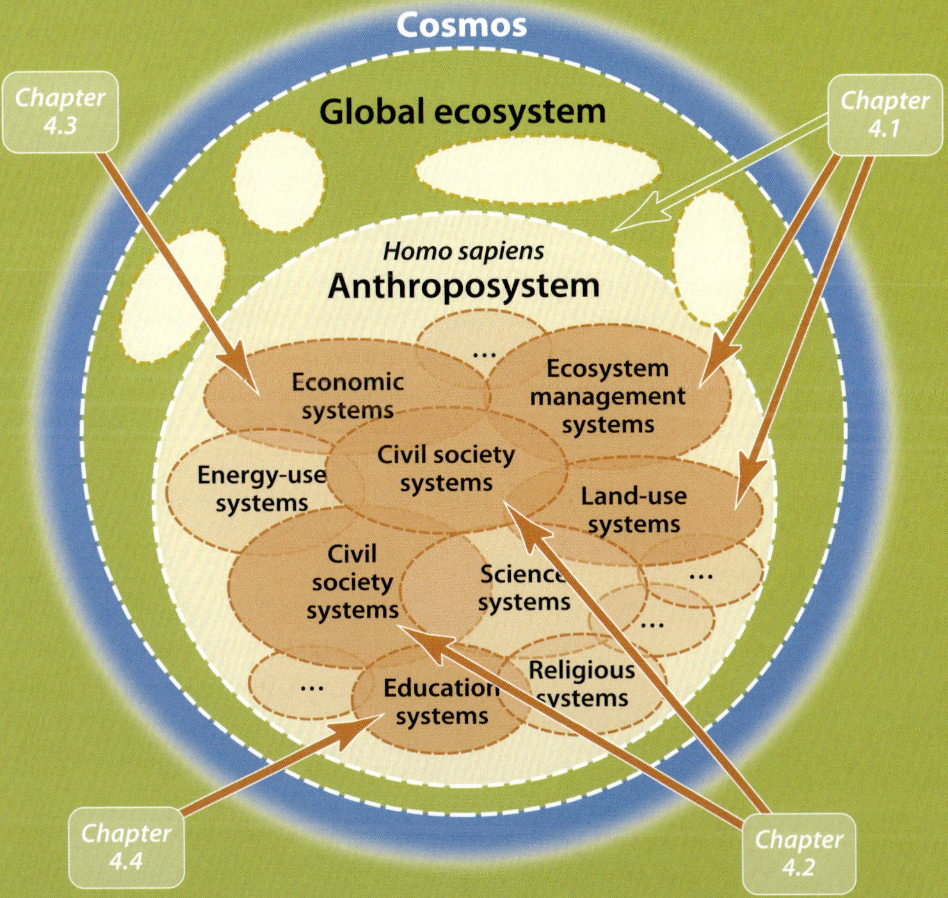

4.1

Ecosystem-based sustainable development

Pierre L. Ibisch

If the anthroposystem — humanity with its social, economic and political systems — is a dependent element of the global ecosystem, then sustainable development cannot succeed unless the functionality of the ecosystem is maintained or restored. It is not just about 'reconciling' humans with the existential foundations that ensure their subsistence, but consciously and systematically reorganizing all economic activity on the basis of functioning and working ecosystems. This does not require new technological solutions or greater knowledge, but more consistent political decisions based on existing understanding and a new ecosystem ethics. This last element is not compatible with the currently prevailing legal and economic framework.

The rationalization of nature, and the discovery that it is worth protecting

The desacralization of nature

The violence perpetrated by humans in their struggle for survival against other members of their ecosystems — which under agriculture goes far beyond the killing of individual animals by hunters — brings about a change in the attitudes of individuals and societies towards nature. The degree to which communities are embedded in and dependent on local ecosystems shapes their cultural development and religious beliefs — and conversely, cultural change can alter attitudes towards nature. Cultures that subsist on largely self-regulating ecosystems, using a greater variety of 'the gifts of Nature', subject themselves to a greater number of taboos regarding their use and tend to have pantheistic religions — the earth and its constituents are themselves sacred. By contrast, in regions where humans extensively reshaped the ecosystems, oligo- or monotheistic religions tended to develop, in which not all natural elements are sacred (Ibisch et al. 2010a). Rituals involving spiritual contact or connection with nature lost their importance, and ultimately the gods themselves were no longer walking amongst us, but located in heaven or in abstract spheres. While an Andean peasant might still offer sacrifices to Mother Earth and baulk at placing himself — at least spiritually — above nature, the degree to which Christianity subdues and claims dominion over the Earth is proverbial. The spread of the Christian monastic orders throughout Europe went hand in hand with the gradual pushing back of the forests (Williams 2006). And whereas the religions of peasant farmers and herders, with their focus on the hereafter, still maintain some degree of respect for creation, or at least a deep-rooted cultural sense of human responsibility for the preservation and protection of their fellow creatures (▹ Chapter 1.2), in the modern secular societies of today's megacities even the last vestiges of a spirituality linked to nature have been mostly lost. (That said, such spirituality is also being rediscovered in a new form in so-called 'postmodern' societies, see Griffin 1988; Taylor 2001.)

The growing importance of social systems at the expense of nature

The more all-encompassing and complex social systems become, and the more they not only serve the needs of individuals but also require their care and attention, the more they also provide a sense of identity and belonging. In doing so, they replace the identity-creating function which in less differentiated societies is largely fulfilled by nature. The emergence of so-called social media and networks, which require considerable time and attention to deal with their constant offers and demands (▹ Chapter 1.1), represents a high point in this process of increasing domination through social systems. Many members of post-industrial societies no longer feel a desire to interact with nature; or rather, they satisfy their corresponding, more or less unconscious, needs virtually, or through contact with isolated components of nature (such as pets). Modern societies enjoy expressing their superiority over nature through cultural artefacts: for example, kilometre-high buildings, artificial islands and air-conditioned indoor beaches[1] (Ibisch et al. 2010a).

The antagonism between nature and culture

In other areas outside religion, too, attitudes towards the natural world have become rational. This means that living things are seen as renewable resources or biomass, and ecosystems become service providers (▹ Chapter 3.1: ecosystem services approach). With the emergence of agriculture, non-cultivated plants become weeds, wild animals and fungi become pests or pathogens. Competition for space arises with wild ecosystems not under human control, and with their constituent parts and processes. In Europe especially, where modern economic activity and the prevailing economic-political globalization culture arose, there developed a veritable antagonism between culture and nature which is still deeply embedded in language and thought (Haila 2000; Ibisch et al. 2010a) (▹ Chapter 3.1).

Conservation as a cultural task

In modern, highly urbanized cultures, a romantic yearning for unspoiled nature can of course break through. Yet it is ultimately only another facet of the nature-culture antagonism. An idealized version of nature is contrasted with culture as an object of yearning (▹ Chapter 2.2).

---

1 'Chill out, you beautiful people, the Versace beach is refrigerated' (J. Leake, The Sunday Times, 14 December 2008).

This means that for modern-day society, nature is often a destination for day trips to a particular site — for example, a national park — that is visited because it is so beautiful and relaxing; but it is also a vulnerable object to be guarded and preserved on account of its singularity and rarity, both for its own sake and for the pleasure of the public. Interestingly, nature conservation and the establishment of modern **conservation areas** were initially promoted notably by creative artists. Conservation became a cultural movement which, at the end of the 19th century, could be understood as a response to the progressive rationalization of ecosystem use and the corresponding transformation of the landscape. The early nature conservation movement in Germany, for example, was principally concerned with the preservation of natural monuments. In addition, however, a rational justification soon arose, resulting from the perception of the threats to species survival and to other elements of the natural world, as well as from the growing understanding of natural processes (Kreft et al. 2012). Increasingly, biology and ecology also produced arguments for the protection of ever more objects for which dedicated spaces needed to be reserved. This led to the corresponding invention of many categories such as areas for the protection of birds or wildlife, or for resource or landscape conservation.

Conservation areas

The establishment of areas where limits are placed on the use of resources, and/or human activities of various kinds are constrained, has become the central and probably most effective instrument of nature conservation. The founding of Yellowstone National Park in 1872 is regarded as the birth of modern nature conservation. There are now about 209,000 protected areas around the world; they cover 15.5 % of the continents and 8.4 % of the sea surface (Juffe-Bignoli et al. 2014). The IUCN *(International Union for Conservation of Nature)* has defined various management categories that are more or less reflected in different national legislations: Category Ia/Ib: *Strict nature reserve/Wilderness area*; Category II: *National park*; Category III: *Natural monument*; Category IV: *Habitat/species management area*; Category V: *Protected landscape/seascape*; Category VI: *Managed resource protected area*.

Protected areas represent, to varying degrees, the different biomes and ecosystem types (Juffe-Bignoli et al. 2014). For example, almost

40 % of the forests in the tropics of Central and South America are in protected areas, but only about 18 % of those in the African tropics. According to the Biodiversity Convention's global nature conservation objectives (or 'Aichi targets'), 17 % of the land surface and 10 % of the oceans are to be protected by 2020. However, it should be noted that the existence and management of protected areas do not automatically guarantee, for different reasons, the level of protection desired. Some are no more than 'paper parks', existing only in theory; ensuring effective management is therefore particularly important (see Bruner et al. 2001). The importance and effectiveness of protected areas may continue to decline in the future, above all if climate change is not adequately taken into account in their management (Geyer et al. 2017a).

Figure 1: Nature conservation area in Brandenburg, Germany: protection of an ecosystem that has been modified and continues to be influenced by human use, with the aim of conservation of rare species or preservation of a culturally-influenced ecosystem state. In this case, the ecosystem management seeks to counter the natural processes of succession, which would lead to reforestation, as indicated by the pine trees. This is a static protection approach; the functionality and adaptability of the system are ignored. (photo: P. Ibisch)

From the ecosystem approach to ecosystem law and ecosystem ethics

Segregative nature conservation vs. the ecosystem approach

Many protected areas embody a segregative nature conservation approach: nature should be preserved apart from and independent of the spaces used by humans. Such areas separate conventionally used land from protected land. More integrative approaches involve combining ecologically-founded restrictions with land use (e.g. rules relating to good practice in agriculture or forestry).

The ecosystem approach

A major problem with segregative protected areas is that they are often located in the midst of a used landscape, and the ecosystems or species they are intended to protect lose their functionality there. The problem is exacerbated by climate change, as protected areas are often too small to allow adaptation processes (Ibisch et al. 2012). Even more problematic are static approaches that try to preserve defined conditions or states within ecosystems (see Fig. 1). Since the 1990s especially, it has been increasingly recognized that protected areas and nature conservation do not work in isolation — that what matters is the larger whole. A deeper, ecosystem-based understanding of nature enabled the establishment of important principles for an ▹ **Ecosystem approach**. A corresponding conceptual framework was officially adopted in 2000 at the 5th Conference of the Parties to the Convention on Biological Diversity (also known as the Biodiversity Convention, or CBD), and termed the 'ecosystem approach'.

The reintegration of humans into nature

The ecosystem approach addresses (among other things) the management of protected areas, and is relevant to nature conservation strategies. Its importance for different ecosystem uses has also been recognized (e.g. forestry). Remarkably, it was the ecosystem approach, originally motivated by ecological concerns, that 're-integrated' humans into nature. It might be seen as ironic that scientific, systemic and ecological research has served to illustrate the degree to which people, as individuals and societies, are only a dependent part of the global ecosystem (▹ Chapter 1.3) — indeed, to an even greater degree than is suggested in the Biodiversity Convention's ecosystem approach. This has important conceptual and ethical implications. Among other things, it means nature conservation is now no longer the preserve of lovers of rare birds and romantics in love with beautiful landscapes.

If all human beings are part of the global ecosystem and depend on its proper functioning for their survival, then the protection of nature is also the protection of humanity. Furthermore, if there exists a universal

The ecosystem approach

'The ecosystem approach is a strategy for the integrated management of land, water and living resources that promotes conservation and sustainable use in an equitable way. [...] It recognizes that humans, with their cultural diversity, are an integral component of many ecosystems. [...] The ecosystem approach requires adaptive management to deal with the complex and dynamic nature of ecosystems and the absence of complete knowledge or understanding of their functioning. Ecosystem processes are often non-linear, and the outcome of such processes often shows time-lags. The result is discontinuities, leading to surprise and uncertainty. Management must be adaptive in order to be able to respond to such uncertainties and contain elements of "learning-by-doing" or research feedback. [...] Ecosystem management should be envisaged as a long-term experiment that builds on its results as it progresses' (CBD 2000).

A voice in the management of ecosystems as a human right

right to life, liberty and security (see the Universal Declaration of Human Rights, United Nations 1948), then there must also be a right for people to have a say in the management of ecosystems; which then fully deserves the status of a human right (Ibisch 2015). It was identified as an important principle of the ecosystem approach that the management of ecosystems should be a matter of social choice, precisely because humans are a part of ecosystems. Accordingly, it is particularly important, and represents a distinct and discrete human right, that people are informed about the state of the ecosystems in which or from which they live. The world's most far-reaching body of legislation in this context is the **Aarhus Convention** (▷ Chapter 3.5), now ratified by 47 European countries (including all EU member states), which follows new legal and socio-philosophical approaches and grants citizens the right to the relevant information.

The Convention not only calls on the government and other public administration bodies to provide information, but also 'natural or legal persons in private law, insofar as they perform public functions or provide public services related to the environment, in particular those of environmental public services in the general interest, and are thereby under the control of the Federal Government or of a legal person governed by public law under the supervision of the Federal Government' (§2(1) UIG).

Aarhus Convention

Convention on Access to Information, Public Participation in Decision-making and Access to Justice in Environmental Matters

The Convention draws on the Stockholm Declaration on the Human Environment, the Rio Declaration on Environment and Development, several UN Resolutions and the European Charter on Environment and Health. It was passed in 1998:

'Recalling the European Charter on Environment and Health adopted at the First European Conference on Environment and Health of the World Health Organization in Frankfurt-am-Main, Germany, on 8 December 1989; affirming the need to protect, preserve and improve the state of the environment and to ensure sustainable and environmentally sound development; recognizing that adequate protection of the environment is essential to human well-being and the enjoyment of basic human rights, including the right to life itself; recognizing also that every person has the right to live in an environment adequate to his or her health and well-being, and the duty, both individually and in association with others, to protect and improve the environment for the benefit of present and future generations' (UNECE 1998, p. 1 f.).

In Germany, for instance, implementation is governed by the Environmental Information Act (Umweltinformationsgesetz, UIG). All citizens, in accordance with §2(3), have the right to obtain information on the environment — not only on 'the condition of environmental elements such as the air and atmosphere, water, soil, landscape and natural habitats including wetlands, coastal and marine areas, biodiversity and its components, including genetically modified organisms, and the interactions between these elements; factors such as substances, energy, noise and radiation, all types of waste as well as emissions, discharges and other releases of substances into the environment [...] that affect or are likely to affect [...] or are intended to protect its elements', but also on 'cost-benefit analyses or other economic analyses and assumptions used to prepare or carry out measures or activities, [...] and on the state of human health and safety and human living conditions, as well as on cultural sites and structures, insofar as they are affected or may be affected by the condition or state of elements of the environment [...] or by factors, measures or activities' (§2(3) UIG).

The (not entirely new) **realization that humanity is part of the global ecosystem** has other far-reaching ethical consequences. It follows that the man-made culture-nature antagonism is not valid. An anthropocentric ethics with humanity at its core cannot be played off against an ecocentric one. The global ecosystem includes all life, meaning that any differentiation between ecosystem ethics and an ethics of life (see 'Oikos Ethics' and 'Bios Ethics', Münk 2006, pp. 141 f.) is thus also void. A 'land ethic' as proposed by the American philosopher and nature conservationist Aldo Leopold (1887–1948) (today we might call it an 'ecosystem ethics') encompasses concern for human well-being. It follows from this that the destruction of ecological systems should be regarded as a crime against humanity. In fact, for some time, there has been an initiative under way to have ecocide recognized internationally as a potential crime.[2]

Ecosystem ethics as the basis for the definition of ecocide as a crime against humanity

Historical views on the position of humanity in nature

'Man lives on nature — meaning that nature is his body, with which he must remain in continuous interchange if he is not to die. That man's physical and spiritual life is linked to nature means simply that nature is linked to itself, for man is a part of nature.'

Karl Marx 1844

'[A] land ethic changes the role of Homo sapiens from conqueror of the land-community to plain member and citizen of it. It implies respect for his fellow-members, and also respect for the community as such.'

Aldo Leopold 1949

2 European Citizens' Initiative *End Ecocide on Earth*: 'Ecocide is the destruction of the global environment. End Ecocide on Earth is therefore defining Ecocide crime as "an extensive damage or destruction which would have for consequence a significant and durable alteration of the global commons or Earth's ecological systems" — upon which rely all living beings in general and humankind in particular — and in compliance with the known planetary boundaries.' (https://cop21.endecocide.org/en/examples-of-ecocide-2/, accessed 20/02/2019).

A licence to kill: altering and destroying ecosystems

One important specific question is: who has the right to change, disturb or destroy ecosystems for their own benefit? In general, nation states have the sovereign right not only to dispose of land, but also to distribute concessions for the extraction of natural resources. However, if all human beings are part of the global ecosystem and ecosystems generally operate across national boundaries, then global rules and conventions that take account of this are essential — and have in fact been drawn up, especially since the Earth Summit in Rio de Janeiro (▹ Chapter 1.2).

Can an ecosystem be owned?

In this context, the problem arises whether land ownership also includes ecosystem ownership, or whether ownership of ecosystem components has to be steered towards supporting their conservation. In the German Basic Law (Article 14), the so-called social obligations attached to property ownership are of great importance:

> »Property entails obligations. Its use shall also serve the public good. (Art. 14 para. 2 German Basic Law)

Landowners ultimately can also freely dispose of organisms that occur on their land (e.g. trees), but on the other hand may not remove or damage specimens of protected species. So they do not actually own them. This applies also to mobile organisms such as resting birds, and even to ecological processes such as pollination. However, strictly speaking, the issue is not simply whether ecosystem components can be owned. Rather, what would be needed is a counterpart to social obligation: an 'ecological obligation' linked to property would mean that land ownership and use must serve to benefit the greater ecosystem.

Participation in ecosystem management and in its control

The monitoring and control of the landowners who manage and alter the ecosystems, and checking whether they are complying with their social obligation, is the responsibility of the state, which also draws up the legal provisions regarding what is prohibited or permitted in specific ecosystems.

A more direct and active involvement of citizens in ecosystem management has so far only been possible in a few instances. There is a substantial need for more to be done in this area (Ibisch 2015). Theoretically, it is easier when government agencies are entrusted with the management of forest or protected areas, for example. In such cases, however, the relevant actors also supervise their own management bodies (e.g. forestry ministries), so that independent monitoring and evaluation are not possible. In the case of private land ownership, not only is citizen participation difficult, but monitoring and control are only possible by means of legal provisions to that effect.

Permitted ecosystem interventions

Especially with regard to the primary economy, which supplies society with natural goods, the laws in most countries are restrained. For example, in the German Federal Nature Conservation Act (BNatSchG) there is an intervention regulation intended to ensure that nature does not suffer substantial damage:

> »Interventions in nature and landscape, as defined in this Act, shall refer to any changes affecting the shape or use of areas, or changes in the groundwater level associated with the active soil layer, which may significantly impair the performance and functioning of the natural balance or landscape appearance (§14 para. 1 BNatSchG).

At the same time, it is stated that agriculture, forestry and fisheries are usually not considered to be interventions if general nature protection and soil protection requirements are met. In order to maintain the competitiveness of these sectors, legislators and governments have even felt obliged to defend or promote forms of industrialization of ecosystem use with significant environmental impacts. In doing so, they are one-sidedly following the stipulation that certain immediate needs of the present generation must be met. This means that even indirect health damage suffered by the current population—for example as a result of the use of pesticides, or nitrate levels in drinking water—is deemed acceptable. Clearly, neither land nor ecosystem ethics, nor concern for future generations, is being applied here.

Can ecosystem destruction be offset?

Economic sectors which intervene in nature for the supposed well-being of society (e.g. through the expansion of settlements, or traffic and energy infrastructure) can make up for this through offsetting or substitution measures (at other locations) (§15 para. 2–4 BNatSchG). If necessary, the following rule applies: 'If an intervention [...] is permitted or carried out, even though the relevant adverse effects cannot be avoided or cannot be offset or substituted within an appropriate period of time, the intervening party shall provide monetary substitution. The amounts of such substitution payments shall be determined in accordance with the average costs of the unfeasible offsetting and substitution measures, including the necessary average costs for planning and maintaining such measures and for providing the necessary areas, also taking account of relevant personnel costs and other administrative costs. If such costs cannot be determined, the amount of a substitution payment shall be determined in accordance with the duration and severity of the intervention, taking account of the resulting advantages accruing to the intervening party' (§15 para. 6 BNatSchG).

The precautionary principle

Although modern principles such as 'prevention is better than cure' (the precautionary principle) are embedded in national and international environmental law,[3] in practice things are often completely different. On the basis of a mechanistic and economistic model of nature, lawmakers assume that regular disturbances to, or even destruction of, ecosystems cannot be avoided, and that they can in principle be offset or at least justified by cash payments. This illustrates how far removed German nature conservation legislation is from any kind of ecological or ecosystem ethics. Conservation is often regarded — not just in Germany — as just another land-use sector in competition for space. Similarly, there is a strong public perception of nature conservation as a barrier to construction and progress, slowing economic development through too many rules and prohibitions. The underlying misunderstanding needs to be tirelessly addressed and explained.

The burden of proof

The problem of the burden of proof is also relevant. As a rule, those who are engaged in nature conservation (associations, scientists) must prove that interventions and ways of using nature cause damage to the ecosystem. Ecosystem users are under no obligation to prove that their practices do not cause lasting damage to the ecosystem. However, for the state to take action, in theory 'the conviction that a risk is actually real and present is not required. Plausible or serious indications of a risk to the environment is instead sufficient. If this exists, it is then incumbent on the causer to refute the substantiated indications of specific cause-and-effect relationships and to shatter the assumptions which have given rise to the concern' (Umweltbundesamt 2015b).

Taking into account the preceding observations, we can ask: what would be the essence of an ▹ **ecosystem ethics** conceived as an updated revision of Leopold's land ethics?

3 See, for example, Article 34 (1) of the German Unification Treaty as an example of self-obligation on the part of the legislature, Article 191 of the Treaty on the Functioning of the European Union, the United Nations Framework Convention on Climate Change, the OSPAR Convention for the Protection of the Marine Environment of the North-East Atlantic (Umweltbundesamt 2015b).

Ecosystem ethics

Ecosystem ethics is defined here as a normative basis for the task of not disrupting the dynamically changing structures of the global ecosystem in such a way that its functionality and reparability might be irreversibly damaged, which would also threaten the subsistence and habitats of humans — of present and future generations.

It involves due respect for the emergent properties of complex systems,[4] including their ability to self-organize (▹ Chapter 1.3), which can only be foreseen to a limited extent, let alone sustainably controlled.

It recognizes that ecosystems — given their functionality, which is based on emergence and ongoing development, and their temporal-spatial dynamics — cannot be owned by individuals or by specific generations. Accordingly, it is also impossible to determine what value parts of nature have for humanity. On the contrary, land ownership within given ecosystems implies an obligation to contribute to the functioning of the greater whole.

Furthermore, ecosystem ethics is rooted in an awareness that humans are a dependent component of the global ecosystem, and that adequate information about its condition, as well as participation in ecosystem management, is therefore a human right.

All ecosystem management must adhere to the precautionary principle and must be adaptive in order to take account of changes or of new knowledge within appropriate time frames.

4 Properties of complex systems that arise only from the interaction of their components.

Ecosystem-based sustainable development

A question of survival: making economic activity ecosystem-friendly

Significant progress has been made in the scientific understanding of our place on Earth, and in a corresponding development of environmental ethics and environmental law. The discourse is under way and growing, but at the same time very far from entering into general public knowledge. Many of the spokespersons for ecology and nature conservation have begun to speak the 'language of the economization of nature' in order to get their voices heard in politics and business. The crucial question, however, is how to achieve ecologization of the economy (▹ Chapter 4.3). This involves some simple but fundamental insights:

1. We humans have emerged from the global ecosystem as a result of evolution.
2. We can liberate ourselves from local ecosystems by obtaining the ecosystem services we need from other systems, which involves expending huge amounts of energy.
3. However, for all our economic activity and for our physical existence, we remain dependent on the functioning of the global ecosystem — which is rapidly deteriorating as a result of our economic activity.

It follows inescapably from this that we have to learn to conduct our lives and above all our economic activity in a way that is ecosystem-compatible. This is a question of survival, if not for humanity as such, then for industrialized civilization at the current levels of population density. This question of survival has now taken on a new dimension, for it is no longer just about tempering our aggressiveness towards the ecosystems that support us. It also means coping with the consequences of the climate changes we have caused. We would be aided in this by the ecosystems if only we could maintain them, or at least allow them to do their work unhindered. Conservation efforts should focus on those ecosystems which currently still function best and which provide life-sustaining services not only for humans but for the global ecosystem as a whole (Freudenberger et al. 2012; Freudenberger et al. 2013). Of central importance is the question of how many largely self-governing wild ecosystems are required to maintain key ecological functions, and how ecosystems can be used intelligently without losing too much functionality. Non-conventional uses must also be taken into account (e.g. the application of paludiculture in peatlands; ▹ Digression 2 Paludiculture).

There is a need for new complex, multi-dimensional development models, including the spatial, economic and social dimensions, which can guide us towards how to interact more sustainably with the ecosystems. For this purpose, ecosystem 'living labs' should be created around the world, in which not only gentler practices of resource use could be tested but also innovative approaches to management and governance (▹Chapter 4.1). The **biosphere reserves** created by UNESCO could play such a role.

Ecosystems as 'living labs'

Biosphere reserves as model regions and ecosystem living labs

An outcome of the Man and the Biosphere programme (MAB) under the auspices of the United Nations' cultural organization, UNESCO, the network of biosphere reserves—now around 670, in 120 countries—provides a unique platform for testing new paths towards sustainable development. Biosphere reserves serve to further international (often cross-border) cooperation, the exchange of experience and the creation of model areas in all biomes of the Earth to show that humans can live and work in an ecosystem-compatible way, for example through being empowered to participate in ecosystem management.

The MAB programme was set up in 1971 following the Biosphere Conference (officially, the Conference on the Rational Use and Conservation of the Resources of the Biosphere, 1968) and launched in 1974—long before the first major milestones of the international sustainability agenda were reached (▹Chapter 1.2). In 1976, the first 58 biosphere reserves were established. From the outset, the sustainable management of land and genetic resources in case study ecosystems has been one of the principal aims. Initially, the strategic focus was on conservation and research. Ecosystem use and aspects of human development were not on the agenda at first.

From early on, the overall concept included graded zoning, so that nature prioritization areas, in so-called core zones, could be established within the biosphere reserves, as well as buffer zones and transition zones. After the first few decades, it was found that many actors and countries had not really implemented the original concept. Often, biosphere reserves are regarded as just another protected area category, but one designated as such by the United Nations, which helps with marketing for tourism. In this context, it was not especially helpful that several countries (including Germany) included biosphere reserves in their protected area legislation, which often — with regard to institutional embedding and personnel capacities — places little focus on working with citizens and has serious deficits in the area of stakeholder participation. For a comprehensive assessment of biosphere reserves and their history, see e.g. Coetzer et al. (2014).

Figure 2: Organic free-range cattle farming as an element of sustainable ecosystem use in the development zone of the Schorfheide-Chorin Biosphere Reserve, Brandenburg, Germany (photo: P. Ibisch)

As part of the development of the so-called Seville Strategy in 1995, the concept of biosphere reserves (UNESCO 1996/2018) was modernized, and one of the outcomes was a stricter application of the criteria for expulsion and evaluation.

'Biosphere reserves are designed to deal with one of the most important questions the World faces today: How can we reconcile conservation of biodiversity and biological resources with their sustainable use? An effective Biosphere reserve involves natural and social scientists; conservation and development groups; management authorities and local communities — all working together on this complex issue' (UNESCO 1996/2018).

The Madrid Action Plan of 2008 represented a new update, highlighting key issues of the 21st century such as climate change and globalization (of environmental problems). And in 2015, the MAB programme took a more strategic direction with the Lima Action Plan (see also UNESCO Man and the Biosphere Programme 2018), aiming for a stronger integration into the global sustainability agenda. Now, biosphere reserves are to become, by 2025, model regions for the implementation of the Global Sustainable Development Goals (SDGs) (▹Chapter 1.2). The objectives of the programme explicitly include not only maintaining biodiversity, but also restoring and supporting ecosystem services and advancing the sustainable use of natural resources. Biosphere reserves should help to create sustainable, healthy and equitable societies, economies and settlements. They should also play a special role in Education for Sustainable Development (▹Chapter 4.4) as well as in climate change management. The participation approach receives even greater emphasis here.

Nature-based solutions taking us towards ecosystem-based sustainable development

The challenge for our modern societies is to achieve truly ecosystem-based sustainable development. The appropriate principles have been proposed for some sectors at least, such as the ecosystem approach (the Biodiversity Convention), ecosystem-based management (especially in the marine area, see e.g. NOAA — the National Oceanic and Atmospheric Administration of the US Department of Commerce) and ecosystem-based adaptation to climate change. For many environmental and development actors, however, these seem to be 'solutions for poor countries' which (still) cannot afford advanced technologies, even though there have been efforts for some time now to illustrate their relevance for rich and supposedly 'developed' countries (Doswald & Osti 2011). Nature-based solutions have even found their way onto the political agenda of the European Union, where there is now also discussion of re-naturing cities (European Commission 2015b; 2017). From a scientific and systems theory perspective, it would be preferable to talk about ecosystem-based solutions. At the same time, efforts are being made to make such ideas catch on with the general public using 'snappy' words and phrases. Thus, ecosystems and their components are also referred to now as *green infrastructure*, as they not only deliver ecosystem functions but also provide ecosystem services to humans (e.g. Tzoulas et al. 2007). The

Figure 3: Ecosystem-based sustainable development must anticipate and take account of future changes in ecosystems. The challenge is illustrated by the Californian Sierra Nevada. The massive dieback of coniferous trees in a redwood pine/Douglas fir forest beginning in the 2010s is probably ultimately due to climate change, triggered more immediately by a variety of factors such as fungal diseases (Sequoia National Park, top). In addition, extensive forest fires are increasingly frequent (as here in the oak forests on the lower elevations, Sequoia National Forest, bottom). In the near future, important ecosystem functions such as the generation and 'trapping' of rainfall, water retention and soil conservation are endangered. It is unclear which tree species might form a new forest in the future—or whether the forest will be replaced by shrub or grassland. (Sequoia National Park and Sequoia National Forest, USA, photos: P. Ibisch)

term first emerged in urban planning, where there was a growing recognition of the importance of urban green space for human well-being (Sandström 2002). The European Union regards green infrastructure as a network of healthy ecosystems, an alternative to so-called grey infrastructure, and has even proposed a corresponding strategy in 2013 (European Commission 2013). The US Environmental Protection Agency has also taken up this concept.

Ecosystem-based potential decreases as ongoing damage grows

In the further development of ecosystem-based solutions, a naive neo-romantic approach must be avoided. Nature is not an ever-caring mother to human beings, but has a neutral 'attitude' towards them. Since humanity has severely damaged many ecosystems, their functionality, performance and ecological carrying capacity are all changing or decreasing. This also means that the potential for ecosystem-based sustainable development is progressively diminishing. Of particular concern is the fact that anthropogenic climate change is imposing deep changes on many ecosystems (Fig. 3). Local ecosystem protection measures alone can no longer guarantee their proper functioning. However, local management measures which run counter to ecosystem processes, or which make healing and adaptation carried out by ecosystems more difficult, can generate additional risks.

Ecosystem design and geo-engineering

The more intervention has taken place in an ecosystem, the more likely it is that costly measures will be needed to maintain a level of functionality. This is particularly evident in the case of artificial agro-ecosystems such as irrigated plantations, for which water has to be procured from ever-greater distances. Since the feeding of a large part of the world's population depends on industrialized monocultures far removed from nature, where the conditions for production can only be ensured by means of a considerable input of technology and energy, vulnerability to environmental change is much greater than is currently perceived. Ecosystem management has often turned into ecosystem design, whereby complex interactions within and between ecosystems are frequently underestimated. A big issue for humankind in the future is likely to be what is called geoengineering. This refers to suggestions for limiting (mainly, but not only) climate change by manipulating the atmosphere (for example, by releasing particles to filter solar radiation). Since the recent past has shown that, even in comparatively small local ecosystems, engineers often fail to foresee the unexpected consequences of their actions, extreme caution is called for with regard to technical intervention at the global level. Renewed effort is also required under international law to prevent countries hit hard by climate change and at the same time technologically advanced — such as India, for example — from forging ahead with their own measures without broader consultation or agreement. Current proposals and demands are already causing great uncertainty and sparking a variety of conspiracy theories (Tingley & Wagner 2017).

Econics: social systems learning from ecosystem husbandry and functioning

Many actors have set out on the journey. If enough knowledge about ecosystems could be integrated into political decision-making processes, and if a systemic problem-solving competence could be developed in many societies based on an ecosystem ethics that no longer set humanity in opposition to nature, then there might be grounds for hope. Given that social systems exist as dependent holons in the global ecosystem and in the superordinate anthroposystem (▹ Chapter 1.3), they are subject to the same conditions governing functionality and sustainable development. Accordingly, it should be possible to learn lessons for human development from the study of husbandry in long-term successful biological and ecological systems. *Bionics* or *biomimicry*[5] searches nature systematically for inspiration for use in construction and technology (see e.g. Cerman et al. 2005). Systemic biomimicry focuses on aspects of self-organization and self-regulation (Küppers 2015). *Econics* emphasizes the emergent property of husbandry and development in efficient, resilient, sufficient, and coherent natural and man-made systems (Ibisch et al. 2010b).[6] Examples include industrial ecology, which places great emphasis on the circular economy; adaptive management (Holling 1978), which was inspired by adaptive cycles; systemic management (e.g. Malik 2006); and holacracy, which has learned from the functioning of natural holarchical systems that it is not leaders and leadership that counts, but governance principles and mechanisms (Robertson 2016) (▹ Chapter 4.2).

5 See e.g. Benyus 1997.

6 The term econics ('Ökonik' in German) was first used by Dirk Althaus in 2007 (Althaus 2007) and introduced into the international debate in 2010 (Ibisch et al. 2010a). The concept was 'discovered' independently by a number of research groups.

The threefold task is easy to describe, but not easy to master:

The task: for, with and from ecosystems

1. **Looking after ecosystems:** ecosystems should above all be ensured the space and the framework conditions that will allow them to work for themselves and for humanity.
2. **Using ecosystems:** ecosystem usage must be based on an ecosystem ethics within the framework of careful and appropriate ecosystem management, so that it works not counter to but rather in harmony with ecosystem processes so that they retain as much functionality as possible.
3. **Learning from ecosystems:** societies must be so constructed that they can allow nature to inspire them to better ways of conducting economic activity and better governance.

Sustainability transformation

4.2

Heike Walk and Pierre L. Ibisch

Many of those actively working for sustainability often feel a sense of impatience and impotence, because their efforts and achievements do not seem adequate to the task and require much too much time in view of the accelerating crises. At the same time, it is becoming ever clearer that the required switch to sustainability will lead to deep structural changes in society affecting culture, values, modes of production and consumption, and politics — ultimately, almost all areas of human coexistence. We are currently facing a 'Great Transformation'. But what exactly does that mean? What is a 'Great Transformation' in the 21st century? What is the relationship between a transformation to sustainability and the functional capacity of democratic systems? What should social transformation processes involve in detail? Who are the actors involved in these processes? And how should politics itself be transformed in order to effectively govern the sustainability transformation?

The need for radical changes

The sustainability discourse as an extraordinary achievement

The many and diverse joint efforts of regions, communities, and business and civil society initiatives show that humanity has initiated an unprecedented process, one which became possible only in the wake of globalization and recent developments in (information) technology. It is an attempt to change direction towards sustainable development at various levels of the complex global social system in a coherent, integrated way. It addresses the enormous challenge of questioning an economic rationale with roots going back thousands of years which billions of people still hope — without thinking too much about the well-being of future generations — will improve their personal situation in the short term. Looked at in this way, the growing discourse on sustainability, the many global and regional environmental conventions and agreements, and the worldwide 2030 Agenda for Sustainable Development (▹ Chapters 1.2 and 3.5) — following a century which saw the first global wars — are an extraordinary historical achievement.

The search for socio-ecological governance mechanisms

Nevertheless, many of those working for sustainability share a sense of impatience and powerlessness, as their efforts and concrete achievements seem either inadequate or too slow in the face of accelerating and concatenated crises (Biggs et al. 2011). Given that a comprehensive pool of knowledge regarding the challenges and their causes — but also their solutions — is available, the problem now is primarily of a political nature: what governance and regulatory mechanisms can be constructed that are not only socially acceptable and politically attractive, but also effective in counteracting the crises of the ecosystems, the economic and financial system and many social systems? Can societies (at least in the Western industrialized countries) politically decree a transformation and then implement it without risking dramatic ruptures and conflicts?

The need for a systemic perspective

There are widely differing pictures of societal change, which must also be understood as a kind of collective process of inquiry, but there is no common strategy. Even the global sustainability goals (▹ Chapter 1.2) stand in part in direct contradiction to each other (e.g. ecosystem conservation vs. economic growth). There are also different views on the question of how sustainable development can serve as a normative frame of reference, in other words as a fundamental value, in the context of this societal process of inquiry. Experience shows that democracy and sustainability are not self-generating. And there are more and more studies pointing to the fact that societal change requires a systemic perspective (▹ Chapter 1.3). A systemic perspective here means not looking at individual, more or less isolated measures, programmes and innovations, because such isolated approaches and activities are unlikely to be sufficient to promote sustainable change. Instead, as many social areas as possible must be taken into account, as well as the interactions between individual policy decisions. It is the unintended consequences or negative interactions of individual activities which show most clearly that linear chains of cause and effect often do not work; instead, systemic causes and levers need to be identified and taken into account (for example, the deliberate cultivation of energy crops has taken up valuable land which is then no longer available for the cultivation of food). Above all, it is a significant insight from systems science, and one that politics finds difficult to accept, that—regardless of the degree of political control—the actual development of any complex system, and which unexpected processes will promote or inhibit the desired change, remain unpredictable.

A transformation guided by solidarity and democracy

The required transformations to sustainability will lead to very far-reaching structural changes in society, affecting culture, values, modes of production and consumption, politics—and ultimately almost all areas of human coexistence (Simonis 1989; Umweltbundesamt 2015c). These transformational requirements can trigger a very wide range of social initiatives and movements. These may be supportive protest movements seeking to promote greater fairness and a transformation that is socially just and ecological. However, reactionary protest movements can also emerge that oppose the transformative processes, incite nationalist-racist sentiments and seek to push back changes. A politics of sustainability therefore requires not less, but more policymaking (▷ Chapter 3.4). At the same time, the transformation must not only serve democracy and social solidarity, invoking human rights; it must also pacify reactionary counter-movements.

What does transformation mean?

Origins of the term

The Term ▷ **transformation** was originally used mainly in the natural sciences (e.g. physics/electrical engineering, soil science, genetics), but from the middle of the 20th century it became a political term with great resonance and appeal. This is mainly due to the idea of the **Great Transformation** coined in 1944 by the Hungarian-Austrian economic sociologist Karl Polanyi (1886–1964) to help make the processes of industrialization and nation state formation understandable.

> **Political and social transformation describes a comprehensive change in the structure and functioning of decision-making and discourse-dominating systems within a state, one that is not only fundamental but often irreversible. The term emphasizes a structural metamorphosis, a radical change from one form to another. Transformations can take place at very different speeds: they can take anything from a few years to several decades or even centuries. They can be peaceful and barely perceptible, but they can also be extremely violent and conflictual. They can also unfold in a completely unexpected and unplanned way, or they can be deliberately initiated. In the case of a planned transformation, however, a social consensus is needed on its overarching goals, as well as the appeasement of oppositional forces (vested interests, diffuse fears among the population, short-term solutions, etc.).**

Karl Polanyi's 'Great Transformation'

Polanyi's study dealt with the social and political upheavals that took place in England during the rise of the market economy—from the 16th century to the 1930s. Polanyi deconstructs the myth of a natural and spontaneously developing free market economy. The market, he wrote, was 'the outcome of a conscious and often violent intervention on the part of government which imposed the market organization on society for noneconomic ends' (Polanyi 2001, p. 258). In this sense, the analysis underlying Polanyi's 'Great Transformation' is evidence for the possibility that radical and profound change can be triggered politically (even if the outcome remained unforeseeable for those triggering it).

Transformations are profound changes

Transformation is used in the literature when—as indicated above—the changes in culture, technology, consumption and politics are not only profound, but also interlinked and mutually reinforcing. These interactive systemic processes mean that the changes cannot be stopped or reversed.

Transformations are conflictual and provoke resistance

There are several historical examples of transformations, including the shift from a feudal agrarian society to a capitalist and urban industrial society. The transition of the Central and Eastern European states from socialist systems to democratically constituted market economies is also described as a transformation. Transformations are usually beset by conflict and provoke resistance, because prevailing values, traditions and lifestyles are called into question. In addition, there are often distributional conflicts between different sectors of the population, and new power constellations emerge. Sudden ruptures and revolutions can therefore be part of a longer-term transformation process, and either expedite it or slow it down.

Dimensions of transformation

In one of the reports of the German Advisory Council on Global Change (WBGU 2011), transformations are differentiated according to their dimensions, among other things. We are currently facing a 'Great Transformation' and a new social contract that combines responsibility for future generations with a culture of democratic participation and a sustainable form of economic activity. But what exactly does that mean? What does a 'Great Transformation' mean in the 21st century? Some authors have tried to offer more concrete answers. In their report to the Club of Rome, for example, Graeme Maxton and Jørgen Randers identify 13 tax and trade policy measures that they believe will help curb global resource depletion and — with the help of political restrictions — move people toward sustainable economic behaviour (Randers & Maxton 2016).

Transformations emerge from niches

Even big transformations often arise from niche developments, which then spread into the regime (i.e. the actors and institutions of the prevailing political system) (Geels & Schot 2007). Which ideas, innovations or policies break through is often determined in individual cases by an interaction of some intensity with other sub-elements of the regime. A system transformation is not made up solely of technical innovations. Innovations have to come together from different areas of technology, business, politics and culture.

Like innovations, transformations follow a typical course (at different speeds) (Ilten 2009). In an initial phase, numerous experiments take place, often in niches. Few people are involved, with hardly any sup-

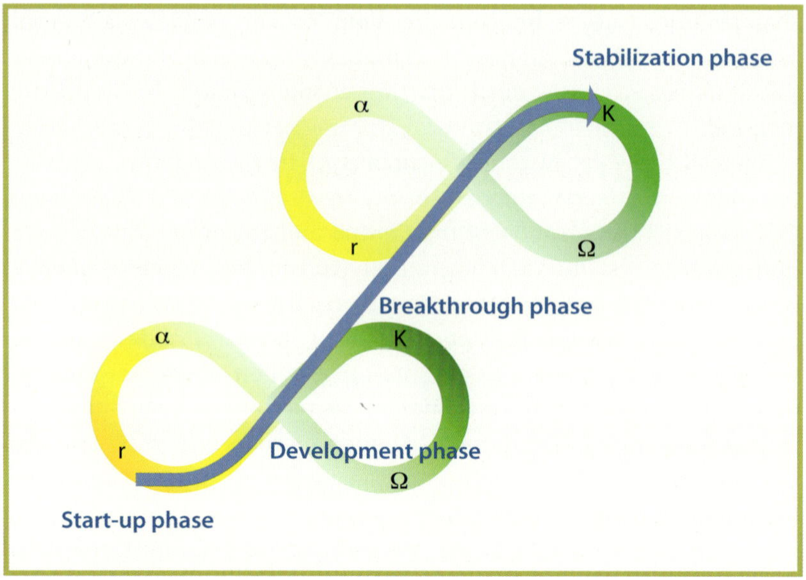

Figure 1: Transformation as a non-linear transition from one system state to another. Transformation is often characterized by an S-curve. However, in the course of the transition following phases of rapid growth and the change of key attributes (r-phase) and subsequent stabilization phases (K-phase), greater or lesser breakdowns of subsystems (Ω-phase) and the rearrangement of system components can often occur (α-phase). Stabilized states are often only temporary until further adjustments are required. This dynamic and adaptive system change occurs across all interconnected system levels. This is why the model is also called a 'panarchy' (from the Greek: pan = everything, archie = rule). For a discussion of the panarchy model in the analysis of world systems, see Gotts (2007). (author's own illustration with acknowledgements to Rotmans et al. 2001; Gunderson & Holling 2002)

port from the political system. In the second phase, the experiments are increasingly taken up by prominent actors in the political system. Conflicts arise, but also the first synergies or links with dominant current technologies. In the third phase, the breakthrough phase, the experiments and niche innovations win out over the 'old' dominant actors and models of the political system. The fourth phase is described as a period of stabilization in which the political systems return to equilibrium at a different level. This ideal-typical course for transformations is often represented by an S-curve (Fig. 1).

Transformations proceed in phases

After democratization and modernization processes faltered or even failed in the so-called transition countries which emerged from the dissolution of the Soviet Union, the political belief which had prevailed in the 1990s in the inevitability of a transformation 'for the better' has now given way to a greater realism. Here, too, a systemic perspective can help, one that has taken on board that in complex systems there can be no certain and predetermined path, and that the integrity and functionality of systems is constantly being challenged by external and internal factors. Ideally, political systems will develop both resistance and resilience, enabling them to cushion disturbances, but also to emerge from crises stronger than before by virtue of flexible adaptation and restructuring processes (Gunderson & Holling 2002) (▹ Chapter 1.3).

Transformation is not automatically and inevitably 'for the better'

New forms of cooperation for transformation

To date, there are no theory-based approaches to explaining the interaction of the (sub)systems in transformation processes, and there is a lack of detailed strategies for an overarching structuring of transformations across the borders of state, market economy and civil society systems. The relevant literature (Umweltbundesamt 2015c) draws attention to the importance of so-called active stakeholder cooperation and to a new quality of political design and social communication.

A sustainable society, according to a report by the *German Advisory Council on Global Change* (WBGU 2011), requires not only a post-fossil economy, but also 'a new social contract'. This 'new social contract' involves a social transformation through **novel forms of cooperation** between governments and citizens. But what can such new forms of actors and of cooperation look like? Some lessons can be learned from successful networks and from best practice. Many NGO and corporate networks are of great importance for disseminating new ideas and innovations, and especially in the context of sustainable development.

The significance of networks

Successful examples of novel forms of cooperation

The *International Council for Local Environmental Initiatives* (ICLEI), a city network, is active in a very wide variety of cities and communities around the world, helping local authorities move toward greater sustainability. It is a worldwide association of cities and municipalities committed to and active in sustainable development. ICLEI started in August 1991 in Toronto, Canada; its European Secretariat was established simultaneously in Freiburg im Breisgau, Germany. ICLEI now has twelve offices in ten locations, and is represented on every continent. It acts as a link between the local and global levels. On the global stage, ICLEI represents the municipalities at UN conferences and in many international bodies. At the same time, ICLEI promotes municipal sustainability by offering consultancy on innovative relations between public administration and the wider society.

UnternehmensGrün (the Federal German Association of Green Business) supports responsible business practices. It is an ecologically-oriented association in which, since 1992, companies have been engaged in taking responsibility for the economy, the environment and society. UnternehmensGrün is supported by more than 220 member companies and is active in a wide variety of fields, such as the financing of educational projects and the drafting of position statements, studies and surveys on ecological-social impact assessments of policy decisions (such as agricultural reform, the sectoral impacts of free trade agreements on small and medium-sized enterprises, the German Renewable Energy Sources Act [EEG], ecological tax reform, etc.).

It is clear from the preceding section that it is the combination of organizational and individual competencies that determines the success of cooperations. The competencies are located in the willingness to innovate, the capacity for dialogue and persuasion, and the ability to push forward change in process management, and are applied by individual actors. These are referred to as change pioneers, or *change agents*, among other terms. They use targeted *change management*, focusing primarily on coordinating processes and activities and formulating visions and mission statements that help convey new values (David & Leggewie 2015).

Change agents

Change agents play an important role in transformation processes—they can become promoters, identify opportunities, and develop new options. It is useful to differentiate between specialist promoters and relationship promoters (Umweltbundesamt 2015c, p. 16). While specialist promoters are important in terms of specialized expertise, the identification of problems and possible alternatives, relationship promoters can be important for communication and interaction processes.

The distinction between specialist and relationship promoters

With targeted support of this kind from the promoters, transformative ideas, values and innovations can be widely disseminated. The inclusion of a large number of individual activists and organizations is of great importance for the success of a transformation. Citizens are required to bring in more independent initiative, and politics to provide appropriate discussion forums. Especially in the municipalities, independent initiative can be promoted through targeted support; a familiarity with, and readiness to apply, different participation procedures are often very helpful here (▷ Chapter 3.5).

The importance of citizen participation has already been discussed in detail in ▷ Chapter 3.5; here, the systemic perspective needs to be emphasized once again. A transformation policy with a systemic perspective cannot focus solely on individual (e.g. social) innovations, but must seek to combine different innovations and to transform several social subsystems simultaneously. The sustainable development of societies requires an ecological economy that respects the limits of the earth system and avoids irreversible damage. This requires, firstly, many different innovations, for example in agricultural and industrial production, in the power supply and in mobility. In addition, a new understanding of prosperity is essential: social innovations and new models of participation and of solidaristic cooperation at the local, national and international level, so that the changes to the economy, to the forms of production and to lifestyles are supported across society.

Linking innovations and (sub)systems

Daring to seek greater solidarity – solutions for a systemic transformation

Although the need for a societal transformation has been widely recognized and accepted, questions regarding its detailed content and its socio-political organization are only beginning to be addressed. For a changeover to sustainability, the prevailing ways of thinking and acting which have shaped us and which determine our everyday lives — the patterns and routines we have 'automated' and internalized (Haderlapp & Trattnigg 2013, p. 113) (▹ Chapter 1.2) — have to change. This change can be understood as a learning process of and in democracy and social solidarity, initiated by politics and promoters (or change agents). These actors serve to open doors and, as mediators, to link local debates and needs to the global sustainability agenda.

Real world experiments and living labs

However, change agents need time and support to initiate such learning and experimenting processes and to try out new design options. For this, some scientists propose the use of 'real world experiments' or 'living labs' (Schneidewind & Singer-Brodowski 2013). A characteristic of such **living labs** is the possibility of experimentation, as well as intensive and constructive cooperation between science and practice.

The Transition Town movement has created living labs in some cities, notably in their place of origin in southern England (▹ Chapter 4.3). UNESCO Biosphere reserves, which have been established in almost 700 model areas across the world, are also well-suited as living labs on account of their underlying structural conception (▹ Chapter 4.1).

Living labs

So-called 'living labs' are places where practitioners from local authorities, social and environmental associations and companies are involved in the research process and in support of the protection of natural resources, landscape planning and maintenance, or the introduction of new mobility and energy systems. The concerns and interests of an environmental association, a cultural association or a cycling club can be heard in just the same way as those of a technology company. In open-ended processes, knowledge is generated that promotes sustainable lifestyles and sustainable economic models in practice (Schäpke et al. 2017). What is key here is that this modern form of 'knowledge creation' is understood as a co-productive process in which local knowledge is also valued and integrated (see Jasanoff 2006).

Participation is a necessary condition

In addition to the scope for experimentation, another element is essential for socio-ecological transformation: the support of different social groups and their active participation in transformative processes. Linked to this is the need to enable the design of our immediate living spaces to be based on responsibility, mutual trust and solidarity. This implies a greater commitment from society, more participation on the part of the political system, and a greater emphasis on social solidarity in work and business processes. Although policy can provide appropriate incentives, an opening up on the part of the political system is usually not sufficient. We need more sustainable models of participation on the economic side, too, based on the democratic values of a republic or commonwealth (in the original sense of the term). The economic doctrine of the primacy of individual economic benefit and of competition, which has been dominant for decades now, stands in stark contrast to the values-based discourse of a community based on social solidarity (Altvater 2005).

Fundamental changes to work and the economy

There are many approaches to a sustainable economy (▹ Chapter 4.3). Paech (2015) points out that our ideas about the future of work, the economy and political communities would have to change radically, and argues that cooperatives should become the dominant form of economic enterprise because they could curb the current pressures on the use of capital through democratic forms of corporate control and governance. Products could be more easily reparable and more durable, and service companies could maintain, optimize or repurpose the existing stock of goods. Initial steps towards such a change of mindset can be seen in resurgent initiatives like community gardens, repair cafes and multi-generation houses.

Cooperatives as important actors

▹ **Cooperatives** could play an important role in developing new forms of cooperation paving the way for the Great Transformation. Cooperatives require their members to pursue their interests and goals jointly under their own responsibility, and to actively contribute to their local environment and community. Cooperative principles and values such as solidarity, democracy and sustainability mean that they will be key actors in the transformation.

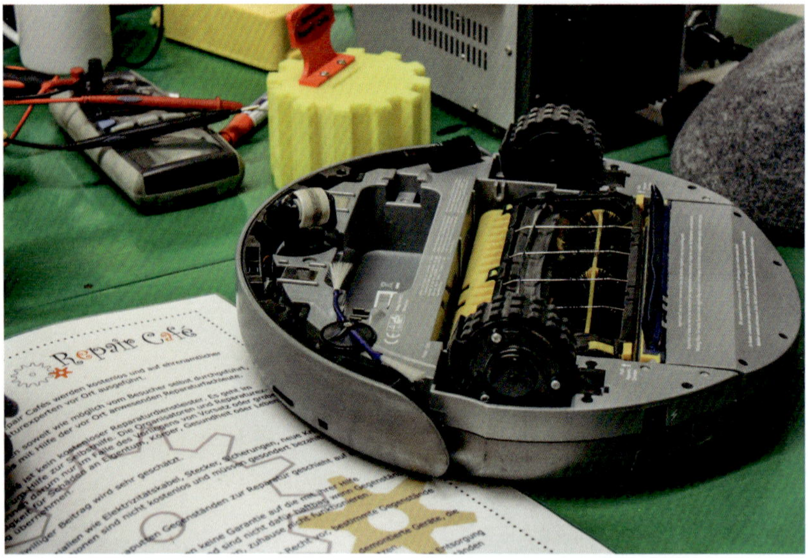

Figure 2: Resource conservation and communal rethinking in the Hebewerk e. V. Repair Café (top), and a cargo bike construction workshop run by the Transition Town initiative 'wandelBar', in the open transformation workshop of Hebewerk e. V. in Eberswalde, Germany (photos: C. Gäbler)

Cooperatives are voluntary associations of people who want to pursue common goals (e.g. in consumption, mobility, housing, health) on a basis of social solidarity, and who set up an economic enterprise to that end. In this form of organization, trust is regarded as an important prerequisite for joint self-help, or economic association. Cooperatives differ from other business models, such as the private limited company, in that no minimum capital is necessary and the share of votes is not proportionate to the capital invested (that is, each member has one vote—regardless of the amount of capital they put in). Moreover, cooperatives are not exclusively dedicated to the pursuit of economic goals; as a rule, social, democratic and sustainability goals are equally important for their members.

The cooperative as a place of learning

Cooperatives represent an economic model for eco-social transformations, and at the same time act as schools of civic and democratic values. As a rule, cooperatives arise out of specific needs in a civil society context (Atmaca 2014). While cooperatives primarily pursue economic goals, they also have a social and cultural dimension and are not dominated by capital and growth interests. On account of this distinctive rationale (and their involvement in civil society), they open up new possibilities in people's day-to-day lives that can lead to market innovations in fields such as recycling, organic farming and the energy sector.

'Our village shop' – a cooperative in Grambow, Germany

Since October 2014, the cooperative 'Our village shop' has been running a village shop in Grambow in Mecklenburg-Vorpommern (a village of some 680 inhabitants). The shop offers a wide range of groceries and other goods with a focus on regional products. Many customers are also members of the cooperative and volunteer in the shop; there are also some employees. Because of their democratic participation, members identify with their shop. This in turn strengthens customer loyalty. In addition, the village shop serves as a meeting place for communication among the villagers.

Figure 3: Village shop in Grambow, Germany (photos: 'Unser Dorfladen' Grambow eG)

The democratic framework for the 'Great Transformation'

A 'politics of sustainability' (Loske 2016) has to redirect the complex social system towards sustainability by changing the supporting framework in terms of education, science and social solidarity, and by incentivizing more sustainable modes of production. A central question is whether the existing democratic systems are sufficiently prepared for this or whether they too need to be changed first.

In Germany, one of the most modern of democracies was created, with a wide variety of checks and balances which can to a certain extent protect society against recklessly putting that democracy at risk. However, this system is poorly equipped to incorporate the interests of future generations into current decisions. Also, for many democratic systems, the path to maturity went hand in hand with the liberalization of the economy and the promotion of growth and competition. In the course of economic globalization and the emergence of influential lateral world systems—such as international financial and economic institutions and transnational companies—the scope for control of the political systems at the national level shrank.

Important new impulses from the EU level

Confederations of states—such as the European Union—initially emerged as economic communities. Aspirations towards wielding similar weight and influence politically were long ignored or resisted. In the mid-2000s, the initiative towards an EU constitution failed; the vision of a 'United States of Europe' once again disappeared from the agenda. At the EU level, there were important new impulses for promoting the sustainability transformation. In 2001, the Commission proposed an EU sustainability strategy. Evidence of how the discourse on sustainability transformation is advancing, and how it can also influence state institutions, is provided by France's Ministry of Ecology and Sustainable Development, which in 2017 became the Ministry for the Ecological and Inclusive Transition (Ministère de la transition écologique et solidaire).

Ombudspersons for future generations

Various NGOs have long called for ▹ **ombudspersons** with the power to influence politics in the interests of future generations. The UN started to look at the issue in 2006, and could later point to states that had introduced similar political roles, for example Hungary, from 2008 onwards (United Nations 2013, p. 29). Ultimately, however, the proposal for a High Commissioner for Future Generations was not successful.

An ombudsperson acts as an impartial arbitrator (usually on an honorary basis) to prevent the unjust treatment of particular groups or to ensure the consideration of the concerns of marginalized groups. They act as a mouthpiece for interests that would otherwise be ignored (e.g. those of children, the victims of violence, the environment).

Sustainability councils and future councils

In Germany, Councils (or Advisory Councils) for Sustainability were introduced at federal and state level from 2001 onwards, but they have no political power (and some have been abolished, e.g. the Advisory Council for Sustainable Development in Brandenburg in 2015). (More on the politics of sustainability ▷ Chapter 3.4.)

Nanz and Leggewie started a new initiative with their idea for 'future councils' (Nanz & Leggewie 2016), a kind of permanent 'bottom-up' institution to identify future issues affecting the local district and to come up with solutions that would be firmly anchored in local authority legislation. A future council might include up to 15 selected individuals broadly representative of the local population in their generational mix. The members would meet regularly and receive, like lay judges in Germany, a modest expenses allowance. The term of office might amount to two years, and municipal councillors and magistrates would be obliged to take account of submissions from future councils in the decision-making process.

The underlying system and conventional ways of working have not yet been called into question

The example of Germany shows that not only can democratic systems be relatively resistant to challenges, but they can also exhibit a certain capacity for learning and adaptation. New challenges are met with new structures that follow the underlying system logic and do not present a direct challenge to conventional ways of working. The establishment of advisory councils, participatory bodies and strategies does not amount to a true transformation.

Some important questions include: will an ecologically effective and socially acceptable transformation of society succeed without new institutions having been added to the political system? What mechanisms might enable the democratic state to pay much more attention to the interests of future generations and at the same time to suppress more short-term and selfish needs and desires?

It remains unclear whether the acute and intensifying social and political crises, as well as the web of physical constraints and path dependencies, will leave enough room for a sustainability policy designed for the long term. Moreover, the surfeit of media and information seems to foster ambiguity rather than generating unanimous understanding (Ibisch 2016b). And the recent past teaches us that, perhaps surprisingly, 'new forms of authoritarian governmentality'[1] are spreading (e.g. in the Balkans or in Turkey; Džihić 2017), and that supposedly successful democracies can come under considerable pressure in divided societies (e.g. Poland, Hungary, USA).

Whether and how new concepts founded on systems theory and emphasizing non-hierarchical self-organization and the wider distribution of power — for example, **sociocracy** or **holacracy** — will establish themselves in the existing political systems may become clear in the not too distant future.

Sociocracy and holacracy: decision-making without hierarchies

Sociocracy is a systemic approach that seeks to override — at least temporarily — hierarchical structures in organizations. This involves representatives of a subsystem in an organization, such as a group of employees who meet up on a regular basis, being temporarily delegated to higher hierarchical levels of an organization, thereby influencing decisions at a higher system level (De Florio 2015).
Holacracy is when decision-making processes no longer take place in rigid hierarchies, but instead people are assigned specialist roles and given responsibility for autonomous decisions (Robertson 2016; HolacracyOne 2017). Other models of shared leadership are also currently being discussed more intensively (e.g. Mohammed & Thomas 2014).

A 'Great Mindshift' is needed

In order for a 'Great Mindshift' to be achieved, it is perhaps not necessary for every part of the current system to be called into question (Göpel 2016). Transformative literacy can result from a better understanding of the system, enabling actors to recognize and abandon path dependencies at an early stage and to change systems and their functions (Göpel 2016, p. 155). New paths can ideally be constructed so as to be recogniz-

1 The concept of governmentality *(gouvernementalité)* comes from the French philosopher Michel Foucault, who in the 1970s sought to develop a systematic theory of governance and analysed the connection between the ruling techniques of governance *(gouverner)* and the 'self-governing' mindset of subjects *(mentalité)*.

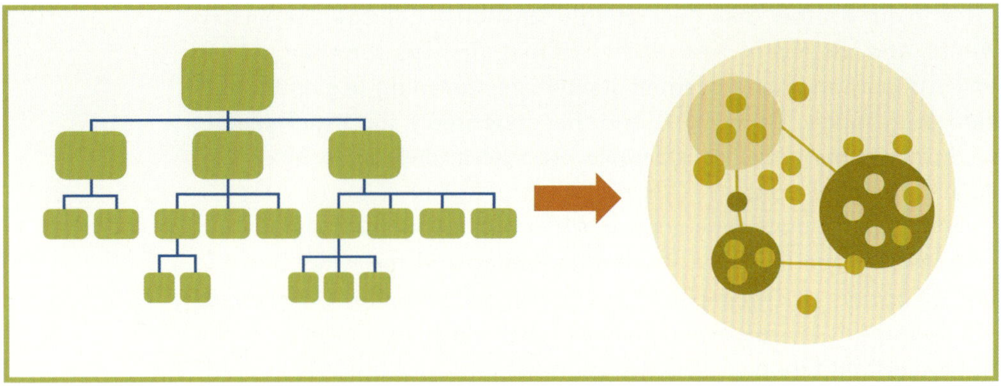

Figure 4: The development could lead from rigid, hierarchical institutional models to more organic and systemic structures, for example to structures with greater orientation toward the self-regulation of adaptive ecosystems. This is the direction taken by holacratic organization models.

ably 'pleasurable and better ways'; and if not, then growing awareness and increasing suffering will also play their part (Loske 2016). In other words, the possible options for promoting and influencing a systematic transformation are manifold, but two further aspects are necessary. First, politics must be open and provide social spaces where it can happen. Secondly, the impulses for change must come from a multitude of different actors—not just from the state, but from civil society and from thought leaders and practitioners.

A sustainable economy

4.3

*Alexander Conrad, Jan König
and Hans-Peter Benedikt*

Since the invention of the steam engine, capitalism — more than any other system — has delivered prosperity for broad sections of the population, especially in the Western world. As the economic historian Werner Plumpe notes, capitalism has triggered 'the greatest increase in prosperity in world history' (Plumpe 2010). Nevertheless, capitalism, and the particular type of prosperity it creates, do not come without crises — economic crashes such as speculation bubbles, and ecological crises caused by the overuse of natural resources and the pollution of the environment. It is like a predator that has turned our economy into a casino where the basest instincts rule, such as greed and lust for profit. But it also promotes innovation, creates technical progress and 'contributes to the peaceful coexistence of peoples, because it excludes ideological and religious differences and places in the foreground the common interest in material improvement' (Miersch 2009). Since the great financial crisis, however, capitalism has become suspect for many people, and this often hinders — even in the academic and research community — a rational and sensibly-conducted search, without angry accusations and ideologically-based anti-capitalism, for a better economic framework. Yet despite this, or perhaps precisely because of this, it is important to ask and to discuss what a sustainable economy might look like.

Economics in the context of weak and strong sustainability

Chapter 1.2 discusses the evolution of various sustainability concepts and illustrates how, depending on the concept, the economic dimension of sustainability can be seen as the basis for the development of a sustainable economic system. In the three-pillar model, the economy is an independent pillar alongside society and ecology, and neither the relative importance of, nor the links between, the individual pillars are specified. Environmental damage, social problems, etc. were thus either neglected by economists or seen as ▹ **externalities**.

A sustainable economy could thus be succinctly defined as a system that continues to focus on growth and profit maximization while taking into account certain ecological and social aspects.

If one of the three pillars were weakened, this could theoretically be compensated by the extension or development of another pillar (▷ Chapters 1.2 and 1.3). If it is possible to replace natural capital with physical capital (machinery, technologies, etc.), there are no limits to growth; or alternatively, the limits could be constantly extended, and business as usual would remain possible.

Externalities, or external effects: In economics, this refers to the uncompensated effects of economic decisions on uninvolved third parties. They can be positive or negative. A negative externality exists when a factory, for example, feeds harmful effluents into a river in the course of its production processes, thereby jeopardizing the income of local fisheries.

The results of a 'business as usual' scenario for Germany

In a wide-ranging empirical work, Gran (2017) shows how certain economic policies might affect the economic, social and ecological spheres in Germany. As a reference point, he constructs a business as usual scenario (BAU scenario), based on historical data and taking into account existing trends (including among others efficiency gains and a reduction in GDP growth rates as a result of demographic trends and the pursuit of environmental targets, etc.), enabling a forecast for the years up to 2040. The relevant results from the BAU scenario are summarized in Table 1. It turns out that in the BAU scenario not only the economic but also the social and ecological indicators improve: per capita economic output and the employment rate both rise despite a continuing decline in growth rates (see Fig. 1). On this basis, the debt-to-GDP ratio can be significantly reduced. Income distribution is improved, education levels rise, and life expectancy (measured by the probability of reaching the age of 60) is also increased. Finally, the ecological footprint per capita shows a clear improvement in this scenario. However, this improvement is nowhere near enough to bring it down to the theoretical limit of 1.7 global hectares (gH) (Global Footprint Network 2017), or alternatively, allowing for productivity increases on existing bioproductive areas, of approx. 2.2 gH (scenario!) by 2040.

Table 1:
Results of the BAU scenario for Germany up to 2040 (Source: Gran 2017, p. 207)

Indicator	Change from 2007 to 2040	Absolute value 2040
GDP	+24 %	2,961 billion euros
GDP per capita	+34 %	38,787 euros
Debt ratio	−32 %	44 %
Unemployment rate	−29 %	6 %
Ecological footprint per capita	−21 %	3.8 gH
Gini*	−7 %	0.27
Education**	+8 %	78.3 %*
Mortality***	−83 %	1.4 %**

* The Gini coefficient is used as an indicator of income distribution. The closer the value to zero, the more even the distribution of incomes in the national economy.

** The level of education is measured by the proportion of 20 to 24-year-olds with an educational attainment of between 3 and 6 in the German system (3 = dual system vocational education, 6 = completion of university undergraduate studies).

*** Mortality is measured by the probability of not reaching the age of 60.

Figure 1:
Change in GDP growth rates under the BAU scenario (with acknowledgements to Gran 2017, p. 209)

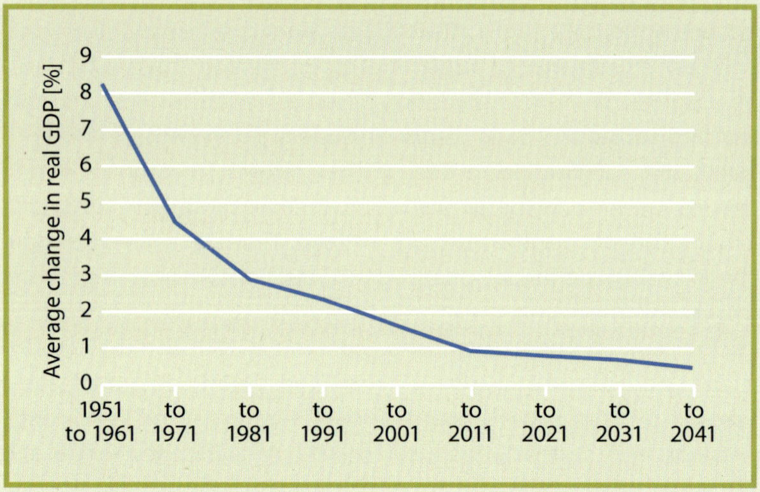

However, (complete) substitutability is not possible (▹ Chapters 1.1, 3.1), which is one of the arguments for limits to growth. The systemic understanding of sustainability incorporates these limits (▹ Chapter 1.2). A sustainable economy could therefore be described as follows:

» The economic system is embedded in these systems [the ecosystem and the social system] to preserve the natural foundations of life for humans, animals and plants. […] [Sustainable economics] therefore includes precautionary resource conservation and responsibility with respect to possible interventions, as well as a careful and future-oriented approach to people, who are the agents of economic activity. This understanding is based on the insight that a single-minded focus on growth — even with proper concern for ecological and social issues — no longer offers a realistic prospect for the future (HNEE 2016a).

Or more succinctly (based on Rogall 2013, p. 135): a sustainable economy is a 'long-term economy […] that learns to respect the limits of natural carrying capacities and the principles of fairness [to humans, animals and nature].' Since the economy is understood here as a subsystem of the ecological and social system, there are no externalities. All the impacts of economic activity on society and the environment would always be 'priced in', i.e. would be economically effective.

This perspective reveals various requirements for a sustainable economic system which will be discussed in further detail in the course of this chapter. Specifically, the focus will be on the following questions:

1. How should an economic system be structured that is not based on growth, but instead takes its orientation from the limits to the ecological and social systems?

2. How can we ensure that the economic system can continue to serve human needs by preserving the natural foundations of human, animal and plant life and not destroying them for the enrichment of individuals, countries or economic areas (etc.)?

Economics under strong and weak sustainability

If the ecological system is finite in its usability, then from an economic point of view there is a scarcity and allocation problem. The central question is how to manage the given natural resources, and how to distribute them over time.

One answer to this is the concept of strong sustainability, which leans heavily on the ideas of Herman E. Daly. Here, the stock of natural resources (or natural capital, e.g. the atmosphere, forests, fish stocks, air) is seen as having intrinsic value and as not substitutable by physical reproducible capital (infrastructure, factories, etc.), and therefore as requiring protection or conservation. A radical interpretation of this gives rise to the Constant Natural Capital Rule, which effectively means the non-use of fossil, non-renewable resources. The concept of strong sustainability therefore demands the preservation of natural capital. Building on this and on the arguments put forward by Daly (1974) and by the German Advisory Council on the Environment (SRU 2002), it is possible to formulate the basic rules for a sustainable economy as follows (▹ Section 3.1):

1. Preservation of the functionality of the global ecosystem; avoidance of irreversible damage, such as the extinction of species
2. Creation of adequate environmental resilience, i.e. of capacity for resistance and adaptability
3. Renewable resources may only be used within the limits of their ability to regenerate
4. Non-renewable resources may only be used within the limits to which they can be replaced by renewable substitutes

A key element of strong sustainability is thus the complementarity between physical capital and natural capital. In contrast, the advocates of weak sustainability regard substitutability as a viable option (Solow 1974, p. 11). In this view, economic growth is possible as long as limited natural resources can be replaced by machinery, technology, and so on. The only question is whether natural resources can be replaced at all without impacting the complex ecosystem.

The theoretical framework – structure and content of a sustainable economic system

Subsystems

Economic systems depict complex economic processes involving different actors and their interactions. For the purposes of distinguishing or describing economic systems, it is useful to look at their subsystems. According to Sik (1987, p. 17 ff.), the following five interacting subsystems are relevant (see also Fig. 2):

1. *The distribution of goods and property.* This is about asking, among other things, who owns the means of production and goods.
2. *Economic interests and motivation.* The central issue here is the question of who does what, and why; for example, why someone takes on an economic risk.
3. *Economic leadership and coordination.* Who decides what is produced, and how, when, where and for whom?
4. *Relationships between economic agents.* Who is in control of a company and appoints the management?
5. *Information processing and knowledge creation.* Finally, how is knowledge about mismanagement or malfunction (e.g. externalities) handled?

Figure 2:
Sub-systems which distinguish economic systems. Arrows indicate originating sources of factors which influence individual subsystems (and their associated economic agents).

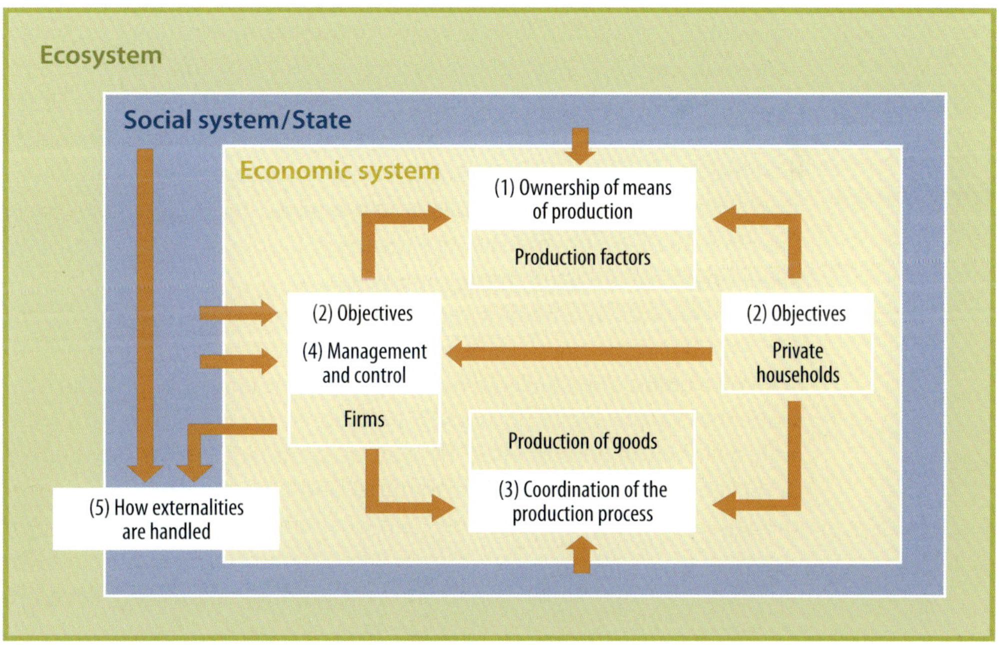

The following outline of a sustainable economic system could thus emerge from the contrast between the two opposing poles — the capitalist economic system and the planned economy:

Ownership of the means of production

In the capitalist system, the means of production are private property, and in the planned economy system they are state property. Both systems, which can be viewed as opposite poles, demonstrate functional deficiencies in operation. The capitalist system is based on inequality (win-lose). This can lead to a such a heavy concentration of the means of production that it becomes a danger to a democratic social system (the key term here is 'defective democracy'). The planned economic system, on the other hand, is based on the idea of complete equality. The means of production belong to, or serve, everybody equally. A state planning authority — which ideally is democratically legitimized (theory), but in the worst cases is the organ of a dictatorship (practice) — concerns itself with maintaining the means of production and ensures that scarce resources are put to the most efficient use, i.e. that nothing is wasted. This results in a monopoly of control over the means of production.

A sustainable economic system, on the other hand, could have 'numerous forms of ownership' (Rogall 2013, p. 159) and, in addition to private and state ownership of the means of production, could also belong to actors from a strong 'third sector', the 'social economy' (e.g. non-profit cooperatives, charitable foundations, non-profit organizations, social entrepreneurs and the like). This diversity would make the economic system more resilient by preventing concentration, and the resultant abuse of power, in the hands of only a few (capitalism) or only a single entity (planned economy). Ownership of the means of production by a cooperative (▹ Chapter 4.3), for example, would ensure that the members have a say in management, and that the means of production associated with the cooperative are used in their interests.

Economic interests, motivation and the focus on growth

The economic agents in a capitalist economic system have different interests. Businesses aim to maximize profits, and private households to maximize utility. The consequences of these aims correspond to the consequences of the focus on growth (▹ Chapter 3.2). What does the centrally planned economy offer, by way of contrast? It aims (in theory!) to maximize the satisfaction of material and cultural needs, and operationalizes these goals with the help of production targets. The main objective is not the generation or maximization of profits from individual production units; the goal is to meet production targets which are set centrally, top-down. The consequences can be:

- demotivation, if targets are set too high or too low
- production errors, if goals are not formulated specifically enough
- planning errors, if individual production units feed incorrect information about their performance into the planning process

The focus on growth arises from the pressure to constantly raise the targets — not least because of competition with the capitalist system.

Alternatively, a sustainable economic system could be designed to 'promote personal initiative [...] and put self-interest at the service of the common good [...] to ensure the well-being of current and future generations. It should be organized in such a way that it preserves at the same time the superordinate interests [protection of the ecosystem]' (Lexikon der Nachhaltigkeit [Dictionary of Sustainability] 2015), and thus also defines the limits to growth. More specifically: 'Cost recovery and reasonable profits [the level and distribution of which are decided on by the owners], sufficiently high ecological, economic and socio-cultural standards within the framework of the natural carrying capacity' could, according to Rogall (2013, p. 161), be the concrete objectives of a sustainable economic system built on market economy principles.[1]

Management and coordination of business entities

In capitalist systems, the owners themselves determine the control and coordination of their businesses. In owner-managed businesses (e.g. owner-managed private limited companies), owners can freely decide on the use of the means of production and how to maximize profits, within the legal framework. Alternatively, management and coordination can be delegated to a management team. This is what always happens when the owners themselves do not have sufficient knowledge or interest in running the business, or when the interests are very heterogeneous. Public limited companies are a good illustration. Management then has the task of satisfying as far as possible the individual interests of the owners (shareholders). These individual interests can usually be reduced to a common denominator, namely the maximization of the return on capital invested. This means the management is focused on the objectives of stakeholders who are not directly involved in the actual business activity — at the expense of the objectives of other stakeholders in the performance and profitability of the enterprise, such as employees, customers, suppliers, etc. As a result, conflicting goals can arise:

1 The extent to which a sustainable economic system needs the market as a coordinating mechanism can be summarized, following Corneo (2014), as follows: all previous planned economy systems have failed due to problems of either cooperation or allocation, so markets are indispensable for the management of a complex economy.

while shareholders demand cost savings to maximize their own profits, employees may be keen on higher, performance-related pay. Delegation itself can also lead to problems, as described in the Principal-Agent Theory. According to this theory, the management (the agent) can create an information advantage over the shareholders (the principal) and by this means pursue its own interests (maximization of self-interest). Management may then take excessive risks and endanger the existence of the company. Although there are regulatory frameworks to prevent principal-agent problems, experience regularly demonstrates their limitations. The same problem exists in centrally planned economies. Here, a state body delegates corporate governance, and, just as in the capitalist market economy, has to contend with the moral hazards facing the appointed management teams.

What would control and coordination look like in a sustainable economic system? There are very different views on this in the literature and practice. Some approaches envisage restricting the control and coordination of enterprises to those actors who are directly involved in producing the goods and services (see Felber 2012, p. 89 ff.[2]). This means that employees would have the same power to determine the business activity of the company as committed, long-term, risk-bearing shareholders. In this vein, Rogall (2013, p. 159) calls for new company law that would give owners/shareholders with a long-term focus greater voting powers than those with short-term, more speculative reasons for investing. In the 'self-managed economy' approach, the management and coordination of companies would even lie exclusively with the employees. Corneo formulates this principle, which defines the self-managed economy, as follows:

» Self-management here means that authority in the company comes from all its employees. The people who work there are not under the control of a capitalist or a bureaucrat, but decide for themselves about their own specific productive activity and about the use of the products of that activity. The employees of a self-managed company form a permanent community of workers who participate as equal members in the management of their business (Corneo 2014, p. 191).

2 English edition: *Change Everything: Creating an Economy for the Common Good* (Felber 2015).

Relations between economic agents, and decisions on production programmes

It is also necessary to look at the relationship between economic agents and how decisions are made on production programmes in a sustainable economic system. In a centrally planned economy, a distinction is made between producing (state enterprises) and consuming (private households) economic agents. The aim is, as described above, to meet the needs of private households as far and as well as possible. For this purpose, those needs are centrally recorded and form the basis for a central planning process. This procedure is enormously complex, and as a result it often led to planning errors under actually existing socialist economies, and thus to a waste of resources. In the former GDR, around 4,500 individual plans were drawn up in the course of central planning, with roughly 100,000 entries in the area of generation and performance planning and around 82 million entries in the central article catalogue (see Gutzeit 2006, p. 114).

In contrast to central planning systems, in the capitalist market economy planning is decentralized. All economic agents make their own best possible plans, taking into account their own needs, goals and constraints (budget, costs, etc.). The individual plans form the basis for negotiations in markets, which serve to reconcile or align those plans. The theoretical outcomes of decentralized market coordination are, once again, the already familiar concepts of utility maximization for private households and profit maximization for enterprises. But what if the market mechanism does not function properly? What is the outcome of a situation where one negotiating party has sufficient market power not to need to reconcile interests? Further, what is the outcome if decentralized plans do not take into account all the consequences of their implementation—for example, if they lead to externalities?

In view of the imperfections of both central (state) and decentralized (market) coordination processes, the question arises as to how planning should be structured in a sustainable economic system. Stiglitz says in this context:

> »I believe that markets lie at the heart of every successful economy but that markets do not work well on their own [...]. Economies need a balance between the role of markets and the role of government—with important contributions by nonmarket and nongovernmental institutions (Stiglitz 2010, p. xii).

A sustainable economic system could then look something like this: 'Competitive framework conditions must be designed in such a way that

Economic system Subsystem	Capitalism	Centrally planned economy	Sustainable economy
Ownership of the means of production	Private individuals, companies	State	Variety of ownership forms (private, public, mixed)
Objectives of the economic agents	Companies: profit maximization Private households: utility maximization	Meeting targets set in economic plans	Promoting personal initiative and individual responsibility combined with a focus on the common good
Coordination of the production process	Markets	State	Efficient markets operating within socially defined limits ensuring e.g. the internalization of externalities
Corporate governance	Owners and shareholders (not necessarily directly involved in the work of the company)	State	The actors directly involved in the production of goods and services (employees, long-term committed investors)
Ways of dealing with externalities	Limited or no attempts to deal with environmental damage (e.g. through legislation); economic growth as the basis for solutions to environmental problems	Lack of innovation, no incentives for more efficient behaviour, so that environmental problems are not resolved	Total internalization of environmental costs (inclusion of environmental costs in pricing)

Table 2: Summary of proposals for the design of a sustainable economic system, contrasted with the design of the capitalist and planned economy systems (based on Royal 2013, p. 145)

functioning markets are created and maintained, innovations are stimulated, an orientation to the long term is rewarded and the social change necessary to adapt to future requirements is supported. [...] Prices must [in a sustainable economic system] always carry out [...] the essential [i.e. efficient] steering function. To do so, they should as far as possible reflect the scarcity of resources, sinks, production factors, goods and services [and thus internalize or price in all external effects]' (Lexikon der Nachhaltigkeit 2015).

Rogall (2013, p. 161) suggests as supplementary measures the 'use of instruments (guard rails) for achieving the system goal' of a sustainable economy. Examples might be: supplementing the provision of services through the market by state-owned and municipal enterprises (re-communalization instead of privatization), effective laws and regulations for the protection of humans and nature (ecological tax reform, nature conservation certificates), and so on.

Dealing with negative external effects

Finally, the question of how externalities should be dealt with in a sustainable economic system must be addressed. The net environmental (and social) balance (▹ Chapters 1.1, 3.1) for both capitalist and planned economic systems[3] is poor. Consequently, it has been proposed that the design of any sustainable economic system should require the total internalization of all external effects. There is a need for economic instruments to enable such internalization to work. These tools would have to be used at both the microeconomic level, i.e. individual economic agents, and at the macroeconomic level, i.e. throughout the economy and/or across economic sectors. What this means in concrete and detailed practical terms will be examined next in a consideration of tried and tested approaches to sustainable economic activity.

Implementation approaches for a sustainable economy

Pluralist economics and the economics of weak and strong sustainability

Although today a large proportion of economics students are still confronted with the so-called standard models of mainstream economics (consumer/producer theory in microeconomics, and the neoclassical and Keynesian schools in macroeconomics), more and more teachers and students around the world are working on the creation of a 'pluralist economics' curriculum and on making it freely accessible. The aim is to showcase alternatives to mainstream economics and their contribution to sustainability (Network Plural Economics 2018). This includes approaches which have not been fully elaborated in terms of theory or mathematics and/or which do not aspire to universal validity. Rather, they are often still outline concepts or field trials with limited regional applicability.

Table 3 summarizes some of the relevant approaches and categorizes them under weak and strong sustainability. They will be discussed in more detail below. Particular attention is paid to post-growth economics and the 'economy for the common good', as both approaches contain a comparatively wide-ranging conceptual framework that can be elaborated in more detail by drawing on other approaches.

3 For example, the planning bureaucracy had no knowledge of more efficient production possibilities, and there was little incentive in the factories to develop new products or processes, so that overall innovations were very slow to establish themselves and inefficiencies persisted for a long time (see Gutzeit 2006, pp. 156 and 168).

Table 3: Selection of approaches for a sustainable economy

Approaches	Objectives
(weak[er] sustainability)	
Green Economy	Green restructuring of the economy
Blue Economy	Innovative uses of waste and resources to establish a Zero Emissions Economy
Factor X	Increasing the efficiency of the use of natural resources
(strong[er] sustainability)	
Steady State Economy	Economic development at an optimal physical level
Degrowth and post-growth	Downsizing the economy, and/or creating an economy which enables a high quality of life without growth
'Economy for the Common Good'	An economy based on principles focusing on the common good
Buen Vivir	Balance with nature, reduction of social inequality, including a solidarity-based economy
The Solidarity Economy	Economic activity based on bottom-up democracy and focused on needs
Transition Town movement	Local, resilient and self-sufficient communities

Green Economy, Green New Deal, Green Growth

The term 'Green Economy' is also associated with the terms 'Green New Deal' and 'Green Growth', and is today at the core of the German Sustainable Development Strategy (see Bundesregierung 2017). The goal it encapsulates is the green restructuring of the existing economic system. For this, the political framework needs to be adjusted accordingly, and the financial, economic and ecological crises made into win-win situations. A crisis is hereby understood as an opportunity that can lead to an economy which promotes human well-being and social justice while reducing environmental risks and ecological scarcity. Linked to this is the idea that the transformation of the economy needed from the environmental perspective—to greater energy and resource efficiency, and better management of natural capital—can become a powerful driver of growth: new 'green' markets can be opened up, and competitive advantage can be gained from eco-innovations. In other words, in the Green Economy, growth and environmental protection can go hand in hand. The focus on growth can remain (see Fiorino 2018, p. 130).

The term 'Blue Economy' was coined in 2010 by the entrepreneur Gunter Pauli and is based on the colour of the earth (seen from above). In contrast to the Green Economy, the Blue Economy aims to find approaches that are universally affordable and completely environmentally friendly, that change the entire economic system and that do not lead to supposedly sustainable products that actually continue to harm the ecosystem or that are only available to a few people.[4]

The Blue Economy as a development of the Green Economy

The aim of this initiative is therefore to find the best nature-inspired, climate-friendly, efficient and marketable technologies in the field of natural resource management and make them accessible to the general public. The 'Blue Economy Alliance' collects and publishes relevant ideas and helps translate them into business models. A related aim is to support a new generation of entrepreneurs who only use local resources and who make waste the starting point for their business models (see Pauli 2017, pp. 2, 104 ff.). The Blue Economy is thus based on the way ecosystems function and imitates the 'cascade economy' in nature, where waste (from metabolic processes) always provides the starting material for the next step in the value-added chain. There is no clear statement on the issue of the orientation towards growth; rather, the Blue Economy can be understood as a further development of the Green Economy, one which adds important contributions, such as the concentration on local resources and a consistent focus on the circular economy.

The Factor X movement – growth of the limits instead of limits to growth

The ideas of the Factor X movement can also be found in both the Green and the Blue Economy. It aims at the dematerialization of the economy, i.e. reducing the consumption of materials worldwide to a sustainable level. Factor X represents the factor of the consumption reduction needed to return to within carrying capacity. However, there is disagreement about the figure itself. Factors four, five and even ten have been discussed in the past. 'The term Factor X indicates that there is enormous potential for resource efficiency in many areas of production and consumption which is largely untapped and in some cases not sufficiently recognized' (Umweltbundesamt 2016b). The Factor X movement starts from both the macro and micro levels: at the macro level, it calls for a reduction in the consumption of some national and global resources, and at the micro level, a reduction in the production of some goods. The approach also includes making some forms of consumption measurable, e.g. by using the MIPS model (material input per service unit). Material input is illustrated using ecological 'backpacks' (water backpack, CO_2 backpack, etc.) (see Lebensministerium 2012, p. 29 ff.). At the heart of the concept is the idea that more wealth can be created with less natural

4 Pauli criticizes among other things the poor eco-balance of biofuels and points out that sustainable products are often very expensive and can therefore only be consumed by the privileged (Pauli 2017, p. 14 f.).

capital. Growth is therefore still possible, to the extent that it succeeds in using existing natural capital more efficiently, or in substituting it with new technologies. One weakness, however, is that the Factor X movement does not offer any answers to the rebound effect: resources that are supposed to be saved through efficiency gains are more often switched to new forms of production, so that overall resource consumption continues to increase (see Umweltbundesamt 2014) (▹ Chapter 3.4). This makes it clear that focusing on increasing efficiency alone will not lead to a sustainable economic system. The limits to economic growth can be pushed ever further outwards by efficiency gains. Instead, approaches need to be found which take the limits to the ecosystem as their starting point and put them at the centre of all economic activity — as is the case in the 'Steady State Economy'.

A 'steady state' as a condition of strong sustainability, and thus of a sustainable economic system

The 'Steady State Economy' can be interpreted in different ways. It could mean that all relevant variables, such as population, labour input, resource use, consumption, etc., are constant in relation to each other; alternatively, they can remain at a constant level in absolute terms — the rates of change are zero. In the first case (constant relative to one another), it follows that all the variables can grow at the same positive rate. The economy is then on a steady development path with constant positive growth (for example of 1%, as is called for in one of the Reports to the Club of Rome — see Randers and Maxton 2016). But if this growth is at the expense of a finite resource, because resource use cannot be decoupled from growth, then an end to the economic growth is inevitable. At this point, there is zero growth — all variables then remain at a constant level. The growth of this economy ends at the boundaries of the ecosystem. The 'Steady State Economy' then describes the foundations of an economy that matches the concept of strong sustainability. But how exactly can the 'Steady State Economy' be implemented? In other words, what would it look like in practice, and what consequences would it have for the actors in the economy?

A 'stationary state' as the endpoint of economic growth?

Adam Smith (1723–1790) already referred to the idea of a 'stationary state' as the endpoint of economic growth (see Smith 1776). In addition to Smith, Robert Malthus (1766–1834) also foresaw such a development as inevitable, since exponential population growth is not accompanied by a sufficient growth in food supply (see Malthus 1798) (▹ Chapter 1.2).

However, with the onset of industrialization and new scientific discoveries, the view of the stationary state also changed. Due to more productive methods and technological advances, natural boundaries seemed surmountable and unlimited growth possible. This was also reflected in economistic ways of thinking (see Schumpeter's idea of creative destruction from 1942). Only in the 1970s, with the Club of Rome's 'Limits to Growth' (Meadows et al. 1972), did the critique of growth, and with it once more the idea of the stationary state, become the focus of economic debate. Embedding the economy in the existing ecological system means that unlimited growth and thus constantly increasing prosperity are not possible. On the contrary, the consumption of finite natural resources reduces the capital stock necessary for prosperity. Drawing on this and on the stationary state, Daly (1974) put forward a definition which can be paraphrased and summarized as follows:

A steady-state economy is characterized by constant levels of physical (man-made) capital and a constant population. Both levels should be set so that they are characterized by the lowest possible throughput rates. Throughput begins with extraction from nature and ends with the pollution of nature. Both involve costs for maintaining stocks, and it therefore makes sense to minimize throughput for any given stock. While the classical economists interpreted the steady state as stagnation, and therefore sought ways—such as global trade—to avoid it, Daly sees it as a necessary scenario for survival, because an economy with a finite supply of resources cannot grow indefinitely. And since the steady state, despite what Smith foresaw, does not come about automatically, Daly calls for permanent state restrictions on, and interventions in, resource use in order to protect the fragile ecosystem, which cannot be utilized indefinitely.

Post-growth economics – downsizing the economy

Post-Growth Economics (PGE) tries to combine the requirements of a steady state economy with concrete ideas for its implementation. With regard to ecological limits, PGE proposes 'downsizing' the economy. According to Paech (2011, p. 66 ff.), however, not only have the ecological limits which make a post-growth economy necessary already been reached, but he also points out that the economy itself has reached its limits, and that there is no longer any connection between economic growth and the welfare of the population. Further, with regard to the Green Economy and the Factor X movement, Paech notes that ultimately there is no possibility of easing the burden on the ecosystem, because efficiency gains are always 'eaten up' by the rebound effect as described earlier (rebound effect, ▹ Chapter 3.4) (Paech 2011, p. 93).

According to PGE, the limited resilience of ecosystems and the limited availability of natural resources require an end to growth. The central question is therefore not whether the economy should shrink, but how. Paech seeks ways to ensure a 'soft landing', meaning a way to contract the economy which is ecologically bearable and socially acceptable. Gran's model calculations for Germany (Gran 2017, p. 222) underline the need for a planned and state-directed 'contraction therapy'. Production and consumption must be reduced, but this is not synonymous with a reduction in the quality of life. On the contrary: according to Paech (2011, p. 110), who cites in support findings from research on happiness and satisfaction, quality of life can even be increased if existing excesses are reduced (less work stress, less pressure to consume, etc.). Gran (2017, p. 228 ff.) underpins this line of argument with empirical findings which show that welfare indicators such as education levels, life expectancy and even government debt ratios can improve despite declining economic output.

Results of a degrowth scenario for Germany

The results of the earlier BAU scenario show that 'business as usual' is not possible within the ecological limits. Gran's degrowth scenario therefore presents a prognosis for Germany that is explicitly geared towards 'adjusting' the economy so that it develops within the ecological limits, as measured by the available biocapacity (2.2 gH in 2040). Specifically, this means a sufficient reduction in GDP (on the expenditure side) to bring the ratio of the per capita ecological footprint to biocapacity into balance by 2040, with no overshoot. Table 4 presents the results of the contraction scenario compared to the BAU scenario, using 2007 as the base year. As could be expected, it shows reductions in GDP and GDP per capita compared with the BAU scenario and also compared with the base year. Given that Gran's model involves a controlled process of contraction (accompanied by government intervention), the smaller economy continues to show high levels of employment in 2040, albeit at an average per capita working week no longer of 40 hours but of around 24 hours (see Gran 2017, p. 239). The sovereign debt ratio has also been reduced in the contraction scenario, mainly due to a corresponding reduction in government spending. Income distribution is more equal in comparison to both the base year and the BAU scenario. The indicators for education and for the probability of reaching the age of 60 are better than in the base year, but slightly worse than in the BAU scenario, as public spending on education and health would be reduced. Finally, with regard to per capita ecological footprint, the target of 2.2 gH would be reached by 2040 in the contraction scenario. It turns out, then, that a reduction in economic output does not equate to a deterioration in people's life conditions. However, this is subject to those life conditions and people's overall social well-being not being measured solely by the established economic indicators such as GDP or GDP per capita.

Indicators	Absolute values 2007	Absolute values 2040 BAU	Absolute values 2040 Degrowth
GDP	2,387 billion euros	2,961 billion euros	2,039 billion euros
GDP per capita	28,945 euros	38,787 euros	26,710 euros
Debt ratio	65 %	44 %	42 %
Unemployment rate	9 %	6 %	6 %
Ecological footprint per capita	4.8 gH	3.8 gH	2.2 gH
Gini*	0.29	0.27	0.23
Education**	72.5 %	78.3 %	77.5 %
Mortality***	8.2 %	1.4 %	2.2 %

* The Gini coefficient is used as an indicator of income distribution. The closer the value to zero, the more even the distribution of incomes in the national economy.

** The level of education is measured by the proportion of 20 to 24-year-olds with an educational attainment of between 3 and 6 in the German system (3 = dual system vocational education, 6 = completion of university undergraduate studies).

*** Mortality is measured by the probability of not reaching the age of 60.

According to Paech (2011, pp. 131 and 150), the shrinking of the economy should be achieved by (among other things) drastically reducing the input of labour as a production factor into the global production process. All in all, in future only about 50 % of current total working time should be spent in classical paid employment, i.e. in the monetary or commercial sector. The remaining 50 % can, in line with the sufficiency strategy (▹ Chapter 1.2), be 'saved' or used in the subsistence economy (i.e. for self-provisioning).[5]

Paech (2011, p. 125 f.) explains the savings as follows. If the liberation from 'the burden of prosperity' succeeds, then less paid work will be necessary to earn the money that is still used today to buy products that don't make us happier, but instead weigh us down. And if more time is then used to make some of our consumer goods ourselves or to extend the lifespan of consumer goods we have bought, then the need for money and thus the need for commercial employment will be further reduced.

Paech also makes concrete proposals (2011, p. 130 ff.) for the remaining 50 % of classical paid employment. As in the Blue Economy, his focus is on a strong regional economy with deglobalized value chains and diverse approaches taken from the solidarity economy. This means that as much as possible is produced locally, within the community. The use of regional currencies should also strengthen the regional economy and protect it from external shocks from the global financial economy. Paech acknowledges in this context that sufficiency, self-provisioning and regionalized economic cycles will not be able to cover all needs. Some of the necessary provisioning will continue to be dependent on the global division of labour. For this element, Paech (ibid) proposes the following: (1) increasing the efficiency and consistency of the technologies used, while at the same time preventing rebound effects through appropriate political support measures; (2) dismantling old industries and 'mining' them for usable resources — this will also ensure that old, environmentally damaging technologies and industries do not coexist alongside new, efficient and consistent technologies.

In general, products that continue to be manufactured through the global division of labour will need to have greater durability, reparability and modularity. Obsolescence should therefore be avoided, and the focus should be on maintenance, renovation, upgrading and redesign

5 According to Gran's model (2017), in a degrowth scenario that brings economic activity back within the limits of our ecosystem, commercial working time would have to be reduced to around 60 % of current levels by 2040.

rather than new production. Finally, Paech also addresses the question of the measurement of prosperity. He notes the problems associated with GDP in this regard and calls for a broader metric which includes ecological and social dimensions (see *Measuring welfare – alternatives to GDP*).

PGE thus represents a complex economic approach based on strong sustainability, involving concrete proposals for the application of the strategies (or measures) arising from a systemic concept of sustainability (efficiency, sufficiency, self-provisioning, consistency and education, ▹ Chapter 1.2). Paech's approach calls for conviction and commitment to enable a bottom-up approach[6] that could gradually take hold in business and politics. The approaches described next — the 'Economy for the Common Good', *Buen Vivir*, the solidarity economy and the Transition Town movement — show how the individual pillars of PGE can be combined with specific measures to enable their implementation (Fig. 3).

Felber (2012, 2015) outlines the three central features of his concept of an 'Economy for the Common Good' (ECG), which could make a significant contribution to the detailed elaboration of PGE as sketched by Paech, as follows.

Christian Felber's 'Economy for the Common Good'

In the first place, the ECG aims to 'dissolve the conflict of values between the economy and society by rewarding and promoting in economic activity the same behaviours and values that enable our interpersonal relationships to succeed: confidence building, appreciation, cooperation, solidarity and sharing' (Felber 2012, p. 12). Instead of competition and selfishness, the ECG aims to promote cooperation and solidarity. Competition means: I win only when others lose. The fear of losing and the desire to win are what shape the existing economic order. The prevailing win-lose relations between economic agents should be replaced by a cooperative and solidarity-based economy that leads to win-win relationships.

Secondly, the 'spirit, values and objectives of our constitution […] should be consistently applied in the economy. The actually existing economic system violates the spirit of the constitution' (Felber 2012, p. 12 f.). As an example, Felber cites the constitution of the Free State of Bavaria, which stipulates in Article 151 that 'the entirety of the economic

6 Bottom-up approach: changes take place from 'below'; that is, stakeholders (individuals, associations, communities, etc.) try out solutions independently and in their own interests. With the right communications, these approaches can spread and find application in more and more places, communities, etc. Politicians can then pick them up and integrate them into laws, subsidy programmes, etc., which gives them a broad, supra-regional impact.

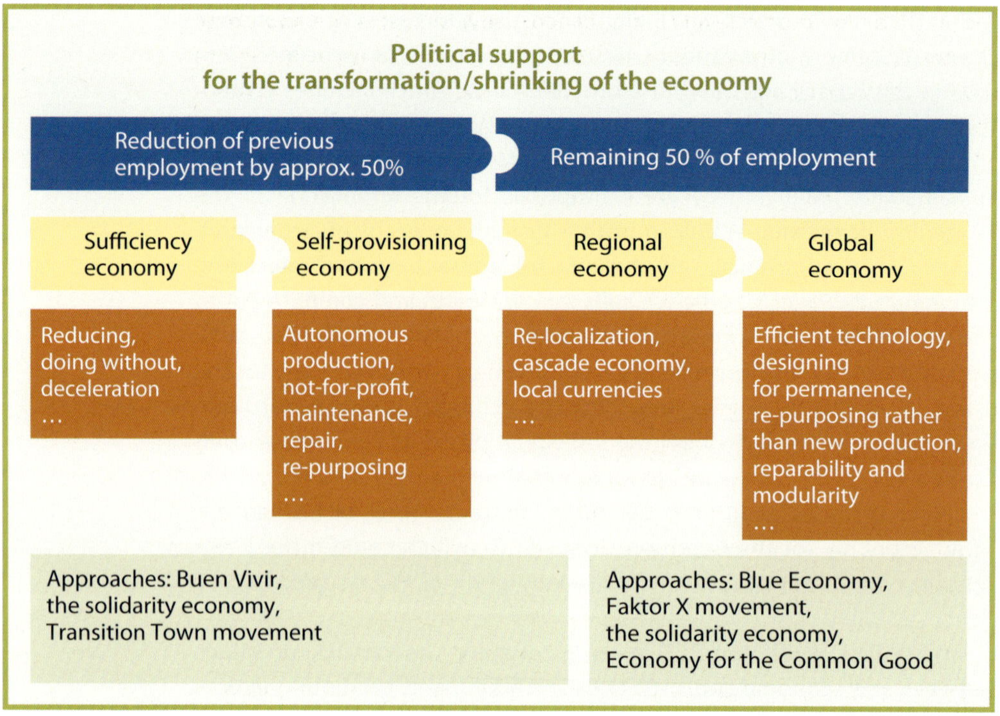

Figure 3: The puzzle of post-growth economics (with acknowledgements to Paech 2001, p. 142)

activity shall serve the public good' (Constitution of the Free State of Bavaria, Art. 151). It follows that economic success would have to be redefined. The 'incentive framework for individual economic actors must be reversed from the pursuit of profit and competition towards the pursuit of the common good and cooperation. […] The new goal for all enterprises is to make the greatest possible contribution to the common good' (Felber 2012, p. 38).

Thirdly, Felber calls for the 'measurement of economic success [to be converted from] exchange value indicators to use value indicators. [For the] purpose of all economic activity is not the provision of exchange value, but of use value […]. At the beginning of the monetary system it was helpful to translate use value into exchange value. Today the means has become the end, the servant the master' (Felber 2012, p. 13). Today's measure of economic success (profit) must be replaced in the future by a focus on the satisfaction of needs (the well-being of all, that is, the common good).

But Felber does not just call for changes; he also underpins these calls with practical suggestions for their implementation. Thus, as a replacement for GDP, he proposes the introduction of the Common Good Product (CGP) at the macro level (national economy), which is the aggregate of the **Common Good Balance Sheets (CGB)** drawn up at the micro level (individual companies).

The Common Good Balance Sheet (CGB)

The CGB (see Felber 2011, pp. 38 ff.) is an evaluation matrix showing the relevant internal and external stakeholders in an enterprise (such as employees, customers, suppliers, the public, etc.) on the Y-axis. This is in recognition of the fact that without these stakeholders there would be no production and no profits. On the X-axis, the CGB relates these stakeholders to the basic principles/values (e.g. human dignity, solidarity, co-determination rights) of the ECG. The matrix entries illustrate by means of point scores how well these basic principles are being applied in the course of the organization's business dealings. The higher the score, the more closely the principle is being followed. The aggregate of all the matrix entries gives the total CGB score. The result is determined not only by positive scores. Business behaviour which ought to be avoided in an ECG — for example, illegitimate environmental pollution, hostile takeovers, etc. — is 'punished' with high negative points.

Applying the logic of the CGB, companies are promoting the common good if they act in a socially responsible manner; produce and distribute goods and services in an ecologically sustainable way and distribute them equitably; if their jobs are of sufficiently high quality; employee participation is practised in the company; and the company recognizes its political responsibilities.

The community of CGB users provides examples of how these goals can be achieved: for example, employee participation in the company is demonstrated if the employees themselves can determine their incomes, their production and sales goals, and how the company profits should be used. In this way, full participation in the company is promoted.

Felber also sets out other ideas (approximately 20 of them) and instruments for the ECG. Particularly relevant suggestions are briefly outlined below (see Felber 2012, pp. 38 ff., summarized and critically appraised in Leschke 2015, pp. 11 ff.):

- The introduction of a Common Good audit for loans and a Common Good Balance Sheet for businesses, meaning that companies with good CGBs enjoy legal and economic benefits.
- Balance sheet surpluses are possible, and may be used for investments (with social and environmental added value), repayment of loans, limited reserves, limited disbursements to employees, or interest-free loans to other enterprises.
- Democratic public ownership (commons) should be encouraged in the form of social enterprises in the education, health, social, transport, energy, communications and finance (democratic banks!) sectors. These are intended to ensure the provision of essential goods and services in all regions.
- The state should be financed primarily through interest-free central bank lending. The central bank is given the exclusive right to create money and handles cross-border capital movements in order to prevent tax avoidance. At the local level, regional currencies can supplement the national currency.
- Income and wealth inequalities are regulated through democratic discussion and decision-making. In addition, a maximum income is introduced. Private wealth, donations and inheritance should also be limited. Inherited wealth that exceeds these constraints is distributed through a generational fund as a 'democratic dowry' to all descendants in the next generation.
- Representative democracy is complemented by elements of direct and participatory democracy.
- Working time is reduced gradually to the popularly endorsed maximum of 30 to 33 hours per week. This frees up time for other core activities such as voluntary community work.

Overall, the ECG hopes to create a new regulatory framework for economic activity that serves the common good, embedded in a binding legal framework. It proposes concrete instruments for this purpose and applies them: 'The economy for the common good is not just an idea but a reality. Worldwide, a civil society movement has spread to more than 40 states, with 19 associations and more than 100 regional groups. More than 2,000 companies support the model, and 350 of them have already drawn up a Common Good Balance Sheet' (Ecogood 2017). Thus (according to Felber, and as with PGE), implementation requires intrinsic motivation and personal responsibility, legal incentives, a regulatory framework, and awareness raising.

Buen Vivir, or the Good Life – nature as a source of rights

The concept of *Buen Vivir* (▹ Chapter 1.2), which can make a contribution to PGE in the areas of sufficiency and self-provisioning, can today be understood as a system-critical response to Western development thinking and in that sense as an alternative to the capitalist model of life. It involves a return to the philosophies of life of the indigenous peoples of South America, who do not see humanity alone as being at the centre of development, but understand everything that exists as an undivided whole. The aims of the concept go beyond merely increasing individual well-being and a high standard of living. The right to a good life is one element. At the same time, nature is accorded great importance and given rights as well. The state has a prominent role in economic and social policy, and is also expected to ensure the protection of nature (see Acosta 2017, pp. 70 ff.).

The Solidarity Economy – use instead of ownership, contributing instead of exchanging, sharing instead of buying

The Solidarity Economy, which could become important for PGE in the area of self-provisioning and regional economics, is a movement focusing on economic forms that place human needs at the centre. At the heart of the Solidarity Economy is non-profit business. Economic activities are geared towards the needs of those involved (including future generations) and should be of benefit to them. The means of production are collectively owned, and their use is determined through cooperative decision-making. The motto 'use instead of ownership, contributing instead of exchanging, sharing instead of buying' (Lebensministerium 2012, p. 48) highlights the general thrust and at the same time the scope of this approach. A variety of activities and models can be found in practice: self-managed businesses, local direct marketing, barter, fair trade, regional currencies, ecovillages, communal garden and housing projects, open source projects, etc. (see Embshoff et al. 2017) (▹ Chapter 3.5, 4.2).

Finally, the Transition Town movement seeks to lead and shape the transition to a resilient, self-sufficient society. Resilience is understood

The Transition Town movement: creating a resilient, self-sufficient society

as the capacity of the local community to withstand disruption from outside, and the movement addresses the question of how communities can continue to live well or better when fossil fuels have been exhausted. The aim is to do without fossil fuels as far as possible and to minimize the community's carbon footprint. The movement sees itself as a social, learning, proactive community developing concrete solutions and disseminating them through networking. These measures, which could also help in the further development of PGE, include: growing your own food, local currencies, local energy enterprises and energy-saving plans, and education and training projects (see Maschkowski et al. 2017, p. 368 ff.). In Eberswalde and Barnim there has been a Transition Town branch ('wandelBar') for several years now as part of the worldwide transition movement. It 'aims to inspire, encourage and support people to develop a positive vision of the future and to make this change themselves' (wandelBar 2015).

Monitoring the achievement of goals in sustainable economic systems

Measuring welfare – alternatives to GDP

GDP is associated with 'measurement errors', because it ignores relevant social and ecological aspects, and some important economic ones as well (▷ Chapter 3.2). In the course of the description of sustainable economic systems (or more accurately, of potential building blocks towards a sustainable economic system), it has already been observed that tracking the development of an economy based on principles of strong sustainability requires alternative metrics. The Common Good Balance Sheet, which can be aggregated at the national level to a Common Good Product, represents one concrete model of an alternative metric targeting the development of the common good of society. Additional metrics are identified and briefly outlined in Table 5 (see Diefenbacher and Zieschank 2011, p. 39 ff., and on the GNH Index, see Ura et al. 2012, p. 13 ff.).

It is clear that a large number of detailed concepts already exist, some of which are already in use. In contrast to GDP, they are often dependent on qualitative analysis, e.g. measuring people's well-being or satisfaction. The lack of accuracy and the limited transferability (e.g. for international and intergenerational comparisons) of qualitative metrics have to be set against the shortcomings of GDP with regard to measuring progress. More new metrics would therefore be useful.

Table 5: Alternative metrics of social welfare (selection)

Metric	Description	Dimensions of welfare methodology
	Alternative metrics employed internationally	
***Index of Sustainable Economic Welfare* (ISEW)**	Extends GDP to include income distribution, unpaid housework and family work, spending on public health, education, environmental and air pollution, costs of global warming	Economic and social dimensions, and some ecological data Quantitative/objective
***Genuine Progress Indicator* (GPI)**	Builds on ISEW and measures whether economic growth actually leads to increasing prosperity — by incorporating social and environmental dimensions	Economic, social and ecological dimensions Quantitative/objective
***Happy Planet Index* (HPI)**	Developed in the UK in 2006 as an alternative to GDP; identifies countries where the citizens have a good life — by incorporating measurement of subjective well-being, life expectancy and ecological footprint	Social and ecological Qualitative/subjective
***Human Development Index* (HDI)**	UN progress indicator which takes into account three dimensions: education (years of education), health (life expectancy at birth), and income (Gross National Income per capita); no ecological dimension	Economic and social Quantitative/objective
***Canadian Index of Well-Being* (CIW)**	Canadian national measure of quality of life; eight dimensions, 64 variables: standard of living, health, education, environment, vitality of the community, functioning of democracy, culture, time use, etc.	Economic, social and ecological Qualitative/subjective
***National Accounts of Well-Being* (NAWB)**	Two main dimensions: personal well-being (emotions, satisfaction, vitality, resilience, self-esteem, positive functioning) and social well-being (supportive relationships, trust)	Principally social Qualitative/subjective
Gross National Happiness (GNH)	Indicator developed in the Buddhist Kingdom of Bhutan, covering four areas: (1) sustainability and equitable socio-economic development, (2) conservation of the natural environment, (3) preservation and promotion of culture, and (4) good governance.	Economic, social and ecological Qualitative/subjective

►

Metric	Description	Dimensions of welfare methodology
	Alternative metrics used in Germany	
National Welfare Index ***Nationaler Wohlfahrtsindikator*** **(NWI)**	Indicator that expands GDP to include previously neglected welfare contributions such as housework and voluntary work and reduces GDP to exclude environmental damage and the diminution of natural capital. It also includes social factors such as distributive justice, public spending on health and education, crime prevention.	Economic, social and ecological Quantitative/objective
Environmental accounts ***Umweltgesamtrechnung*** **(UGR)**	German national accounting data supplemented by an ecological dimension; environmental impacts, environmental conditions and environmental protection measures are recorded, as well as how they are affected by economic activities. Based on gathering data on raw materials, energy, water and land use, waste, sewage and air pollution.	Economic and ecological Quantitative/objective

Conclusion

This chapter has looked at the foundations for a sustainable economic system. Following a brief glance at actually existing (and in parts idealized) economic systems, the focus turned towards the design of a sustainable economic system, which to begin with was rather theoretical. This showed that any reform of the current capitalist economic system (weak sustainability strategy), or of its alternative models, would have to pass a double test: it would need to ensure efficient cooperation between the various actors involved in the economy, and it would need to ensure fair allocation, without waste. Following on from this, some concrete implementation models were outlined. There was a particular focus on Post-Growth Economics and on the Economy for the Common Good.

PGE represents a conceptual framework that can be filled out in detail using ideas from, for example, ECG. The ECG, in turn, incorporates ideas from the areas of the Solidarity Economy and the Transition Town movement. It emerges that a combination of different economic approaches can add up to a comprehensive overall concept which — compared with the standard formal theoretical models that dominate mainstream economics — demonstrates a striking degree of practical relevance, and from the outset incorporates knowledge from neighbouring fields of science and research (sociology, psychology, the natural sciences in general). Thus, the approaches to sustainable economics emerge clearly defined from the sphere of the abstract, and interested actors are offered concrete fields for their implementation in the real world.

Education for sustainable development

4.4

Heike Molitor

Sustainable development as a process includes learning for sustainability. A reorientation of education towards sustainable development is very important as a strategy for the creation of a sustainable society. Education for Sustainable Development (ESD) is an international value-based approach that has a special role to play in the process of implementing sustainable development. Educational institutions, such as universities, are being encouraged to implement ESD, both in teaching and at the organizational level. Education for sustainable development can be implemented in various formats in higher education.

Background and political history

Chapter 36 of Agenda 21: reorienting education

Following the first mention of sustainable development in the Brundtland Report (1987), the largest ever United Nations Environment and Development Conference was held in Rio de Janeiro in 1992, on the recommendation of the Brundtland Commission. In addition to other documents and conventions, this event produced Agenda 21, a package of measures for public action (▷ Chapter 1.2). Chapter 36 called for a reorientation of education, and for educators to implement sustainable development through different programme areas: reorienting education towards sustainable development; increasing public awareness; and promoting training (United Nations 1992a).

» Both formal and non-formal education are indispensable to changing people's attitudes so that they have the capacity to assess and address their sustainable development concerns. It is also critical for achieving environmental and ethical awareness, values and attitudes, skills and behaviour consistent with sustainable development and for effective public participation in decision-making. To be effective, environment and development education should deal with the dynamics of both the physical/biological and socio-economic environment and human (which may include spiritual) development, should be integrated in all disciplines, and should employ formal and non-formal methods and effective means of communication (BMU no year, p. 261).

Here was the first official linkage between education and sustainable development. This reorientation of education has been conceptually developed since the mid-1990s as Education for Sustainable Development (ESD).

United Nations Decade of Education for Sustainable Development

Ten years after Rio (2002), the world community met in Johannesburg for the *World Summit on Sustainable Development* to clarify and advance the plans decided on at the first Earth Summit (Rio + 10). As part of this conference, the United Nations General Assembly decided to give education a special role in the development towards sustainability, and the UN Decade of Education for Sustainable Development was proclaimed for the period 2005–2014. Education actors were thereby called upon to actively implement ESD. All member states undertook to make special efforts to anchor ESD in all areas of education. They were asked to create their own structures for the achievement of the goals of the decade in their own countries. Over the decade, nearly 2,000 projects in Germany were recognized as UN Decade projects, and around 50 official measures and 21 municipalities were commended for good ESD practice. However, it did not prove possible to anchor ESD in educational planning (e.g. in framework curricula).

The Global Action Programme on ESD 2015–2019

For this reason, the Global Action Programme on Education for Sustainable Development (2015–2019) has been under way since 2015, building on the UN Decade of ESD to take a number of projects forward to structural implementation (from project to structure). The Global Action Programme (GAP) is aimed at actors from politics, civil society, the private sector, the media, science and research, organizations that enable and support education, as well as teachers and students (Deutsche UNESCO-Kommission [German Commission for UNESCO] 2016).

Implementing the Global Action Programme on ESD in Germany

The Federal Ministry of Education and Research (BMBF) is responsible for the implementation of the Global Action Programme in Germany.[1] To this end, a National Action Plan with concrete proposals and implementation plans was adopted by the National Platform under the chairmanship of the State Secretary in the BMBF. The National Platform consists of members drawn from politics, research, business and civil society, and draws on international expert advice. In addition, expert forums were established for specific educational areas, bringing together relevant specialist knowledge in that area and contributing to the development of the National Action Plan. In this way, the expertise built up

1 This internet portal provides comprehensive information on ESD and the Global Action Programme: www.bne-portal.de

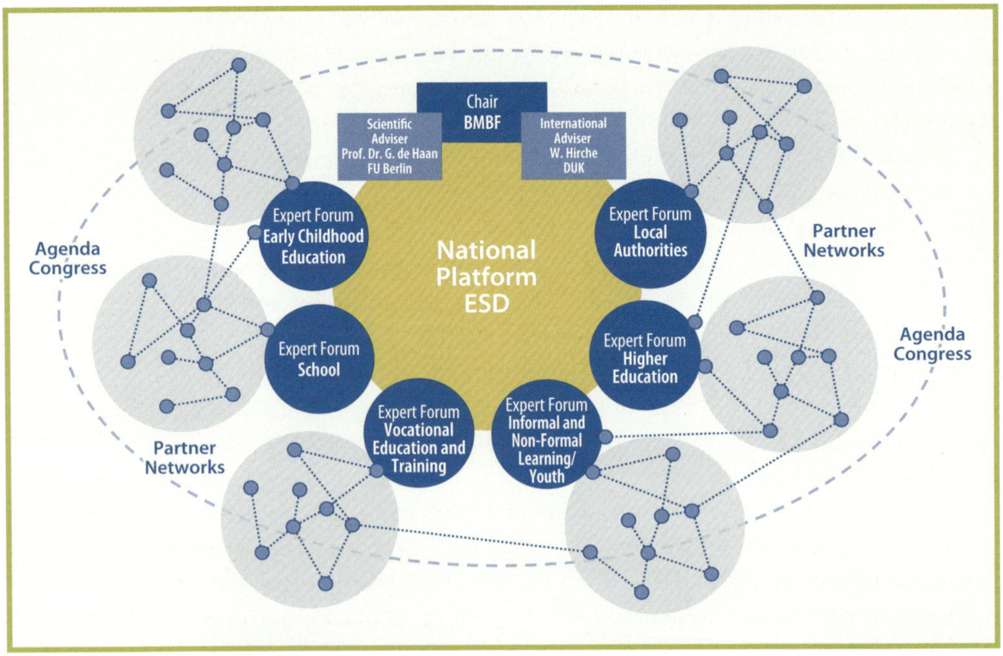

Figure 1: National Action Plan for Education for Sustainable Development: structure of the advisory bodies involved in the UNESCO Global Action Programme on Education for Sustainable Development (Bundesministerium für Bildung und Forschung 2017)

Figure 2: Places of learning honoured for the year 2017/2018 at the Agenda Congress 2017 in Germany. Over the course of the Global Action Programme, the German UNESCO Commission, in cooperation with the Federal Ministry of Education and Research (BMBF), honours places of learning, networks and local authorities for a particularly successful and high-quality implementation of Education for Sustainable Development. As a place of learning, the Eberswalde University for Sustainable Development was honoured as a source of inspiration for sustainable development at the Agenda Congress 2017. (photo: Thomas Köhler/photothek.net)

over the ten years of the UN Decade will feed into the Global Action Programme. Partner networks are affiliated to the forums, some of which were established during the UN Decade and now continue to work as part of the Global Action Programme (Deutsche UNESCO-Kommission 2017; Deutscher Bundestag 2017).

Another political process to promote education for sustainable development is the 2030 Agenda, a global 'plan of action for people, planet and prosperity' that 'seeks to strengthen universal peace in larger freedom' (DGVN 2018) (▹ Chapter 1.2). The 17 Sustainable Development Goals (SDGs) have established verifiable targets. The fourth Goal is particularly relevant for the education sector, aiming to 'ensure inclusive and equitable quality education and promote lifelong learning opportunities for all' (ibid.):

The Sustainable Development Goals (SDGs), sub-goal 4.7

» By 2030, ensure that all learners acquire the knowledge and skills needed to promote sustainable development, including, among others, through education for sustainable development and sustainable lifestyles, human rights, gender equality, promotion of a culture of peace and non-violence, global citizenship and appreciation of cultural diversity and of culture's contribution to sustainable development (United Nations 2017, SDG Goal 4.7).

Educational institutions (especially universities; ibid.) are called upon to implement education for sustainable development in their institution and at the same time to transpose the contents of the SDGs into pedagogical activity (Engagement Global 2016, 2017).

Education (for sustainable development)

Education for sustainable development

Education for sustainable development (ESD) is an international values-based concept with a special role to play in the process of implementing sustainable development. The purpose of ESD is to empower people to shape their lives and their environment so that they contribute to sustainable development. The concept starts from a general understanding of education.

Link to values

In the international context of Education for Sustainable Development, there is an emphasis on the link to values and respect for others, including future generations, as well as respect for diversity and difference, other communities and the resources of our planet. These values should be integrated into all aspects of learning in order ultimately to enable behaviour and action towards a fairer and more sustainable society (UNESCO 2006, p. 4).

ESD as a holistic development of environmental education

From a national perspective, education for sustainable development has evolved from discussions around environmental education. In contrast to environmental education, ESD does not focus solely on ecological aspects, but also covers social and economic aspects of sustainable development, and thus addresses the topic holistically. What is new is the way it addresses fundamental issues of justice both within the current generation and between this and future generations.

> »Education for Sustainable Development […] aims to enable people to actively contribute to the creation of an ecologically sustainable, economically productive and socially just environment, taking global issues into account (BMBF 2002, p. 4).

Theories of education

Education as a process of personality development or of 'self-formation' clearly involves much more than just the acquisition of knowledge (von Hentig 1999). In the search to define what education means, we can distinguish between material and formal approaches (in educational theories). In material approaches, education is understood as knowledge and the exploration of specific, apparently quantifiable content ('the greater the knowledge, the better the education'). In formal approaches, the focus is on the human being, the subject whose individual abilities and skills are to be developed. Education means the development of the subject (Fromm 2015).

In his theoretical discourses on the question of what constitutes education, the educationalist Klafki brought together both approaches in the notion of categorical education. Good education goes beyond pure factual knowledge; it imparts skills on the basis of specific values. Education represents a lifelong process and a task that has to be mastered again and again by humans interacting with the world (Klafki 1991). According to Klafki, education is 'a historically mediated awareness of the key problems of the present day and, as far as can be foreseen, of the future', combined with 'the realization of the co-responsibility borne by all in the face of such problems, and a willingness to contribute to overcoming them' (Klafki 1991, p. 56). For Klafki, key problems of the present day include issues around the environment and peace, which should be addressed in educational contexts (ibid.). This means that already since the 1990s, dealing with the topic of sustainable development in pedagogical contexts is justified on both political and pedagogical grounds.

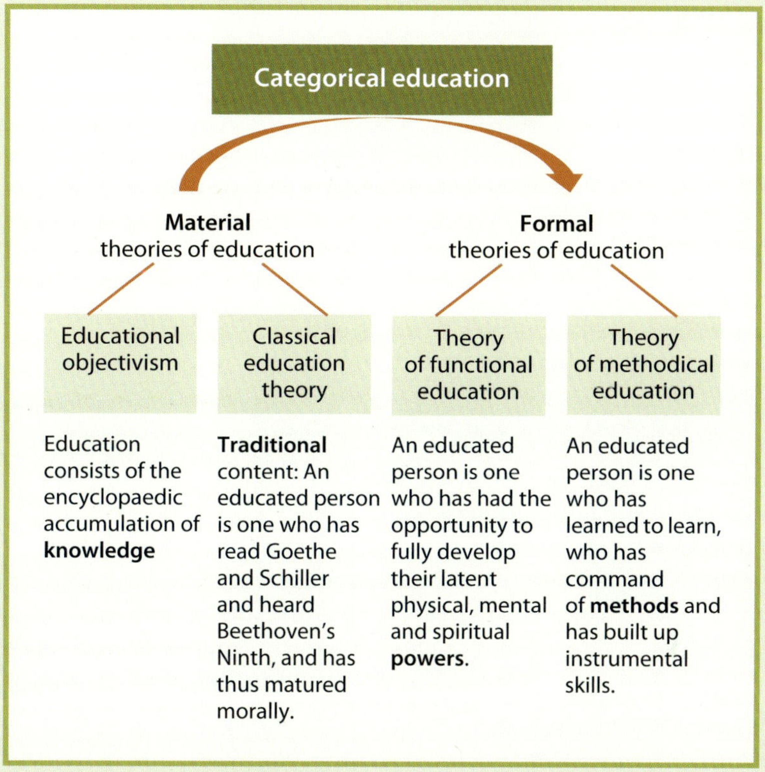

Figure 3: Material and formal theories of education (with acknowledgements to Jank & Meyer 2005)

Important features of education for sustainable development in higher education

Against the background of the political and educational framework, higher education institutions are called upon to conceptually implement education for sustainable development. The following features are particularly relevant:

1. Addressing relevant topics or areas within sustainable development — with reference to the SDGs
2. Competence orientation in teaching
3. Self-determination – co-determination – participation
4. The 'Whole Institution Approach'

Feature 1: Addressing relevant topics or areas within sustainable development – with reference to the SDGs

Topics for ESD

Topics within sustainable development are socially relevant, wide-ranging and diverse. They should:

- concern a key local and global problem
- be of long-term significance
- be based on broad and differentiated knowledge of the subject
- offer as much scope for action as possible

Since the world can be described as a complex, interconnected ecological, psychological, social and economic system, this requires a multi-perspective systemic approach (▹Chapter 1.3). Topics or areas such as biodiversity, climate change, poverty reduction, health, natural resource management, and production and consumption are closely linked and interconnected. They need to be addressed from different perspectives (such as ecology, psychology, sociology, political science, economics, i.e. in an ▹ **interdisciplinary** and ▹ **transdisciplinary** way) so that solutions can be found which contribute to a transformation towards a fairer and more sustainable society. Addressing topics relevant to ESD means at the same time engaging intensively with the SDGs (Overwien & Rathenow 2009; Rieckmann 2018a). Topic areas within sustainable development (especially in higher education institutions) require a transdisciplinary approach, as they affect people at a local level who are experts in their own living and working environment, and who can therefore offer practical and specific answers to key questions (Graumann & Kruse 2008).

An interdisciplinary approach in science and research is a working method that uses theories and methods from various disciplines (subjects/specializations) to solve problems or to answer questions (Graumann & Kruse 2008).

An transdisciplinary approach includes the interdisciplinary, but in addition brings in perspectives and knowledge from non-scientists (stakeholders) for a process of joint problem-solving. The boundary between science and practice is thereby blurred (Graumann & Kruse 2008).

Table 1: Examples of sustainability topics and their links to the SDGs (▻ Chapter 1.2)

Topic	SDG	Relevance of ESD
Biodiversity	SDG 14 Life below water	Conserve and sustainably use the oceans, seas and marine resources for sustainable development
	SDG 15 Life on land	Protect, restore and promote sustainable use of terrestrial ecosystems, sustainably manage forests, combat desertification, halt and reverse land degradation and halt biodiversity loss
Climate change	SDG 13 Climate action	Take urgent action to combat climate change and its impacts
Reducing poverty	SDG 1 No poverty	End poverty in all its forms everywhere
Social equality	SDG 10 Reduced inequalities	Reduce inequality within and between countries
Health	SDG 2 Zero hunger	End hunger, achieve food security and improved nutrition and promote sustainable agriculture
	SDG 3 Good health and well-being	Ensure healthy lives and promote well-being for all at all ages
Management of natural resources	SDG 6 Clean water and sanitation	Ensure availability and sustainable management of water and sanitation for all
	SDG 7 Affordable and clean energy	Ensure access to affordable, reliable, sustainable and modern energy for all
Production and consumption	SDG 12 Responsible consumption and production	Ensure sustainable consumption and production patterns

Feature 2: Competence orientation in teaching

Competence-oriented approach

In addition to dealing with sustainability issues, Education for Sustainable Development aims to support the education process for individuals through competence-oriented methods. Central to this is the promotion of ▹ **key competences**, such as:

- systemic and forward-looking thinking, because issues relevant to sustainability are complex, often unclear, and in some cases subject to uncertainty—much is interlinked or systematically meshed
- the ability to cooperate, negotiate and participate, because sustainability-oriented action requires participation in the negotiations over sustainable development issues and shared decisions
- the ability to reflect on one's own values and those of other people in the context of sustainable development, and on one's own role in local communities and (as a citizen of the world) in a global context, as issues and problems relevant to sustainability do not stop at national borders
- a capacity for empathy, which includes understanding of and respect for the needs, ideas and actions of other people (Michelsen 2009; Deutsche-UNESCO-Kommission 2016; Rieckmann 2018b)

Competences are the abilities people need in order to find their way in open and unmanageable, complex and dynamic situations, and can thus be described as qualities required for self-organization (Lehner 2009).

In German discourse, the concept of ▹ **Gestaltungskompetenz** ('shaping competence') has been developed to make the above-mentioned key competences more concrete and to divide them into 12 **sub-competences**. These are required to enable people to address complex societal, local, global, but also personal challenges of (un)sustainable development and to contribute to a good life with a focus on fairness. The sub-competences were drawn from, or normatively based on (e.g. fairness as a basis for decision-making), the sustainability sciences (e.g. interdisciplinarity), social practice (e.g. planning and acting collaboratively), and futurology (e.g. anticipatory analysis and assessment) (de Haan 2008).

'Gestaltungskompetenz (shaping competence) refers to the forward-looking faculty of being able to modify and model, through active participation, the future of the society in which one lives in the direction of sustainable development' (de Haan & Harenberg 1999, p. 62).

The 12 sub-competences of Gestaltungskompetenz (shaping competence):

- Building up knowledge in a way that is open to the world and integrates new perspectives
- Predictively analysing and evaluating developments
- Gaining knowledge and acting in an interdisciplinary way
- Identifying and weighing up risks, dangers and uncertainties
- Planning and collaborating with others
- Taking conflicting goals into account when reflecting on action strategies
- Taking part in collective decision-making
- Motivating oneself and others to action
- Reflecting on one's own guiding principles and those of others
- Using fairness as a basis for decision-making and action
- Planning and acting independently
- Showing empathy for others

(Programm Transfer-21, undated)

Feature 3: Self-determination – co-determination – participation

Participation

Theoretical considerations regarding how to teach and learn in ESD are based on didactic principles. Central to the implementation of ESD in higher education is the principle of participation, in self-organization, self-determination, co-determination and other forms. This means that students especially should have the opportunity to determine or co-determine their learning process, or at least be able to participate in determining it. Various teaching-learning settings meet this requirement. It requires special attention to conceptual and methodological approaches to teaching. The degree of participation can vary (Hart 1992), from open questions to 'murmuring groups' (two-person dialogues lasting a few minutes as an introduction to a topic), or from small groups to fully self-organized events (project workshops). In the last part of this chapter, two formats are described with different levels of student participation. Firstly, a strongly input-oriented lecture with a large number of students, in which the technical foundations of sustainable development will be communicated. An open discussion event is planned during the semester, at which students can contribute on issues relating to the sustainable development of their own university. The degree of participation is relatively low in this format. By way of contrast, a format is presented which has been developed by students for students and which has a high degree of participation. The students independently choose a sustainable development topic and work on it under their own direction for one semester, accompanied by two members of the teaching staff.

Alternative teaching-learning settings

Students have an important part to play in shaping sustainable development at their own educational institutes. Involvement and participation in higher education processes is crucial for the success of sustainable development. Round tables, open forms of teaching and learning, self-organized events, project work, simulation and role-playing games and so on are particularly suitable as teaching-learning settings for promoting sustainable development in educational institutions (Nationale Plattform Bildung für nachhaltige Entwicklung [National Platform for Education for Sustainable Development] 2017) (▹Chapter 4.5).

In order to help shape the transformation of society towards sustainable development (▹Chapter 4.2), projects that enable real-world problems and situations to be tackled in cooperation with partners from practice (e.g. from the local region) are particularly suitable teaching-learning formats for students. These learning experiences are of great importance, because experiences are acquired from practice and can be worked and reflected on against the background of a theoretical framework. The place of learning is then not just the university, but the public sphere or the working and living environment of the partners from practice and of their issues. This format of project-based learning demonstrates a high degree of student self-determination.

For these forms of teaching and learning, which provide specific support for the transfer of sustainability in teaching,[2] the capacity for dialogue, reflection, and interdisciplinary and transdisciplinary thinking and action are essential preconditions for an orientation toward competence within sustainable development education.

2 Sustainability transfer in teaching means teaching-learning processes that are embedded in a practice setting and in which a genuine exchange between students, teachers and transfer partners is supported. The learning process is jointly determined.

Feature 4:
The 'Whole Institution Approach'

A transformative approach

For the implementation of ESD, not only is the teaching-learning situation crucial, but so too is the organizational level that deals with the relationship between the educational institution and society, as well as for the organization of the educational institution itself (*Whole Institution Approach*; ▹ Chapter 4.5). In this transformative approach, the entire educational institution is guided by the model of sustainable development (Michelsen et al. 2011; Mathar 2014). It involves '[...] the total reorganization of educational institutions. This applies to material flows (material procurement, supply and disposal, resource consumption), cooperation with local sustainability actors (environmental organizations, development aid organizations, citizens' initiatives, sustainable enterprises), staff qualifications, the use of local resources and the strengthening of ESD within the curriculum' (Erben & Haan 2014). This means that sustainable development cannot simply be taught as a subject, but that a university, for example, has to adopt it as a guiding principle and take the following steps (UNESCO 2014):

- the inclusion of sustainability in the curriculum as core content and as a cross-cutting issue
- reducing the environmental footprint of the educational institution
- strengthening students' participation in sustainability activities in the institution and in society (with different stakeholders)
- improving relations between higher education institutions and the region, but also in international relations on all issues related to sustainable development
- support for the school leadership team via an overarching financial policy covering grants, recruitment, funding and purchasing etc.

Models for the implementation of ESD in higher education institutions

In order to add concrete detail to the creation of teaching-learning-settings in higher education institutions, two formats with differing levels of student participation will now be described (see above).

The classical lecture format leaves little room for students to participate via their own proposals, ideas and questions. Lectures often provide the foundations for further courses of teaching. The first example shows that even in such input-oriented events (in some cases, mass events with hundreds of people), a degree of student participation is possible.

Model I: Introductory lecture on sustainable development with student participation

In the context of the cross-faculty introductory lecture on sustainable development, students at the *Eberswalde University for Sustainable Development* have the opportunity, in a very open format, to reflect on the further development of the university. What picture do students have of how their university will have developed in terms of sustainable development by the end of their own studies? What are they expecting after three years of undergraduate study?

The outcome is that visions are formulated that have their starting point (in this example) in everyday life, e.g. in communication with each other, with a preference for regional and seasonal fruit and vegetables, in 'swap shops' or with CO_2-free mobility (cycling, CO_2-neutral student passes for local public transport). The vision for the organizational and societal level is that the conclusive shift from fossil energy sources to green electricity should be complete. And the era of disposable cups will be over. Some of the students assume that compost toilets will be available as an alternative to the water toilets. Rotary taps will have been replaced, the heating system improved, less paper will be used for copies and the sustainability concept will be passed on to other universities. Others dream of revolutions…

Integration of results into sustainability management

The results are collected by the sustainability officer, who is involved in the event, and used for the further development of the university. In this way, students can contribute to the development of their own university (low degree of participation), learn about the broader topic of sustainability, and gain practical experience themselves in the ongoing transformation of their own university.

Figure 4:
Discussion among Bachelor's students, facilitated by Master's students, during a course on the implementation of sustainable development at the University (photos: top left and below: U. Wessollek, top right: K. Kräusche)

Model II: Project workshops

Personal responsibility and self-organization

The second format demonstrates and requires a very high level of student participation. Adaptive and open-ended project workshops enable students to work independently and under their own management on self-selected topics and goals.

An important element in this is the consensus principle (Project Workshop Council at HNEE 2016). Project workshops have full status as recognized course elements, in which professorial staff, though they can contribute in terms of advice or guidance on request, do not play the driving role.

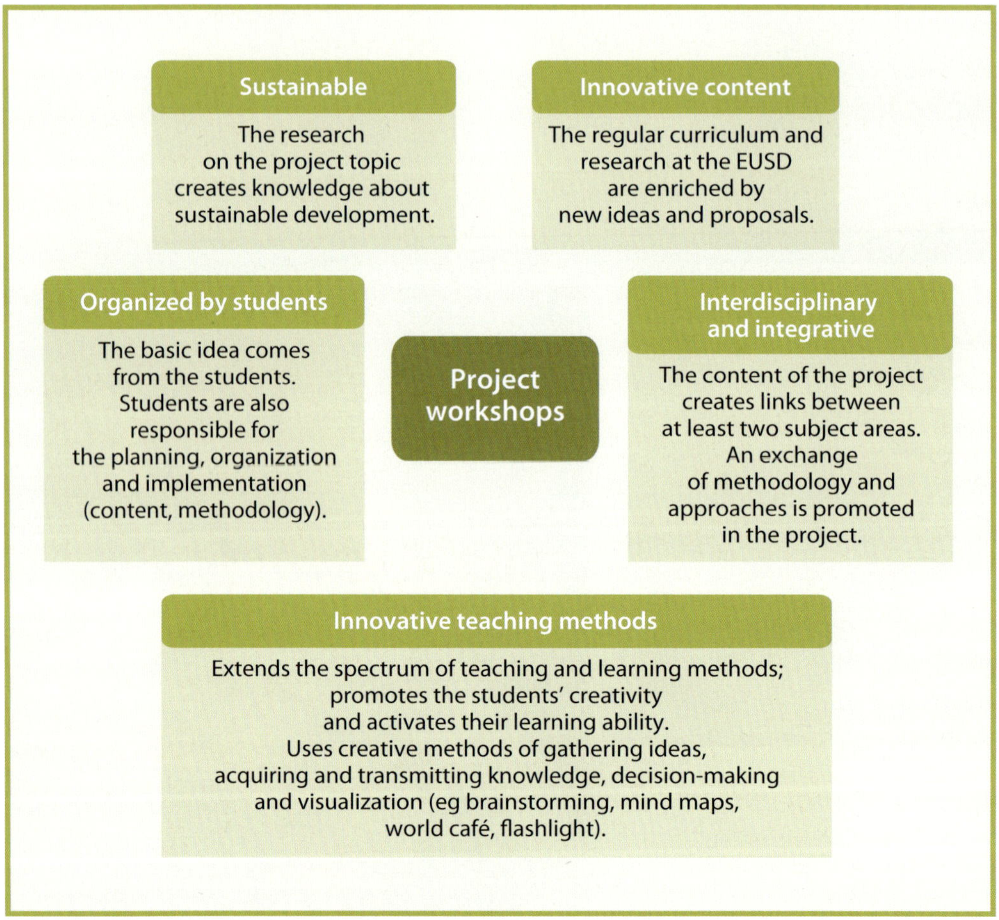

Figure 5: Quality indicators for a project workshop at EUSD (EUSD Project workshop council 2016)

Project Workshop on Permaculture – a participant's report

'At the Eberswalde University for Sustainable Development there are project workshops on various topics, organized by students. All departments can participate. The very limited prescriptive framework for these workshops involves an examination at the end of the semester, which consists of an academic paper to be submitted and the presentation of the results in a format that can be freely chosen. Two student tutors and the supervising professors are on hand to help with questions. Otherwise, the participants are completely free, as a group, to design the module on their selected topic entirely according to their own wishes.

I took part in the project workshop on the fairly broad topic of "permaculture". The two tutors organized a voluntary introductory weekend on the fundamentals of permaculture as a limited form of factual/technical input.

Permaculture, as a concept which covers almost all areas of life, opened up a huge field of possible topics for us. Suddenly we were faced with the question of precisely which topics we wanted to address from this vast range. "Wanted" — not "had to"! There are many ideas, but where should we start? We only have limited time, we can't deal with everything. What is our highest priority? And how do we achieve the goals we have set ourselves? How do we find the right information in a mountain of knowledge, and how do we internalize it without a lecturer preparing it all and teaching us everything? How do we organize ourselves as a group? We found ourselves confronted with these questions, and were initially at something of a loss.

Together at first, and later in self-selected small groups, we sought out topics, created schedules and finally sat ourselves down to the work itself.

Four of us as a small group wanted to tackle the topic of "adobe construction". We developed a "building block" (as used in the sphere of the commons: a transferable "guide" to a specific topic, a collection of knowledge, applicable to everyone, freely available, e.g. on the Internet) as a guide to the construction of a clay oven for baking pizzas and bread, and a short play as a way of presenting the results. Unfortunately, we could not build an actual clay oven in our "laboratory" during a workshop because there was not enough time. The other groups came up with a water use plan for the site, instructions for building a composting toilet, a children's book and other exciting projects.

Organizing ourselves as a big group, and bringing together and aligning different interests and expectations was a challenge and a learning experience for us.

Personally, I realized how motivating it is to master a topic for myself because I want to learn it, rather than having something put in front of me and having to learn it. Finding out what actually interests me, what it makes sense to learn in terms of my personal goals, and how to go about setting targets and planning, all of this is a process that has to be learned. Following the plan, implementing it and finally achieving the goals that were set requires a lot of self-discipline, and above all practice. The project workshop provides the space to learn — in the spirit of permaculture — in these ways in addition to the more obvious acquisition of knowledge.'

(A student in her third semester at the Eberswalde University for Sustainable Development, studying for a Bachelor of Science degree in Wood Science and Technology.)

Conclusion

The path to a holistic approach in higher education

Education for Sustainable Development can be implemented at higher education institutions in many ways. The integration of specific methods in individual courses can be the start of an all-encompassing transformation process. In order for it to become a place of learning for ESD following the Whole Institution Approach, a comprehensive approach is required that covers everything from the teaching-learning setting to the organization and the mission statement. Teaching, research, knowledge transfer and administration at the university must all be purposefully and actively designed along sustainable lines. Holistic thinking and truly sustainable action in all areas is a challenge for the organization, but also for the university's personnel and partners. But at the same time they can provide the criteria of success for sustainable development (▷ Chapter 4.5).

4.5

Institutional transformation to sustainability: the example of Eberswalde University for Sustainable Development

Wilhelm-Günther Vahrson and Kerstin Kräusche

It is not just the economic system or political systems that require transformation to achieve and support sustainable development. Individual institutions, too, can and must become (more) sustainable, within the framework of a 'Great Transformation'. Higher education institutions have a particular responsibility for the sustainable development of our society, not least because future decision-makers are being trained there. Higher education institutions are becoming increasingly aware of this responsibility, and are being called upon by external actors as well as by their own students and staff to contribute to sustainable development, to a just coexistence, and to the common good. The development of the former University of Applied Sciences at Eberswalde into the University for Sustainable Development can serve as an example to demonstrate the constraints, options and risks, but above all the opportunities associated with institutional transformation towards sustainability. How can such a transformation succeed? How can the standard-bearers for such a system change be found? And how is the idea lodged in their heads? Were there any special initial conditions that helped this endeavour? And is such a transformation a tightly controlled process, or rather a sequence of random events?

Historical background

In 1992, the World Summit took place in Rio de Janeiro — more precisely, it was the United Nations Conference on Environment and Development. It achieved a great visionary step forward, one that ultimately led to the call at Rio+20 for the 'Great Transformation' to a climate-friendly society without the use of fossil fuels and to the climate agreements of Paris and Marrakech (▷ Chapter 1.2). Simultaneously — but only temporally connected, and by no means on account of Rio — on April 1, 1992, following the political change in Germany, the University of Applied Sciences at Eberswalde in Brandenburg opened its doors, taking in 45 forestry students. The founding senate, establishing a strong and clear public image from the start, decided that a 'green thread' was to link and draw together all of the institution's departments and courses. This 'green thread' picked up on a long-established tradition essentially based in forestry.

The 'green thread' developed out of a tradition of forestry

In 1804, the forester Georg Ludwig Hartig (1764–1837) formulated the first guide to implementing the principle of sustainability in practical forestry. The second edition of his 'Guide to the Evaluation of Forests' states that

»[...] no long-term forestry can be imagined or expected unless the timber yield is calculated on the basis of sustainability. Wise forestry management must therefore involve making an early and precise calculation of the extent of the state forests, and while exploiting them as much as possible, must do so in such a way that the next generation can derive at least as much benefit from them as does the generation now living (Hartig 1804, p. 1).

In 1821, at the instigation of Georg Ludwig Hartig at the University of Berlin (today's Humboldt University of Berlin), the Royal Forestry Academy was established, and Senior Forestry Advisor Wilhelm Pfeil (1783–1859) was appointed as its Director. Pfeil soon sought to relocate the training for forestry, and in 1830 he founded, in Neustadt-Eberswalde—close to the forests, and with a clear orientation toward practice—the Higher School of Forestry. Sustainability, then, was already an important topic in forestry (▹ Chapter 1.2): Pfeil studied the subject closely, and mercilessly attacked anyone he suspected of deviating from the path of sustainability in his feared and respected 'Critical Journal of Forestry and Hunting' (Pfeil 1822–1859).

From the Institute of Forest Sciences to the University of Applied Sciences

Teaching and researching forestry were closely associated with Eberswalde from that time onwards, until in 1963 teaching in Eberswalde was wound down for political reasons by the government of the GDR. Research continued to be carried out at the Institute for Forest Sciences Eberswalde (IFE), which was founded in 1954 and numbered up to 700 staff until its closure in the years 1990–1991. During that time, sustainable was usually equated with ecological, i.e. 'green', and until the mid-1970s it remained a rather marginal topic everywhere in the world. Soon after the peaceful revolution in East Germany, the staff of the IFE lobbied for the (re)opening of a forestry faculty of the Humboldt University of Berlin to be located in Eberswalde, which finally led to the founding of the University of Applied Sciences Eberswalde. From the closure of the IFE there emerged, via several intermediate steps, today's *Thünen Institute of Forest Ecosystems at Eberswalde*, a Federal Research Institute, the *Eberswalde Forestry Competence Centre* (Landeskompetenzzentrum Forst Eberswalde, LFE), a state research institute, and the *University of Applied Sciences Eberswalde*, whose Department of Forestry then recruited numerous colleagues from the former IFE and which, with more than six joint professorships, was closely linked to the predecessor facilities of the LFE. Former staff from the IFE retained de facto control here for a long time.

Back to the present

A period of stagnation 1992–1998

The University of Applied Sciences Eberswalde was re-established in 1992 with four faculties and four corresponding diploma courses (Forestry, Landscape Management and Nature Conservation, Wood Technology, and Business Administration), and in the structurally weak north-east of the economically deprived state of Brandenburg the university

Key figures for the Eberswalde University for Sustainable Development for 2017

Finances
- State funds: 12.8 million euros
- Federal funds: Higher Education Pact 1.5 million euros
- Third-party funds: 8.7 million euros (>158,000 euros/professorial post)

3 Locations
- City Campus
- Forest Campus
- Arboretum

4 Faculties
- Forest and Environment
- Landscape Management and Nature Conservation
- Wood Science and Technology
- Sustainable Business

3 Research Foci
- Sustainable Rural Development
- Sustainable Extraction and Processing of Natural Materials
- Sustainable Management of Finite Resources

10 Bachelor's and 10 Master's degree courses
55 professorial posts in total
c. 2,200 students
c. 270 university staff
c. 438 graduating students (in 2017)
6,085 graduates since re-establishment

Eberswalde University for Sustainable Development (EUSD) is currently one of the smaller, more popular and more efficient higher education establishments in Germany; third-party funding per professorial post is 3-5 times higher than the national average.

soon found itself in danger of being trapped in a precarious situation. The original expansion plans were reduced in the mid-1990s from 55 to 40 professorships, of which only 28 were actually taken up by 1998; gaps were provisionally filled with the help of temporary federal programmes; construction projects were delayed, some for more than ten years; the higher education budget stagnated, both in the state of Brandenburg generally and in Eberswalde specifically. There was no tradition of universities in Brandenburg, and public appreciation and support were generally low.

Change Process I: Self-discovery

Establishing the guiding principles

In order to determine a distinctive standpoint for the institution, and to escape from the dominant forestry tradition, but also in order to be able to use the limited resources available in a more targeted way, an intensive discussion on guiding principles was conducted from 1998 to 2000. This resulted in the explicit anchoring of sustainability in the preamble to the charter. There, since 2000, the following statement can be found:

> **With nature for humanity:** teaching and research are committed to sustainability, which consists in the unity of ecology, economy and social responsibility. For us, the conservation and use of the diversity of nature do not represent a contradiction.

Creating a public image based on sustainability

A 'green' or strongly ecological understanding of sustainability remains dominant. In the years that followed, this perspective provided the decision-making framework for the further development of the university and for a public image increasingly based on sustainability. At that time, a majority could not yet be found for the proposal to change the name of the university (the option then under consideration was the 'University of Applied Sciences for Nature, Technology and Business'); the proposal had to be abandoned after several heated meetings of the general assembly. In hindsight, this proved to be a blessing in disguise, as the proposed name lacked the requisite concision and resonance, and its adoption in the year 2000 would likely have pre-empted the later discussion on a change of name.

Change Process II: The winds of change blow in from outside

The Bologna Process

From the year 2000, as part of the Bologna Process, the diploma courses were converted to Bachelor's and Master's degrees. Although the loss of the well-structured and popular diploma courses was a matter of regret to many, the university saw this as an opportunity for greater freedom. The new quality assurance instruments of accreditation and re-accreditation, which applied across the higher education sector, served to counter existing prejudices against universities of applied sciences and at the same time to make possible ambitious Bachelor's and Master's courses.

Political support provides a development boost

At this time — and closely associated with the then newly-appointed Minister for Science, Research and Culture, Prof. Dr. Johanna Wanka — appreciation of the value of higher education institutions began to increase in the state government. Through the so-called 'Overload Programme', about 3,000 new places were to be created in readiness for a predicted 'student bulge', in a competitive procedure involving all of the higher education institutions in Brandenburg. The financial resources were small, but the impact was enormous. Following years of stagnation, there was once again an opportunity for growth and development, and the university's guiding principles set the direction: starting in 2001, the

Figure 1: Faculties and degree programmes at the Eberswalde University for Sustainable Development.

Faculty 1: Forest and Environment	**Faculty 2: Landscape Management and Nature Conservation**
Forestry (B.Sc.) International Forest Ecosystem Management (B.Sc.) Forest Information Technology (M.Sc.) Global Change Management (M.Sc.) Forestry System Transformation (M.Sc.)	Landscape Management and Nature Conservation (B.Sc.) Organic Farming and Marketing (B.Sc.) Organic Farming and Marketing (B.Sc. dual) Management in Organic Agriculture (M.Sc.) Regional Development and Nature Conservation (M.Sc.) Strategic Sustainability Management (M.A., in-service)
Faculty 3: Wood Science and Technology	**Faculty 4: Sustainable Business**
Wood Science and Technology (B.Eng.) Wood Science and Technology (B.Eng. dual) Wood mechatronics (B.Eng. dual, in preparation) Wood Science and Technology (M.Sc.)	Finance Management (B.A.) Regional Management (B.A.) Business Management (B.A.) Sustainable Tourism Management (M.A.) Sustainable Business Management (M.A.) Municipal Economy (M.A., in-service)

programme funds enabled the development of study courses in Organic Farming and Marketing (BSc), Sustainable Tourism Management (MSc), Forest Information Technology (MSc), Management in Organic Agriculture (MSc), and Global Change Management (MSc), and thus the practical implementation of the university's guiding principles and ecologically-oriented sustainability ethos.

Research orientation

In the field of research — the university has always been research-oriented, and (not least due to limited public funds) always very good, too, at attracting third-party funding — the topic of renewable energies was identified as a priority topic.

Introduction of environmental management to EMAS standards

From early on — and this was as much a matter of conviction as one of credibility — close attention was paid to the institution's ecological footprint and associated internal processes, and the appointment of a resource conservation officer was stipulated in the charter. However, it was only from 2007 onwards that this area of the administration gained real momentum with the decision to professionalize it through the appointment of a full-time environmental manager to promote systematic **environmental management** and certification to **EMAS standards**.

EMAS requirements

EMAS *(Eco-Management and Audit Scheme)* is an instrument developed by the European Communities in 1993 for companies seeking to improve their environmental performance. Regulation EC No. 1221/2009 of the European Parliament and of the Council of 25 November 2009, on the voluntary participation of organizations in a Community eco-management and audit scheme, covers:

- the adoption of a corporate environmental policy
- the identification, assessment and monitoring of significant environmental issues for the company in a regular environmental audit
- the continuous improvement of environmental performance over and above the legal requirements, as well as the introduction and achievement of environmental objectives
- access to the resources required (personnel, infrastructure, technical and financial, including an environmental management officer)

- training, information and participation in environmental management for all staff
- transparent documentation of the environmental management system, and the publication of an environmental statement on relevant activities and data (for example, on resource and energy consumption, emissions, waste, biodiversity, etc.)
- the auditing and validation of the management system by an independent, external, qualified and authorized expert

Environmental management at the Eberswalde University for Sustainable Development

The environmental management system at the EUSD involves all members of the university — students, teaching, research and administration staff, as well as cooperation partners and business partners (such as the student union, the municipality and the region).

The objectives of systematic environmental management are:

- transparency of environmental activities
- the optimization of processes in corporate environmental protection
- to continuously reduce negative environmental impacts and improve environmental performance
- to increase efficiency in the use of energy and materials
- to inform, sensitize and motivate university members in order to support environmentally friendly behaviour at the university and in their private lives
- to gain a better understanding of the potential for improvement in the organization of processes, the control of resource consumption and the use of technology

The EUSD uses the EMAS's 'easy' methodology for environmental analysis and preparation for EMAS validation (the Environmental Advisory Committee at the Federal Ministry for the Environment, Nature Conservation, Building and Nuclear Safety provides current guidelines for environmental management officers [Environmental

Verification Committee 2015]). Originally developed for small companies, this 'toolbox' was adapted to the requirements of higher education institutions, in particular for the environmental analysis that accompanies the process. It records objective measures (environmental monitoring, ecomapping), but also the environmental situation as subjectively perceived by students and employees (via an anonymous online survey of all university members). The environmental plan sets out objectives that emerge from the aggregated analysis of the key indicators energy (electricity and heat), water, waste, emissions, occupational safety (especially fire safety), soil protection, and storage, as well as from indirect environmental aspects (e.g. communications). The measures set out in the environmental plan are intended to improve the environmental situation at EUSD. The environmental statement describes progress towards the achievement of the overall objective—the continuous reduction of negative environmental impacts. Every second year, the environmental statement is validated by an independent, certified environmental auditor, and then published and included in the sustainability report.

Change Process III: A new name

From a technical college to a university

A change in German higher education legislation in December 2008 allowed what had previously been known as technical colleges *(Fachhochschulen)* to call themselves 'universities of applied sciences' *(Hochschulen für Angewandte Wissenschaften)*. This was taken as an opportunity by the university management to bring the question of a change of name back onto the agenda for internal debate, with the aim of firmly anchoring the topic of sustainability in the institution's name. It was seen as another way to reaffirm and enhance its public profile. At the same time, there was also a concern that another university might stake a claim to this ground first.

The participatory search for a name

At the beginning of 2009, the Senate commissioned a small working group to analyse the changes required to the charter by the new Higher Education Act, and to suggest some new possible names. Many of the proposals involved the word 'sustainability'. The suggested names were hotly debated in the Senate. While everybody at the university was able to participate in the online debates, the actual process of finding a new name was the responsibility of a small working group authorized by the Senate.

Choosing the new name – a thorough and well-informed process

On one side of the debate, a concern was expressed by some of the students, and some of the professors, too, that a new name could damage graduates' career prospects and the public image of the university (this was allayed by the assurance that certificates could still be issued under the old name on request — though no such request was ever subsequently received). On the other side, accusations of greenwashing, megalomania and lack of credibility were heard. The compromise option that was then decided on — 'the Eberswalde University for Sustainable Development' (EUSD) — after much intensive debate, was able to command majority support, and the process by which it was agreed was characterized by thoroughness and wide participation.

Focus

Whole Institution Approach

Since then, EUSD has systematically pursued a 'Whole Institution Approach' (▷ Chapter 4.4) in teaching, research, knowledge transfer and administration. The name change has become an important driver of a self-reinforcing process that permeates all areas of the university. In all its activities, the question of sustainability is posed both externally and internally. It is a self-imposed yardstick, the measure of the university's credibility. Sustainability is not just another thing to be taught; ways of implementing it have to be demonstrated and exemplified. The aim here is to maximize coherence, but also visibility. The implementation of sustainability via a Whole Institution Approach is promoted and reflected throughout the various areas of the university. One example of this is the bottom-up founding of the open Round Table for the sustainable development of the university, in which every university member can participate and help to shape the ongoing process of the university's sustainable development.

The participatory development of sustainability principles

Communication and participation are crucial in all of this, both within the university itself and in its relations with the wider society. In this, the Round Table has become a very effective bottom-up driver, promoting the topic of sustainability at all levels. For example, through the formulation of **sustainability principles**, a binding framework of action for the university was developed, agreed with the Presidium and then passed by the Senate, and it is continuously updated. In addition, the Round Table concerns itself with the detailed implementation of sustainability at the university, including in social interaction (in forms of

EUSD's sustainability principles

Since February 2013, the principles of sustainability have been the foundation of EUSD's activity and development. They represent what all members of the university (students and staff) aspire to in dealing with each other and in their cooperation with partners from outside the university. They include:

- long-term thinking and action as a benchmark for all university activity
- mindful and economical use of natural resources
- to deal with one another with compassion, justice and fairness
- to inform, communicate and reflect on our own ongoing process of sustainable development

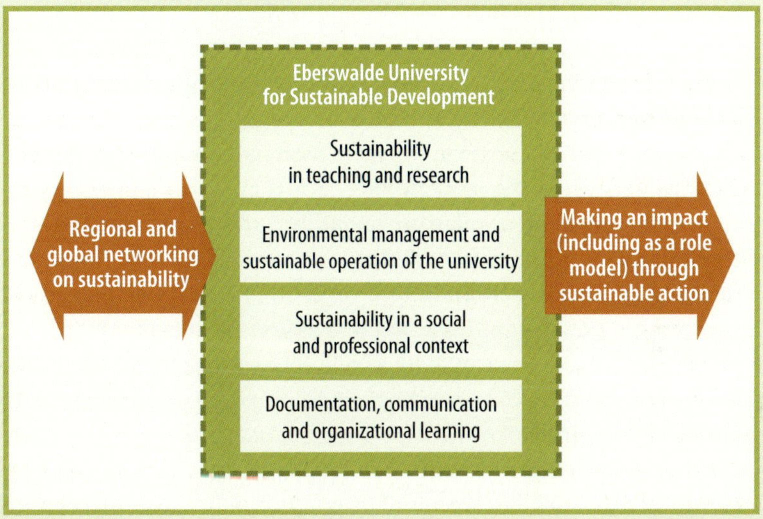

Figure 2: Sustainability principles of the EUSD.

communication, appreciation, and seeking consensus around the common good), but also in terms of more prosaic matters such as sustainable food in the refectories.

Systematic implementation of sustainability in teaching

Sustainability is systematically implemented in teaching. This can be seen in those courses where sustainability is directly relevant, in the compulsory sustainability lecture for all Bachelor's degree courses, and in the project workshops organized by students on sustainability topics (▹Chapter 4.4). In many courses, classes are integrated into the university's own sustainability management process. The aim is to provide students with experiential knowledge, to identify and demonstrate options for action and to involve them in the further development of internal sustainability management (▹Chapter 4.4). The larger goal in all this is for EUSD to train the 'change agents' for the 'Great Transformation' (▹Chapter 4.2). For sustainable teaching and learning, there are certainly no simple paths, but there are good examples for all the different approaches. The project 'Let's do the job properly!' seeks appropriate responses to the increasing diversity of students. The practice module 'Organic businesses as study partners' combines teaching, research and knowledge transfer, and was awarded the *Ars Legendi Prize* of the German Stifterverband and the German Rectors' Conference in 2017. The university received its first prize for its activities in the area of 'Education for Sustainable Development' (ESD) in 2007, and in 2017 it was named educational institution of the year by the German UNESCO Commission (▹Chapter 4.4).

Concentrating research and knowledge transfer on sustainability

The university's research foci — *Sustainable Rural Development*, *Sustainable Extraction and Processing of Natural Materials* and *Sustainable Management of Finite Resources* — turn what was originally a disadvantage — the institution's remote location — into an advantage. They also serve to strengthen the direct cooperation with regional, national and international stakeholders. In the joint project 'HOCH N' — involving 11 German universities researching applied sustainability and transformation — the EUSD leads on the topic of knowledge **transfer**.

The university's systematic orientation toward sustainability

The university has been operating on a CO_2-neutral basis since 2014, with an active **climate protection management** system. Winning the European EMAS Award twice (2009 and 2017) confirmed its Europe-wide leadership position in best-practice environmental management. This systematically recorded knowledge and practical experience is passed on to other institutions via numerous further education and other networks (on topics such as sustainable procurement, sustainable event management, sustainability management, environmental management systems) (HNEE 2018b). Germany's first model project for climate-neutral student travel, the 'Climate-neutral semester ticket', has been under way since September 2017, initiated by the students' union and the university management in partnership with the regional travel association for Berlin-Brandenburg.

EUSD's knowledge transfer strategy

Knowledge transfer (or the so-called 'Third Mission') is a reciprocal institutionalized exchange between universities and society which (according to Berthold et al. 2011, p. 9) can be subsumed under the term 'social engagement'.

The transfer strategy of EUSD states that: 'knowledge transfer at the Eberswalde University for Sustainable Development is understood in a broad sense, and goes beyond the narrow interpretation of it as technology transfer (primarily from science to business). It is the reciprocal exchange, between partners, of knowledge, ideas, services, technologies and experience. It encompasses all forms of cooperation between the university and its external partners in teaching and research — and beyond' (HNEE 2016b, p. 3).

The knowledge transfer strategy describes how the term is understood at EUSD and identifies its objectives. An overview of the structures and actors involved serves to publicize transfer formats and to directly address partners and target groups. The objectives are continuously developed, as part of a separate catalogue of measures.

The transfer strategy at EUSD helps to publicize the variety of transfer activities, to recognize and support commitment and engagement, as well as to facilitate cooperation with business and the wider community. At the same time, it gives the transfer activities a strategic orientation in order to enhance EUSD's public image. The transfer strategy is not static and fixed, but is understood as a dynamic process, triggered by the transfer audit, in which all interested members of the university can and should participate.

EUSD's Climate Protection strategy

The climate protection strategy of EUSD was drawn up by means of an intensive participatory process, and sets out the path by which EUSD can become a climate-friendly university. Central to this are nine guidelines for a climate-friendly university:

1. **Avoidance of CO_2 emissions in higher education operations:** The goal is to have all electricity and heat supplied by renewable energy sources.

2. **Achieving climate neutrality within system boundaries:** From 2014, net greenhouse gas emissions will be offset by climate protection projects that simultaneously promote sustainable development.

3. **Efficiency and sufficiency as guidelines for further action:** Notwithstanding the use of renewable energies and renewable raw materials, energy at EUSD is used sparingly and mindfully.

4. **Climate awareness:** Climate change, climate protection and climate adaptation are addressed both as cross-cutting themes and as specific topics in teaching and research.

5. **Impact as a role model for climate protection:** Effective climate protection management is demonstrated and copied through networked interaction with local and global partners.

6. **Climate protection made visible on the City and Forest campuses:** The measures for climate protection are demonstrated to a wider public, for example during the climate protection week, or through the screens on the PV systems displaying the electricity generated.

7. **Continuous and holistic climate protection accounting:** Within the limits of the system, EUSD practises continuous net accounting in its climate protection efforts, also taking indirect emissions into account.

8. **Communications:** EUSD communicates its climate protection efforts by means of open and regular publications, for example its Climate Change Strategy, EMAS Environmental Statement, and Sustainable Development Report.

9. **Setting up a learning project:** Climate protection is a process closely linked to organizational learning. The idea of the 'climate-friendly university' is developed in close cooperation with all members of the university to integrate existing knowledge and to use and pass on relevant experience.

Building awareness for climate-friendly behaviour is at the forefront. EUSD regards its training of multipliers and its role model function as the biggest contributions it can make to protecting the climate.
It is intended that the further development of EUSD into a climate-friendly university (formulated as a goal in its sustainability principles) should follow a holistic approach. Since EUSD cannot avoid all greenhouse gas emissions, these should be offset by climate protection projects, thus achieving climate neutrality. However, the use of the term 'climate neutrality' can also provoke criticism. How are the system limits for greenhouse gases identified or defined? How important are efficiency and sufficiency? What environmental benefits are ascribed to the use of green electricity? What are the quality criteria applied to the offsetting projects?
Criticism is justified if climate compensation is not at the lowest end of this hierarchy of priorities: efficiency and sufficiency before 'greening' before climate compensation (Harthan et al. 2010). Meeting this criterion is the goal of EUSD's actions here: by implementing a comprehensive energy plan, by saving energy through efficient and sufficient behaviour and appropriate administrative measures, and by supplying all electricity and heat from renewable energy sources (see also guidelines 1–3).

Greenhouse gas accounting for organizations and companies

Standards

The recognized standard for greenhouse gas accounting in organizations, the Corporate Carbon Footprint (CCF), is the Greenhouse Gas Protocol (GHG) or ISO 14064, which is based on it. Emissions are divided into three areas:

Scope 1: all direct emissions from combustion processes as well as emissions from the vehicle fleet

Scope 2: all indirect emissions from supplied energy

Scope 3: all other indirect emissions.

Emissions from Scopes 1 and 2 must be accounted for in accordance with the GHG Protocol. The accounting of emissions from Scope 3 is optional. In addition, not only CO_2 emissions, but all relevant greenhouse gases should be included, using CO_2 equivalents. The calculation is usually made by taking usage figures and multiplying them by recognized emission factors (WWF Germany & CDP 2016).

Comparability of the Corporate Carbon Footprint

Given the room for manoeuvre—in accounting for Scope 3 emissions, in the choice of emissions factors, and in the definition of the applicable system boundaries—it is difficult to compare the Corporate Carbon Footprint of different organizations. Also, some organizational activities which may potentially be relevant to climate-friendly transformations may not be included in these accounts, such as awareness-raising measures. On the other hand, other activities which have a positive effect on the CCF may only lead to displacement effects: for example, outsourcing climate-damaging activity, or purchasing green electricity not linked to the expansion of renewable energies.

The CCF is therefore not unconditionally suitable as an indicator for assessing the climate-friendliness of an organization, and certainly not as a sole indicator. EUSD seeks to resolve this dilemma by making accounting more transparent. In addition, other factors besides the CCF are incorporated into the EUSD climate strategy, as can be seen in the guidelines for the climate protection strategy.

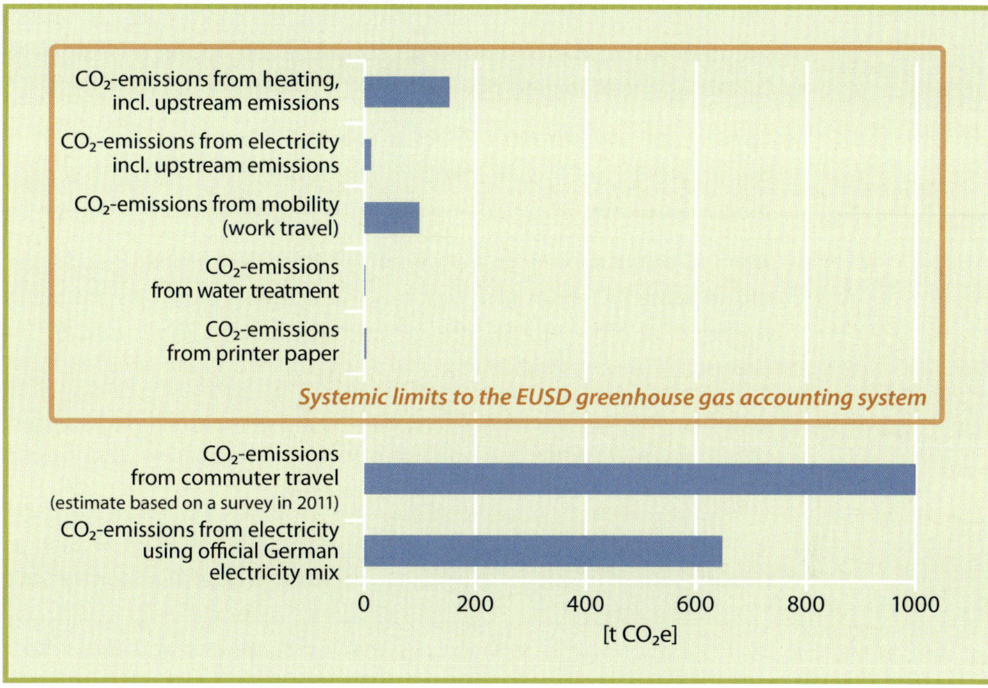

Figure 3: Systemic limits to the EUSD greenhouse gas accounting system, and alternative systemic limits and emissions factors using greenhouse gas emissions for 2016 as an example. Including commuter traffic in EUSD's greenhouse gas accounting, or the alternative of accounting for green electricity consumption according to the Federal electricity mix, would have a dramatic impact on EUSD's CCF. (EUSD calculates its green energy consumption using the trader principle, which includes upstream emissions)

Name change and content restructuring for an entire faculty

The university's new focus on sustainable development also brought with it the renaming and restructuring of the *Faculty of Business* to the *Faculty of Sustainable Business*. This involved replacing the very popular Master's degree in *Marketing Management* with the new continuing education course in *Strategic Sustainability Management*.

Sustainability Report

The Whole Institution Approach finds its most complete expression in the university's integrated biennial Sustainability Report. All the operations of the university, in teaching, research, knowledge transfer and administration, are here presented under the aspect of sustainable development. This involved close cooperation with the German Council for Sustainable Development and controversial discussions over the applicability and application of the German Sustainability Code to university reporting. The report is now based on sustainability principles (HNEE 2018a).

Transformation through diversity rather than consolidation

The organization of the transformation process conforms to an understanding of leadership based not on consolidation but on diversity, and is guided by the following points:

- Sharing a vision but not subject to micro-management, the actors of the university are free to follow different directions, but with a common orientation.
- Joy and disputation: is there anything better than working and arguing together? Only in this way can the shared vision be continually renewed; in other words, internal and external friction is part of the process.
- Patience and decisiveness while waiting for and exploiting 'windows of opportunity': when the guiding principles are clear, sudden opportunities can be taken quickly and decisively.
- Suitability: the administrative and emotional effort must be as small as possible, for all students as well as academic and administrative staff.

Bottom-up approach

It is very important in this respect that prescriptions and ideas from the Presidium are often met initially with suspicion, and often provoke opposition for the sake of it. So leadership can best be implemented in a bottom-up process, in which, to put it bluntly, creative stakeholders are able to wrest innovations from the Presidium who are supposedly in charge of things. It is therefore important to identify and support the people who can provide the drive and impetus.

Achievements and uncertainties

A distinctive public profile in the region and beyond

The distinctive mission of the university has been firmly established at the state level in the current contractual regulations governing higher education institutions and in the higher education development programme. Nationwide and worldwide, EUSD's goals are now extremely topical; this is confirmed by the 2015 UN Sustainable Development Goals (▹ Chapter 1.2) (Goal 4.7):

> »By 2030, ensure that all learners acquire the knowledge and skills needed to promote sustainable development, including, among others, through education for sustainable development and sustainable lifestyles.

What this means for higher education institutions at the national level is set out in the 2017 *German National Action Plan for Education for Sustainable Development* (which EUSD is involved in drawing up):

» Higher education institutions, as sites of research and education, are key to sustainable development. Through research and teaching, they develop and impart knowledge, skills, competences and values, and train multipliers and future leaders (National Platform for Education for Sustainable Development 2017, p. 51).

EUSD's goals could hardly be better described at the moment than in the above quotes from 2015 and 2017.

Losing our USP – institutional risk or societal advance?

The topic of 'sustainable development' has thus clearly moved to the centre of education policy, and is now being taken up in many places, some with significantly greater financial and other resources, so that EUSD's 'unique selling point' could disappear. From the university's perspective, that is certainly a risk. But from a societal point of view, is this not precisely what we should be aiming towards? Certainly, that is what is called for in the United Nations Sustainable Development Goals and the Global Action Programme described above.

Development potential

The university's small size currently makes it impossible to cover the topic of sustainability in all its multiple aspects, so EUSD will have to continue to reinvent itself and set new priorities in the future. The starting point for the sustainable further development of the university consisted primarily of ecological criteria (e.g. the development of the environmental management system and of relevant courses of study). The social factors shaping sustainable development are increasingly becoming the focus for transformation, in addition to being integrated (of course) into EUSD's internal operations. Topics such as the common good, intrinsic value and diversity are under discussion, and health management is being established. This process of updating is also reflected in the adoption (as recently approved by the German Rectors' Conference) of a third research focus, on Sustainable Management of Finite Resources, and in new professorial posts in Transformation Governance.

The never-ending discourse on 'sustainability'

In addition, the growing arbitrariness (as is often pointed out) in the use of the term 'sustainability' ('Everyone's bored with it') is something that could be addressed. However, it is important here to differentiate precisely between oft-misused terminology and content. This discussion is indeed being conducted at the university, and to continue to engage with this discourse remains part of the EUSD's essential mission.

An open system and permanent exchange

Finally, there is a certain risk that EUSD could evolve into a tightly self-referential sustainability community trapped in its own 'echo chamber'. Against this, it can be said that EUSD is in constant dialogue and interaction with many stakeholders in society — not only with its peers, but often with actors who have a different idea of development and who start from different premises. A self-understanding as an (open) system is vitally important, one in which everyone strengthens everyone else with arguments and resources, ensuring that everything works properly, so that all the staff and the graduates are able to act as change agents beyond the walls of the university. Through intensive networking and the numerous transfer activities for the creation of a sustainable society, the EUSD can rebut any accusations that it is an 'echo chamber' or a 'green bubble'.

Sustainable university development as a systemic, self-reinforcing process

The process of renaming and of establishing a unique profile has given a significant boost to the development of the university and notably enhanced its focus on sustainability. This also applies to its external impact and perception, where EUSD — relative to its actual size — has gained wide visibility nationally and to a certain extent also internationally, and is perceived as a credible brand. This benefits the university at all levels, not least in recruiting staff and students, but also with respect to internal morale and academic appointments. Time and again, it thus succeeds in finding new 'drivers' and integrating them into the development process. Within the given guidelines, this systemic process is ultimately an open one: the avoidance of micro-management and the reliance on broad framework precepts is intended to promote a diversity of approaches and emergent properties. The university's name, and the self-reinforcing process of continuous questioning triggered by the adoption of that name ('Is that really sustainable?') form the important and necessary support that renders micro-management unnecessary. Whether this continues to be effective will only become clear in the future.

Conclusion

The institutional transformation at the University for Sustainable Development was certainly tied to a number of specific starting and boundary conditions, and has so far — sometimes only just — had to negotiate some critical cliff-edges. One such tipping point was its renaming, notwithstanding the (tightly contested) decision to do so by the Senate. The overall development of the university could not be foreseen or planned in detail.

As such, the extent to which the transformation might serve as a model for other institutions is quite limited. Higher education institutions are — first and foremost — largely autonomous and democratic institutions; while the state government monitored the transformation process described above in a constructively critical manner, it never attempted to intervene. It was essential to identify and define the core objectives early on, in the form of a clear but open-ended set of guiding principles. This then made it possible to repeatedly interpret and exploit external changes — such as the Bologna Process, the Overload Programme and changes to the Higher Education Act — as windows of opportunity.

Against the background of those guiding principles, the time factor must also be borne in mind. The change process described was neither quick nor overly hasty in its implementation: it took 12 years to go from the first discussion of guiding principles in 1998 to the name change in 2010. Projects that could not be implemented quickly were picked up again when the time was right.

The most important thing in all of this is to keep the discussion going and to support those actors who are driving the process forward.

Bibliography

Acosta, A. (2017): Buen Vivir. Die Welt aus der Perspektive des Buen Vivir überdenken, in: Konzeptwerk Neue Ökonomie; DFG-Kolleg Postwachstumsgesellschaften (Eds.): Degrowth in Bewegung(en) – alternative Wege zur sozial-ökologischen Transformation, München, p. 70–83.

Aiello, L. C.; Wells, J. C. K. (2002): Energetics and the evolution of the genus Homo, in: Annual Review of Anthropology, 31, p. 323–338.

Ajzen, I. (1991): The theory of planned behavior, in: Organizational Behavior and Human Decision Processes, 50(2), p. 179–211.

Alamgir, M., et al. (2017): Economic, socio-political and environmental risks of road development in the Tropics, in: Current Biology, 27(20), R1130–R1140.

Alcántara, S., et al. (2016): Demokratietheorie und Partizipationspraxis. Analyse und Anwendungspotentiale deliberativer Verfahren. Erschienen in der Reihe: Bürgergesellschaft und Demokratie, Wiesbaden.

Alcántara, S.; Kuhn, R.; Renn, O. (2014): DELIKAT – Fachdialoge Deliberative Demokratie. Analyse Partizipativer Verfahren für den Transformationsprozess – Anhang: Verfahren in der Partizipationsmatrix, Dessau-Roßlau.

Althaus, D. (2007): Zeitenwende. Die postfossile Epoche; weiterleben auf dem Blauen Planeten. 1. Auflage, Mankau, Murnau a. Staffelsee.

Altvater, E. (2005): Das Ende des Kapitalismus, wie wir ihn kennen, Münster.

Alvaredo, F., et al. (2018): World Inequality Report [http://wir2018.wid.world/files/download/wir2018-full-report-english.pdf; 25.01.2018].

Andrae, A.; Edler, T. (2015): On global electricity usage of communication technology: Trends to 2030, in: Challenges, 6(1), p. 117–157.

Andreoni, V. (2017): Energy metabolism of 28 world countries: A multi-scale integrated analysis, in: Ecological Economics, 142, p. 56–69.

APA (American Psychological Association) (2016): Stress in America: The impact of discrimination. Stress in America™ Survey [http://www.apa.org/news/press/releases/stress/2015/impact-of-discrimination.pdf; 21.02.2018].

APA (American Psychological Association) (2017a): Stress in America: Coping with change. Stress in America™ Survey. Part 1 [http://www.apa.org/news/press/releases/stress/2016/coping-with-change.PDF; 21.02.2018].

APA (American Psychological Association) (2017b): Stress in America: Coping with change. Stress in America™ Survey. Part 2: Technology and social media [http://www.apa.org/news/press/releases/stress/2016/coping-with-change.PDF; 21.02.2018].

Arepo Consult (2017): Kurzanalyse der nationalen Treibhausgasemissionen für das Jahr 2016. Kurzstudie für die Bundestagsfraktion Bündnis 90/Die Grünen, Berlin.

Arnold, A., et al. (Ed.) (2015): Innovation – Exnovation. Über Prozesse des Abschaffens und Erneuerns in der Nachhaltigkeitstransformation, Marburg.

Arnstein, S. (1969): A ladder of citizen participation, in: Journal of the American Institute of Planners, 35(4), p. 216–224.

Atmaca, D. (2014): Genossenschaften in Zeiten raschen Wandels. Chancen einer nachhaltigen Organisationsform, in: C. Schröder; H. Walk (Eds.): Genossenschaften und Klimaschutz. Akteure für zukunftsfähige solidarische Städte, Wiesbaden, p. 49–72.

Bandura, A. (1991): Social cognitive theory of self-regulation, in: Organizational Behavior and Human Decision Processes, 50(2), p. 248–287.

Barclay, P.; Stoller, B. (2014): Local competition sparks concerns for fairness in the ultimatum game, in: Biology Letters, 10(5), 20140213.

Barne, D.; Khokhar, T. (2017): Year in review: 2017 in 12 charts [http://www.worldbank.org/en/news/feature/2017/12/15/year-in-review-2017-in-12-charts; 25. 01. 2018].

Barnosky, A. D., et al. (2004): Assessing the causes of late Pleistocene extinctions on the continents, in: Science, 306(5693), p. 70–75.

Bateson, G. (1970): Die Wurzeln ökologischer Krisen, in: G. Bateson (Ed.): Ökologie des Geistes. Anthropologische, psychologische, biologische und epistemologische Perspektiven (1985 [1972]), Frankfurt/Main, p. 627–633.

Bateson, G. (1985 [1972]): Ökologie des Geistes. Anthropologische, psychologische, biologische und epistemologische Perspektiven, Frankfurt/Main.

Beck, K.; Ziekow, J. (2011): Mehr Bürgerbeteiligung wagen, Wiesbaden.

Beckerman, W.; Hepburn, C. (2007): Ethics of the discount rate in the Stern Review on the economics of climate change, in: World Economics, 8(1), p. 187–210.

Benyus, J. M. (1997): Biomimicry: Innovation inspired by nature, New York.

Benz, A.; Dose, N. (2010): Governance – Modebegriff oder nützliches sozialwissenschaftliches Konzept?, in: A. Benz; N. Dose (Eds.): Governance – Regieren in komplexen Regelsystemen. Eine Einführung. 2., aktualisierte und veränderte Auflage, Wiesbaden, p. 13–36.

Bertalanffy, K. L. von (1928): Kritische Theorie der Formbildung, Berlin.

Bertalanffy, K. L. von (1940): Vom Molekül zur Organismenwelt, Potsdam.

Bertalanffy, K. L. von (1949): Das biologische Weltbild, Bern.

Bertalanffy, K. L. von (1950): The theory of open systems in physics and biology, in: Science, 111 (2872), p. 23–29.

Bertalanffy, K. L. von (1953): Biophysik des Fließgleichgewichts, Braunschweig.

Bertalanffy, K. L. von (1962): Modern theories of development, New York.

Bertalanffy, K. L. von (1968a): General system theory: foundations, development, applications, New York.

Bertalanffy, L. von (1968b): The organismic psychology and systems theory, Heinz Werner lectures, Worcester.

Bertanlanffy, K. L. von (1937): Das Gefüge des Lebens, Leipzig.

Berthold, C.; Meyer-Guckel, V.; Rohe, W. (Eds.) (2011): Mission Gesellschaft – Engagement und Selbstverständnis der Hochschulen. Ziele, Konzepte, internationale Praxis [https://www.stifterverband.org/mission-gesellschaft; 26. 01. 2018], Essen.

BGBl (Grundgesetz für die Bundesrepublik Deutschland in der im Bundesgesetzblatt Teil III, Gliederungsnummer 100-1, veröffentlichten bereinigten Fassung, das zuletzt durch Artikel 1 des Gesetzes vom 13. Juli 2017 [BGBl. I p. 2347] geändert wurde).

Biggs, D., et al. (2011): Are we entering an era of concatenated global crises?, in: Ecology and Society, 16(2), Art. 27.

Binswangen, M. (2015): Geld aus dem Nichts. Wie Banken Wachstum ermöglichen und Krisen verursachen, Weinheim.

BIS (Bank for International Settlements) (2004): BIZ-Quartalsbericht, Dezember 2004, Internationales Bankgeschäft und internationale Finanzmärkte [http://www.bis.org/publ/qtrpdf/r_qt0412ger.pdf; 29. 05. 2017].

BIS (Bank for International Settlements) (2008): 78. Jahresbericht [https://www.bis.org/publ/arpdf/ar2008_de.pdf; 29. 05. 2017].

Blanchard, O.; Illing, G. (2004): Makroökonomie, 3. Auflage, München.

Blohm, P. (2010): Wie die Moral zur Welt kam, in: Psychololologie heute, 37(11), p. 59–64.

BMBF (Bundesministerium für Bildung und Forschung) (2002): Bericht der Bundesregierung zur Bildung für nachhaltige Entwicklung, Bonn.

BMEL (Bundesministerium für Ernährung und Landwirtschaft) (Ed.) (2014): Der Wald in Deutschland. Ausgewählte Ergebnisse der dritten Bundeswaldinventur, 1. Auflage, Berlin.

BMELV (Bundesministerium für Ernährung, Landwirtschaft und Verbraucherschutz) (Ed.) (2011): Waldstrategie 2020: Nachhaltige Waldbewirtschaftung – eine gesellschaftliche Chance und Herausforderung, Bonn.

BMFSFJ (Bundesministerium für Familie, Senioren, Frauen und Jugend) (2017): Zweiter Bericht über die Entwicklung des bürgerschaftlichen Engagements in der Bundesrepublik Deutschland. Demografischer Wandel und bürgerschaftliches Engagement. Der Beitrag des Engagements zur lokalen Entwicklung. Bundestags-Drucksache 18/1180.

BMU (Bundesministerium für Umwelt, Naturschutz und Reaktorsicherheit) (o. J.): Konferenz der Vereinten Nationen für Umwelt und Entwicklung im Juni 1992 in Rio de Janeiro. Dokumente: Agenda 21, Bonn [http://www.bmub.bund.de/fileadmin/bmu-import/files/pdfs/allgemein/application/pdf/agenda21.pdf: 15.11.2017].

BMUB (Bundesministerium für Umwelt, Naturschutz, Bau und Reaktorsicherheit) (2015): Umweltpolitik in der DDR [http://www.umwelt-im-unterricht.de/hintergrund/umweltpolitik-in-der-ddr/; 21.08.2017].

BMUB (Bundesministerium für Umwelt, Naturschutz, Bau und Reaktorsicherheit) (2017): Die Endlagerung radioaktiver Abfälle [https://www.umwelt-im-unterricht.de/hintergrund/die-endlagerung-radioaktiver-abfaelle/; 20.02.2018].

BMUB (Bundesministerium für Umwelt, Naturschutz, Bau und Reaktorsicherheit) (2007): Nationale Strategie zur biologischen Vielfalt. Kabinettsbeschluss vom 7. November 2007, Berlin.

BMWi (Bundesministerium für Wirtschaft und Energie) (2018): Gesamtausgabe der Energiedaten – Datensammlung des BMWi. Energiedaten und -szenarien [http://www.bmwi.de/Redaktion/DE/Artikel/Energie/energiedaten-gesamtausgabe.html; 20.03.2018].

BMZ (Bundesministerium für wirtschaftliche Zusammenarbeit und Entwicklung) (2017): Die Agenda 2030 für nachhaltige Entwicklung [http://www.bmz.de/de/ministerium/ziele/2030_agenda/index.html; 08.11.2017].

BNatSchG (Bundesnaturschutzgesetz vom 29. Juli 2009 [BGBl. I p. 2542], das zuletzt durch Artikel 1 des Gesetzes vom 15. September 2017 [BGBl. I p. 3434] geändert wurde).

Boesch, C.; Hohmann, G.; Marchant, L. (2002): Behavioural diversity in Chimpanzees and Bonobos, Cambridge.

Bojanowski, A. (2014): Verwirrende Werbefloskel, in: Bundeszentrale für Politische Bildung (Ed.): Nachhaltigkeit. Aus Politik und Zeitgeschichte (APuZ), 31–32/2014 [http://www.bpb.de/apuz/188659/verwirrende-werbefloskel; 22.03.2018].

Bonjean, V., et al. (2018): Gas and galaxies in filament between clusters of galaxies: The study of A399-A401, in: Astronomy & Astrophysics, 609, Art. 49.

Bonneuil, C.; Fressoz, J.-B. (2015): The shock of the Anthropocene, London, New York.

Bottenberg, E. H. (1996): Eine Einführung in die Sozialpsychologie, geeignet für humanistische und ökologische Fragen, Regensburg.

BP (British Petroleum) (2017): BP Statistical Review of World Energy 2017, London.

BP (British Petroleum) (2018): BP Energy Outlook – Edition 2018. Statistical review of world energy 2018; Zusammenstellung von Trends und Zahlen ausgewählter Charts: 1990–2016 [https://www.bp.com/en/global/corporate/energy-economics/energy-outlook.html; 23.03.2018].

BPB (Bundeszentrale für politische Bildung) (2017): Das Bretton-Woods-System [http://www.bpb.de/politik/wirtschaft/finanzmaerkte/54851/bretton-woods-system?p=1; 29.05.2017].

Brand, K.-W. (Ed.) (2002): Politik der Nachhaltigkeit. Voraussetzungen, Probleme, Chancen – eine kritische Diskussion, Berlin.

Brand, K.-W.; Warsewa, G. (2003): Lokale Agenda 21. Zukunftsperspektiven eines neuen Politiktypus, in: GAIA Ecological Perspectives for Science and Society, 12(1), p. 15–23.

Brang, P., et al. (2014): Suitability of close-to-nature silviculture for adapting temperate European forests to climate change, in: Forestry, 87(4), p. 492–503.

Brauer, M., et al. (2016): Ambient air pollution exposure estimation for the Global Burden of Disease 2013, in: Environmental Science & Technology, 50(1), p. 79–88.

Breit, H.; Eckensberger, L. H. (1998): Moral, Alltag und Umwelt, in: G. de Haan; U. Kuckartz (Eds.): Umweltbildung und Umweltbewußtsein. Forschungsperspektiven im Kontext nachhaltiger Entwicklung, p. 69–89.

Brosnan, S. F.; de Waal, F. B. M. (2014): Evolution of responses to (un)fairness, in: Science, 346(6207), 1251776.

Brown, N. J.; Quiblier, P. (Eds.) (1994): Ethics and Agenda 21: Moral implications of a global consensus, New York.

Brümmerhoff, D.; Lützel, H. (2002): Lexikon der Volkswirtschaftlichen Gesamtrechnung, 3. Auflage, München.

Bruner, A. G., et al. (2001): Effectiveness of parks in protecting tropical biodiversity, in: Science, 291(5501), p. 125–128.

Brunnengräber, A.; Klein, A.; Walk, H. (2005): NGOs im Globalisierungsprozess. Mächtige Zwerge – umstrittene Riesen, Bonn und Wiesbaden.

BUND; MISEREOR (Eds.) (1996): Zukunftsfähiges Deutschland. Ein Beitrag zu einer global nachhaltigen Entwicklung. Studie des Wuppertal Instituts für Klima, Umwelt, Energie GmbH, Basel, Boston, Berlin.

BUND (Bund für Umwelt und Naturschutz Deutschland e.V.) (Ed.) (2009): Deutschlands Forstwirtschaft auf dem Holzweg – BUND-Schwarzbuch Wald, Berlin.

BUND (Bund für Umwelt und Naturschutz Deutschland e.V.) (2011): Zukunftsfähige Energiepolitik, BUNDpositionen Nr. 48, September 2011, Berlin.

Bundesregierung (2013): Koalitionsvertrag der Bundesregierung, 18. Legislaturperiode [https://www.bundesregierung.de/Content/DE/StatischeSeiten/Breg/koalitionsvertrag-inhaltsverzeichnis.html; 22.05.2017].

Bundesregierung (2017): Deutsche Nachhaltigkeitsstrategie. Neuauflage 2016, Berlin. [English version online at https://www.bundesregierung.de/resource/blob/997532/188836/7d1716e5d5576bec62c9d16ca908e80e/2017-06-20-nachhaltigkeit-neuauflage-engl-data.pdf?download=1].

BWaldG (Bundeswaldgesetz vom 2. Mai 1975 [BGBl. I p. 1037], das zuletzt durch Artikel 1 des Gesetzes vom 17. Januar 2017 [BGBl. I p. 75] geändert wurde).

Campact e.V. (2016): Transparenz-Bericht 2016. So wirkt das politische Engagement der Campact-Aktiven, Verden.

Carbon Brief (2015): Die Geschichte der Energiewende [https://www.carbonbrief.org/zeitliste-vergangenheit-gegenwart-zukunft-deutschen-energiewende; 01.02.2018].

Carlowitz, H. C. von (1713): Sylvicultura Oeconomica, oder Haußwirthliche Nachricht und Naturmäßige Anweisung zur Wilden Baum=Zucht, Leipzig.

Carson, R. (2007): Der stumme Frühling. 127.–130. Auflage, Nördlingen.

CBD (Convention on Biological Diversity) (2000): COP 5 Decision V/6. Ecosystem Approach, Nairobi.

Cerman, Z.; Barthlott, W.; Nieder, J. (2005): Erfindungen der Natur. Bionik – Was wir von Pflanzen und Tieren lernen können, 3. Auflage, Reinbek bei Hamburg.

Ciompi, L. (1999): Die emotionalen Grundlagen des Denkens. Entwurf einer fraktalen Affektlogik, 2. durchgesehene Auflage, Göttingen.

Clement, R.; Terlau, W.; Kiy, M. (2013): Angewandte Makroökonomie, 5. Auflage, München.

Coetzer, K. L.; Witkowski, E. T. F.; Erasmus, B. F. N. (2014): Reviewing Biosphere Reserves globally: effective conservation action or bureaucratic label?, in: Biological Reviews, 89(1), p. 82–104.

Colditz, G. (1994): Auen, Moore, Feuchtwiesen. Gefährdung und Schutz von Feuchgebieten, Basel.

Core Project (2015): The Capitalist Revolution, Unit 1 [http://www.core-econ.org; 14.02.2018].

Corneo, G. (2014): Bessere Welt. Hat der Kapitalismus ausgedient? Eine Reise durch alternative Wirtschaftssysteme, Berlin, Wien.

Costanza, R., et al. (1997): The value of the world's ecosystem services and natural capital, in: Nature, 387, p. 253–260.

Council of Europe (2017): European Landscape Convention [https://www.coe.int/en/web/landscape/home; 18.01.2018].

Cross, T. (2017): Human obsolescence, in: The World in 2018. The Economist, 144.

Crutzen, P. J. (2002): Geology of mankind, in: Nature, 415(6867), p. 23.

Dalai-Lama XIV (2015): Der Appell des Dalai-Lama an die Welt mit Franz Alt. Ethik ist wichtiger als Religion, Wals bei Salzburg.

Daly, H. E. (1974): The Economics of the Steady State, in: American Economic Review, 64(2), p. 15–21.

David, A., et al. (2017): Heat roadmap Europe: Large-scale electric heat pumps in district heating systems, in: Energies, 10(4), Art. 578.

David, M.; Leggewie, C. (2015): Kultureller Wandel in Richtung gesellschaftliche Nachhaltigkeit. Arbeitspapier, Essen.

De Florio, V. (2015): Reflections on organization, emergence, and control in socio-technical systems, in: R. MacDougall (Ed.): Communication and control: Tools, systems, and new dimensions, Lanham, p. 189–200.

DESA (United Nations, Department of Economic and Social Affairs, Population Division) (2015): World population prospects: The 2015 revision, key findings and advance tables. Working Paper No. ESA/P/WP.241, New York.

Deutsche UNESCO-Kommission (2016): UNESCO-Roadmap zur Umsetzung des Weltaktionsprogramms "Bildung für nachhaltige Entwicklung", 4. Auflage, Bonn.

Deutsche UNESCO-Kommission (2017): Die Gremien der deutschen Umsetzung [www.bne-portal.de/de/bundesweit/gremien-der-deutschen-umsetzung; 03.01.2018].

Deutscher Bundestag (Ed.) (1998): Konzept Nachhaltigkeit. Vom Leitbild zur Umsetzung. Abschlussbericht der Enquete-Kommission "Schutz des Menschen und der Umwelt" des 13. Deutschen Bundestages, Bonn.

Deutscher Bundestag (2013): Bericht der Enquete-Kommission Wachstum, Wohlstand, Lebensqualität – Wege zu nachhaltigem Wirtschaften und gesellschaftlichem Fortschritt in der Sozialen Marktwirtschaft, Bundestag Drucksache 17/13300.

Deutscher Bundestag (2016): Neuregelung beim Fracking-Einsatz. Parlamentsnachrichten (hib 385/2016) [https://www.bundestag.de/presse/hib/201606/-/429076; 15.03.2018].

Deutscher Bundestag (2017): Bericht der Bundesregierung zur Bildung für nachhaltige Entwicklung. 18. Legislaturperiode. Drucksache 18/13665, Berlin.

DGVN (Deutsche Gesellschaft für die Vereinten Nationen e.V.) (2018): Nachhaltig entwickeln. Agenda 20130 für nachhaltige Entwicklung [http://nachhaltig-entwickeln.dgvn.de/agenda-2030-sdgs/agenda-2030-fuer-nachhaltige-entwicklung/; 03.01.2018].

Diamond, J. (2006): Kollaps. Warum Gesellschaften überleben oder untergehen, Frankfurt/Main.

Diefenbacher, H.; Zieschank, R. (2011): Woran sich Wohlstand wirklich messen lässt. Alternativen zum Bruttoinlandsprodukt, München.

Diekmann, P.; Preisendörfer, P. (2001): Umweltsoziologie, Reinbek bei Hamburg.

Dijk, A. I. J. M. van, et al. (2013): The Millennium Drought in southeast Australia (2001–2009): Natural and human causes and implications for water resources, ecosystems, economy, and society, in: Water Resources Research, 49(2), p. 1040–1057.

Dohmann, J. (2016): Thermodynamik der Kälteanlagen und Wärmepumpen. Grundlagen und Anwendungen der Kältetechnik, Berlin, Heidelberg.

Dorsch, F.; Häcker, H.; Stapf, K.-H. (Eds.) (1987): Dorsch Psychologisches Wörterbuch, 11., ergänzte Auflage, Bern, Stuttgart, Toronto.

Doswald, N.; Osti, M. (2011): Ecosystem-based approaches to adaptation and mitigation – good practice examples and lessons learned in Europe. BfN-Skripten 306, Bonn.

Duden (2014): Das Herkunftswörterbuch. Etymologie der deutschen Sprache. Duden Band 7, Bibliographisches Institut Berlin, Berlin.

Dunsworth, H. M., et al. (2012): Metabolic hypothesis for human altriciality, in: Proceedings of the National Academy of Sciences, 109(38), p. 15212–15216.

Džihić, V. (2017): Neue Formen autoritärer Gouvernementalität. Serbien, Mazedonien und die Türkei als Beispiele, in: Südosteuropa Mitteilungen, 57(02/2017), p. 6–17.

Ecogood (2017): Community – Eine weltweite Community entsteht [https://www.ecogood.org/de/community/; 02.01.2018].

EEG (2012) (Gesetz für den Vorrang Erneuerbarer Energien – Erneuerbare-Energien-Gesetz (EEG), 2012, Beschluss des Deutschen Bundestages vom 30. Juni 2011 – BT-Druck. 17/6363).

EEG (2014) (Erneuerbare-Energien-Gesetz vom 21. Juli 2014 (BGBl. I p. 1066), das zuletzt durch Artikel 1 des Gesetzes vom 17. Juli 2017 (BGBl. I p. 2532) geändert wurde).

Ehrlich, P. R.; Ehrlich, A. H. (1981): Extinction: The causes and consequences of the disappearance of species, New York.

Ehrlich, P. R.; Mooney, H. A. (1983): Extinction, substitution, and ecosystem services, in: Bioscience, 33(4), p. 248–254.

Eibl-Eibesfeldt, I. (1997): Die Biologie des menschlichen Verhaltens. Grundriss der Humanethologie, 3. überarbeitete und erweiterte Auflage, Weyarn.

Eichhorn, J.; Guericke, M.; Eisenhauer, R. (2016): Waldbauliche Klimaanpassung im regionalen Fokus, München.

Ellis, E. C. (2015): Ecology in an anthropogenic biosphere, in: Ecological Monographs, 85(3), p. 287–331.

Ellis, E. C., et al. (2013): Used planet: A global history, in: Proceedings of the National Academy of Sciences, 110(20), p. 7978–7985.

Ellis, E. C.; Ramankutty, N. (2008): Putting people in the map: anthropogenic biomes of the world, in: Frontiers in Ecology and the Environment, 6(8), p. 439–447.

Embacher, S. (2009): Demokratie! Nein danke? Demokratieverdruss in Deutschland, Berlin.

Embshoff, D.; Müller-Plantenberg, C.; Giorgi, G. (2017): Solidarische Ökonomie. Initiativen, Ketten und Vernetzung zur Transformation, in: Konzeptwerk Neue Ökonomie; DFG-Kolleg Postwachstumsgesellschaften (Ed.): Degrowth in Bewegung(en) – alternative Wege zur sozial-ökologischen Transformation, München, p. 344–355.

End Ecocide on Earth (2017): Ökozid [https://www.endecocide.org/de/ecocide/; 12.03.2018].

Engagement Global (Ed.) (2016): Orientierungsrahmen für den Lernbereich Globale Entwicklung im Rahmen einer Bildung für nachhaltige Entwicklung. Ein Beitrag zum Weltaktionsprogramm "Bildung für nachhaltige Entwicklung", 2. aktualisierte und erweiterte Auflage, Bonn.

Engagement Global (Eds.) (2017): ESD Expert Net. Teaching the Sustainable Development Goals, Bonn.

Engelmann, J. M., et al. (2017): Social disappointment explains chimpanzees' behaviour in the inequity aversion task, in: Proceedings of the Royal Society B – Biological Sciences, 284(1861), 20171502.

Engl, E.; Attwell, D. (2015): Non-signalling energy use in the brain, in: The Journal of Physiology, 593(16), p. 3417–3429.

Erb, K.-H., et al. (2018): Unexpectedly large impact of forest management and grazing on global vegetation biomass, in: Nature, 553, p. 73–76.

Erben, F.; Haan, G.d. (2014): Nachhaltigkeit und politische Bildung, in: Bundeszentrale für politische Bildung (Ed.): Nachhaltigkeit. Aus Politik und Zeitgeschichte (APuZ), 31–32 [http://www.bpb.de/apuz/188665/nachhaltigkeit-und-politische-bildung?p=all; 26.01.2018].

Eser, U. (2012): Naturerfahrung, -erleben und -beziehung aus naturschutzethischer Sicht, in: N. Jung; H. Molitor; A. Schilling (Eds.): Natur, Emotion, Bildung – vergessene Leidenschaft? Zum Spannungsfeld von Naturschutz und Umweltbildung. Eberswalder Beiträge zu Bildung und Nachhaltigkeit 4, Opladen, p. 33–52.

Estrada, A., et al. (2017): Impending extinction crisis of the world's primates: Why primates matter, in: Science Advances, 3(1), e1600946.

Europäische Kommission (2013): Mitteilung der Kommission an das Europäische Parlament, den Rat, den Europäischen Wirtschafts- und Sozialausschuss und den Ausschuss der Regionen: Grüne Infrastruktur (GI) – Aufwertung des europäischen Naturkapitals. COM(2013) 249 final, 6. Mai 2013, Brüssel.

Europäische Kommission (2015a): Paket zur Energieunion. Mitteilung der Kommission an das Europäische Parlament, den Rat, den Europäischen Wirtschafts- und Sozialausschuss, den Ausschuss der Regionen und die Europäische Investitionsbank. Rahmenstrategie für eine krisenfeste Energieunion mit einer zukunftsorientierten Klimaschutzstrategie. COM(205) 80 final, 25. Februar 2015, Brüssel.

European Commission (2015b): Towards an EU research and innovation policy agenda for nature-based solutions & re-naturing cities. Final report of the Horizon 2020 expert group on "Nature-based solutions and re-naturing cities", Brussels.

European Commission (2017): Nature-based solutions [https://ec.europa.eu/research/environment/index.cfm?pg=nbs; 02.02.2018].

Evelyn, J. (1664): Sylva: or, A discourse of forest-trees, and the propagation of timber in His Majesty's dominions: as it was deliver'd in the Royal Society the XVth of October, [MDCLXII], London [https://openlibrary.org/books/OL13518723M/Sylva; 26.02.2018].

Felber, C. (2012): Gemeinwohl-Ökonomie. Eine demokratische Alternative wächst, Deutike E-Book, Wien.

Felber, C. (2014): Geld: Die neuen Spielregeln. Deuticke E-Book, Wien.

Felber, C. (2015): Change everything: creating an economy for the common good, ZED Books, London.

finanzen.net (2017): Historische Kursdaten für NASDAQ Composite Index [https://www.finanzen.net/index/Nasdaq_Composite/Historisch; 09.12.2017].

Fiorino, D. J. (2018): A good life on a finite Earth – The political economy of green growth, New York.

Fisher, B. (2008): Review and analysis of the peak oil debate. Institut of Defence Analysis, IDA Document D-3542, Alexandria.

Flaxman, S.; Goel, S.; Rao, J. M. (2016): Filter bubbles, echo chambers, and online news consumption in: Public Opinion Quarterly, 80(S1), p. 298–320.

FNR (Fachagentur Nachwachsende Rohstoffe e.V.) (2016): Basisdaten Bioenergie Deutschland 2016, Gülzow-Prüzen.

Folke, C. (2006): Resilience: The emergence of a perspective for social-ecological systems analysis, in: Global Environmental Change, 16(3), p. 253–267.

François, C. (1999): Systemics and cybernetics in a historical perspective, in: Systems Research and Behavioral Science, 16(3), p. 203–219.

Freedom House (2017): Freedom in the world 2018 [https://freedomhouse.org/report/freedom-world/freedom-world-2018; 25.01.2018].

Freudenberger, L., et al. (2012): A global map of the functionality of terrestrial ecosystems, in: Ecological Complexity, 12, p. 13–22.

Freudenberger, L., et al. (2013): Nature conservation: priority-setting needs a global change, in: Biodiversity and Conservation, 22(5), p. 1255–1281.

Fritzsch, H. (2004): Elementarteilchen: Bausteine der Materie (Beck'sche Reihe), München.

Fromm, M. (2015): Einführung in die Pädagogik. Grundfragen, Zugänge, Leistungsmöglichkeiten, Münster.

Frorip, J., et al. (2012): Energy consumption in animal production – case farm study, in: Agronomy Research Biosystem Engineering Special Issue 1, p. 39–48.

Fuhrer, U. (2010): Freizeitumwelten für Kinder und Jugendliche, in: V. Linneweber; E.-D. Lantermann; E. Kals (Eds.): Spezifische Umwelten und umweltbezogenes Handeln, Göttingen, Bern, Toronto, Seattle, p. 209–234.

Fuhrer, U.; Wölfing, S. (1997): Von den sozialen Grundlagen des Umweltbewußtseins zum verantwortlichen Umwelthandeln. Die sozialpsychologische Dimension globaler Umweltproblematik, Bern.

Fukuyama, F. (2012): The origins of political order. From prehuman times to the French revolution, London.

Furuichi, T. (2011): Female contributions to the peaceful nature of bonobo society, in: Evolutionary Anthropology, 20(4), p. 131–142.

Gabler Wirtschaftslexikon (2017): Springer Gabler Verlag (Ed.) Gabler Wirtschaftslexikon, Stichwort: Nutzen, [http://wirtschaftslexikon.gabler.de/Archiv/2440/nutzen-v10.html; 22.05.2017].

Gebhard, U. (2009): Kind und Natur. Die Bedeutung der Natur für die psychische Entwicklung, 3. überarbeitete und erweiterte Auflage, Wiesbaden.

Geels, F. W.; Schot, J. (2007): Typology of sociotechnical transition pathways, in: Research Policy, 36(3), p. 399–417.

Gehlen, A. (2004 [1950]): Der Mensch. Seine Natur und seine Stellung in der Welt, Wiebelsheim.

Generalversammlung der Vereinten Nationen (2015): Entwurf des Ergebnisdokuments des Gipfeltreffens der Vereinten Nationen zur Verabschiedung der Post-2015-Entwicklungsagenda. A/Res/69/315. Resolution der Generalversammlung, verabschiedet am 1. September 2015 [http://www.un.org/depts/german/gv-69/band3/ar69315.pdf; 02.01.2018].

Gerlach, C. (2015): Hunger in der Geschichte des 20. Jahrhunderts, in: Bundeszentrale für politische Bildung (Ed.): Aus Politik und Zeitgeschichte (APuZ), 49 [http://www.bpb.de/apuz/216231/hunger-in-der-geschichte-des-20-jahrhunderts?p=0; 15.08.2017].

Geyer, J., et al. (2017a): Assessing climate change-robustness of protected area management plans – The case of Germany, in: PLoS ONE, 12(10), e0185972.

Geyer, R.; Jambeck, J. J.; Law, K. L. (2017b): Production, use, and fate of all plastics ever made, in: Science Advances, 3(7), e1700782.

Gilbert, D. T.; Wilson, T. D. (2007): Prospection: Experiencing the Future, in: Science, 317, p. 1351–1354.

Global Footprint Network (2017): Ecological wealth of nations. Ecological footprint per capita [http://www.footprintnetwork.org/content/documents/ecological_footprint_nations/ecological_per_capita.html; 08.03.2018].

Global Water Partnership (2017): Bolivia: The water war to resist privatisation of water in Cochabamba (#157) [http://www.gwp.org/en/learn/KNOWLEDGE_RESOURCES/Case_Studies/Americas--Caribbean/bolivia-the-water-war-to-resist-privatisation-of-water-in-cochabamba-157/; 31.01.2018].

Gómez, J. M., et al. (2016): The phylogenetic roots of human lethal violence, in: Nature, 538, p. 233.

Gómez-Baggethun, E., et al. (2010): The history of ecosystem services in economic theory and practice: From early notions to markets and payment schemes, in: Ecological Economics, 69(6), p. 1209–1218.

Goodwin, N., et al. (2014): Macroeconomics in context, 2. Auflage, New York.

Göpel, M. (2016): The great mindshift. How a new economic paradigm and sustainability transformations go hand in hand, Berlin.

Gorke, M. (2000): Die ethische Dimension des Artensterbens, in: K. Ott; M. Gorke (Eds.): Spektrum der Umweltethik, Marburg, p. 81–99.

Gorke, M. (2010): Eigenwert der Natur. Ethische Begründung und Konsequenzen, Stuttgart.

Gotts, N. M. (2007): Resilience, panarchy, and world-systems analysis, in: Ecology and Society, 12(1), Art. 24.

Gramelsberger, G.; Feichter, J. (2011): Modelling the climate system: an overview, in: G. Gramelsberger; J. Feichter (Eds.): Climate Change and Policy: The Calculability of Climate Change and the Challenge of Uncertainty, Berlin, Heidelberg, p. 9–90.

Gran, C. (2017): Perspektiven einer Wirtschaft ohne Wachstum. Adaption des kanadischen Modells LowGrow an die deutsche Volkswirtschaft, Hochschulschriften Band 155, Marburg.

Graumann, C. F.; Kruse, L. (2008): Umweltpsychologie – Ort, Gegenstand, Herkünfte, Trends, in: E.-D. Lantermann; V. Linneweber (Eds.): Grundlagen, Paradigmen und Methoden der Umweltpsychologie, Bern, Toronto, Seattle, p. 3–65.

Griffin, D. R. (Ed.) (1988): Spirituality and society: Postmodern visions (Suny Series in Constructive Postmodern Thought), Albany.

Gritschneder, M. (2005): Der Einfluss der Philosophie Max Schelers auf die Logotherapie Viktor E. Frankls, in: D. Batthyány; O. Zsok (Eds.): Viktor Frankl und die Philosophie, Wien, New York, p. 109–124.

Grober, U. (2012): Sustainability: A cultural history, Totnes.

Grunwald, A.; Kopfmüller, J. (2012): Nachhaltigkeit, 2., aktualisierte Auflage, Frankfurt/Main.

GtV-Bundesverband Geothermie (2012): Hintergrundpapier zur Stimulation geothermischer Reservoire, Berlin.

Gudynas, E. (2009): Politische Ökologie. Natur in den Verfassungen von Bolivien und Ecuador, in: Juridikum, 2009(4), p. 214–219.

Gunderson, L. H.; Holling, C. S. (Eds.) (2002): Panarchy: understanding transformations in human and natural systems, Washington, D.C.

Gutzeit, W. (2006): Wirtschaftssyteme in der Entwicklung – Theorieansatz für die gesamtwirtschaftliche Organisation einer Volkswirtschaft, Volkswirtschaftliche Schriften Heft 546, Berlin.

Haan, G. de (2008): Gestaltungskompetenz als Kompetenzkonzept für Bildung für nachhaltige Entwicklung, in: I. Bormann; G. de Haan (Eds.): Kompetenzen der Bildung für nachhaltige Entwicklung. Operationalisierung, Messung, Rahmenbedingungen, Befunde, Wiesbaden, p. 23–43.

Haan, G. de; Harenberg, D. (1999): Bildung für eine nachhaltige Entwicklung: Materialien zur Bildungsplanung und zur Forschungsförderung. Bund-Länder-Kommission für Bildungsplanung und Forschungsförderung (BLK), Heft 72, Bonn.

Haberl, H., et al. (2007): Quantifying and mapping the human appropriation of net primary production in earth's terrestrial ecosystems, in: Proceedings of the National Academy of Sciences, 104(3), p. 12942–12947.

Habermas, J. (1992): Faktizität und Geltung. Beiträge zur Diskurstheorie des Rechts und des demokratischen Rechtsstaates, Frankfurt/Main.

Haderlapp, T.; Trattnigg, R. (2013): Zukunftsfähigkeit ist eine Frage der Kultur. Hemmnisse, Widersprüche und Gelingensfaktoren des kulturellen Wandels, München.

Haeckel, E. (1866): Generelle Morphologie der Organismen. Band 1: Allgemeine Anatomie der Organismen, Berlin.

Haidt, J. (2001): The emotional dog and its rational tail: A social intuitionist approach to moral judgement, in: Psychological Review, 108(4), p. 814–834.

Haila, Y. (2000): Beyond the nature-culture dualism, in: Biology and Philosophy, 15(2), p. 155–175.

Haines-Young, R.; Potschin, M. B. (2017): Common International Classification of Ecosystem Services (CICES) V5.1 and guidance on the application of the revised structure [https://cices.eu/; 18. 01. 2018].

Hallmann, C. A., et al. (2017): More than 75 percent decline over 27 years in total flying insect biomass in protected areas, in: PLoS ONE, 12(10), e0185809.

Hamann, K.; Baumann, A.; Löschinger, D. (2016): Psychologie im Umweltschutz. Handbuch zur Förderung nachhaltigen Handelns, München.

Hamilton, W. D. (1975): Innate social aptitudes of man: an approach from evolutionary genetics, in: R. Fox (Ed.): Biosocial anthropology, London, p. 133–153.

Handelsblatt (2015): Umstrittene Technik CCS. So wird die Kohle nicht sauber [http://www.handelsblatt.com/technik/zukunftderenergie/umstrittene-technik-ccs-lippenbekenntnisse-der-kohleindustrie/12687956-2.html; 20. 03. 2018], in: Handelsblatt, 07. 12. 2015.

Hare, B.; Wobber, V.; Wranghamb, R. (2012): The self-domestication hypothesis: evolution of bonobo psychology is due to selection against aggression, in: Animal Behaviour, 83(3), p. 573–585.

Hart, R. A. (1992): Children's participation. From tokenism to citizenship. UNICEF Innocenti Essays No. 4, (Ed.), Florence.

Harthan, R.; Brohmann, B.; Fritsche, U.; Grießhammer, R.; Seebach, D. (2010): Positionspapier Klimakompensation, Öko-Institut, Berlin, Freiburg, Darmstadt.

Hartig, G. L. (1804): Anweisung zur Taxation der Forsten, 2. Auflage, Gießen und Darmstadt.

Hartje, V.; Klaphake, A. (2006): Implementing the Ecosystem Approach for freshwater ecosystems – A case study on the Water Framework Directive of the European Union. BfN-Skripten 183, Bonn.

Hartmann, K. (2015): Nachhaltigkeits-Blabla, in: DER SPIEGEL Nr. 38, 12.09.2015, p. 70.

Hasenbein, S., et al. (2016): A long-term assessment of pesticide mixture effects on aquatic invertebrate communities, in: Environmental Toxicology and Chemistry, 35(1), p. 218–232.

Haskell, D. G. (2017): Der Gesang der Bäume. Die verborgenen Netzwerke der Natur, München.

Hauff, M. von; Jörg, A. (2013): Nachhaltiges Wachstum, 1. Auflage, München.

Hauff, V. (Ed.) (1987): Unsere gemeinsame Zukunft. Der Brundtland-Bericht der Weltkommission für Umwelt und Entwicklung, Greven.

Hayek, F. A. von (1979): Die drei Quellen menschlicher Werte, Tübingen.

Heinrichs, J. (2007): Öko-Logik. Geistige Wege aus der Klima- und Umweltkatastrophe, 2. überarbeitete Auflage, München.

Heiss, K. (2014): Finanz- und Realwirtschaft – Gemeinsame Zukunft oder getrennte Wege?, in: Akademisches Forum für Außenpolitik – AFA (Ed.): GAP-Journal 2013/14 [http://afa.at/gap/GAP-Journal-2013-2014.pdf; 29.05.2017], Wien, p. 105–113.

Hellbrück, J.; Kals, E. (2012): Umweltpsychologie, Wiesbaden.

Henning, H.-M.; Palzer, A. (2013): Energiesystem Deutschland 2050. Sektor- und Energieträgerübergreifende, modellbasierte, ganzheitliche Untersuchung zur langfristigen Reduktion energiebedingter CO_2-Emissionen durch Energieeffizienz und den Einsatz Erneuerbarer Energien. Eine Studie des Fraunhofer-Instituts für Solare Energiesysteme (ISE), Freiburg.

Hentig, H. von (1999): Bildung, Weinheim und Basel.

Hepach, R.; Vaish, A.; Tomasello, M. (2012): Young children are intrinsically motivated to see others helped, in: Psychological Science, 23(9), p. 967–972.

Herkner, W. (1993): Lehrbuch Sozialpsychologie, 5. korrigierte und stark erweiterte Auflage der "Einführung in die Sozialpsychologie", Bern, Stuttgart, Toronto.

Hinterberger, F., et al. (2009): Welches Wachstum ist nachhaltig? Ein Argumentarium, in: F. Hinterberger, et al. (Eds.): Welches Wachstum ist nachhaltig? Ein Argumentarium, Budapest, p. 29–94.

HNEE (Hochschule für nachhaltige Entwicklung Eberswalde) (2016a): Grundsätze zur nachhaltigen Entwicklung an der Hochschule für nachhaltige Entwicklung Eberswalde [http://www.hnee.de/_obj/8BC45754-A5A8-47A9-A6BE-24BE804CFCAC/outline/Nachhaltigkeitsgrundsaetze-2016final.pdf; 08.03.2018].

HNEE (Hochschule für nachhaltige Entwicklung Eberswalde) (2016b): Ideen- und Wissenstransfer für eine nachhaltige Entwicklung. Transferstrategie der Hochschule für nachhaltige Entwicklung Eberswalde [http://www.hnee.de/de/Forschung/Forschung-Transfer-E1033.htm; 29.01.2018].

HNEE (Hochschule für nachhaltige Entwicklung Eberswalde) (2018a): Nachhaltigkeitsbericht [http://www.hnee.de/de/nachhaltigkeit/Nachhaltigkeitsbericht; 09.02.2018].

HNEE (Hochschule für nachhaltige Entwicklung Eberswalde) (2018b): Nachhaltigkeitsmanagement [http://hnee.de/nachhaltigkeit; 09.02.2018].

Hohmann, G.; Fruth, B. (2008): New records on prey capture and meat eating by bonobos at Lui Kotale, Salonga National Park, Democratic Republic of Congo, in: Folia Primatologica, 79(2), p. 103–110.

HolacracyOne (2017): Holacracy: How it works [https://www.holacracy.org/how-it-works/; 18.12.2017].

Holling, C. S. (Ed.) (1978): Adaptive environmental assessment and management, London.

Homburg, M.; Matthies, E. (1998): Umweltpsychologie. Umweltkrise, Gesellschaft und Individuum, Weinheim, München.

Huber, J. (1995): Nachhaltige Entwicklung durch Suffizienz, Effizienz und Konsistenz, in: P. Fritz; J. Huber; H. W. Levi (Eds.): Nachhaltigkeit in naturwissenschaftlicher und sozialwissenschaftlicher Perspektive, Stuttgart, p. 31–46.

Hublin, J.-J., et al. (2017): New fossils from Jebel Irhoud, Morocco and the pan-African origin of Homo sapiens, in: Nature, 546(7657), p. 289–292.

Hunecke, M. (2013): Psychologie der Nachhaltigkeit. Psychische Ressourcen für Postwachstumsgesellschaften, München.

Ibisch, P. L. (2015): Bürgerbeteiligung im Ökosystemmanagement, in: J. Sommer (Ed.): Kursbuch Bürgerbeteiligung, Berlin, p. 245–268.

Ibisch, P. L. (2016a): Karbonisierung der Weltumweltpolitik oder ökosystembasierte Nachhaltigkeit?, in: J. Sommer; M. Müller (Eds.): Unter 2 Grad? Was der Weltklimavertrag wirklich bringt, Stuttgart, p. 89–103.

Ibisch, P. L. (2016b): Ökologie und Politik. Die Wachstumskrise entfaltet sich weiterhin, in: M. Richter; I. Thunecke (Eds.): Paradies now. André Gorz – Utopie als Lebensentwurf und Gesellschaftskritik. Sammlung Kritisches Wissen Band 70, Mössingen-Talheim, p. 83–102.

Ibisch, P. L.; Hobson, P. R.; Vega E., A. (2010a): The integrated anthroposystem: globalizing human evolution and development within the global ecosystem, in: P. L. Ibisch; A. Vega E.; T. M. Herrmann (Eds.): Interdependence of biodiversity and development under global change. CBD Technical Series Nr. 54., Montreal, p. 149–183.

Ibisch, P. L.; Hobson, P. R.; Vega E., A. (2010b): Mutual mainstreaming of biodiversity conservation and human development: towards a more radical ecosystem approach, in: P. L. Ibisch; A. Vega E.; T. M. Herrmann (Eds.): Interdependence of biodiversity and development under global change. CBD Technical Series Nr. 54., Montreal, p. 15–34.

Ibisch, P. L., et al. (2016): A global map of roadless areas and their conservation status, in: Science, 354(6318), p. 1423–1427.

Ibisch, P. L.; Kreft, S.; Luthardt, V. (Eds.) (2012): Regionale Anpassung des Naturschutzes an den Klimawandel. Strategien und methodische Ansätze zur Erhaltung der Biodiversität und Ökosystemdienstleistungen in Brandenburg, Eberswalde.

ICIDI (Independent Commission on International Development Issues) (1980): Das Überleben sichern [North-South. A programme for survival, dt.]. Gemeinsame Interessen der Industrie- und Entwicklungsländer. Bericht der Nord-Süd-Kommission, Köln.

IfD Allensbach (Institut für Demoskopie Allensbach) (2013): Motive des bürgerschaftlichen Engagements. Ergebnisse einer bevölkerungsrepräsentativen Befragung. Untersuchung im Auftrag des Bundesministeriums für Familie, Senioren, Frauen und Jugend, Allensbach.

IFPRI (International Food Policy Research Institute) (2017): 2017 Global food policy report, Washington, D.C.

Ilten, C. (2009): Strategisches und Soziales Nischenmanagement. Zur Analyse gesellschaftspolitisch motivierter Innovation, Wiesbaden.

Institut für ökologische Wirtschaftsforschung (o. J.): Toolbox für Unternehmen, Zivilgesellschaft und Politik [http://www.partizipativ-innovativ.de/; 01. 06. 2017].

Issing, O. (1993): Einführung in die Geldtheorie, 9. Auflage, München.

Jank, W.; Meyer, H. (2005): Didaktische Modelle, 7. Auflage, Berlin.

Jasanoff, S. (2006): Ordering knowledge, ordering society, in: S. Jasanoff (Ed.): States of Knowledge – The co-production of science and social order, New York, p. 13–45.

Jensen, K.; Call, J.; Tomasello, M. (2007): Chimpanzees are rational maximizers in an ultimatum game, in: Science, 318(5847), p. 107–109.

Jhingan, M. L.; Girija, M.; Sasikala, L. (2012): History of economic thought, 3rd edition, Delhi.

Joas, H. (1999): Die Entstehung der Werte, Frankfurt/Main.

Joas, H. (2006): Wie entstehen Werte? Wertebildung und Wertevermittlung in pluralistischen Gesellschaften. Gedruckter Vortrag, in: tv impuls: Gute Werte, schlechte Werte. Gesellschaftliche Ethik und die Rolle der Medien. Freiwillige Selbstkontrolle Fernsehen e.V., Berlin, 15.9.06 [http://fsf.de/data/hefte/pdf/Veranstaltungen/tv_impuls/2006_Ethik/Vortrag_Joas_authorisiert_061017.pdf; 06.12.2017].

Jochem, D., et al. (2015): Der Holzeinschlag – eine Neuberechnung: Ergebnisse der verwendungsseitigen Abschätzung des Holzeinschlags in Deutschland für 1995 bis 2013, in: Holz-Zentralblatt, 141(30), p.752–753.

Jonas, H. (1984): The imperative of responsibility: in search of an ethics for the technological age, Chicago & London.

Jørgensen, S. E. (2007): An integrated ecosystem theory, in: Annals of the European Academy of Sciences (2006/07), p.19–33.

Jørgensen, S. E.; Fath, B. D. (2004): Application of thermodynamic principles in ecology, in: Ecological Complexity, 1(4), p.267–280.

Jørgensen, S. E.; Fath, B. D. (2006): Examination of ecological networks, in: Ecological Modelling, 196(3), p.283–288.

Jørgensen, S. E.; Müller, F. (Eds.) (2000): Handbook of ecosystem theories and management, Boca Raton, London, New York, Washington, D.C.

Jørgensen, S. E.; Nielsen, S. N.; Fath, B. D. (2016): Recent progress in systems ecology, in: Ecological Modelling, 319, p.112–118.

Juffe-Bignoli, D., et al. (2014): Protected Planet Report 2014, Cambridge.

Jung, N. (2006): Steine und Brücken auf dem Weg zu ganzheitlicher, nachhaltiger Umweltbildung, in: B. Hiller; M. Lange (Eds.): Bildung für nachhaltige Entwicklung – Perspektiven für die Umweltbildung. Vorträge und Studien, Heft 16, Münster, p.179–192.

Jung, N. (2012): Natur und Entstehung von Werten, in: N. Jung; H. Molitor; A. Schilling (Eds.): Auf dem Weg zu gutem Leben. Die Bedeutung der Natur für seelische Gesundheit und Werteentwicklung. Eberswalder Beiträge zu Bildung und Nachhaltigkeit 2, Opladen, p.113–136.

Jung, N. (2015): Reichen kluge Argumente für kluges Handeln?, in: U. Eser; H. Seyfang; A. Müller (Eds.): Klugheit, Glück, Gerechtigkeit – Warum Ethik für die konkrete Naturschutzarbeit wichtig ist. BfN-Skripten 414, Bonn, p.53–70.

Kaiser, I., et al. (2012): Theft in an ultimatum game: chimpanzees and bonobos are insensitive to unfairness, in: Biology Letters, 8(6), p.942–945.

Kant, I. (2011 [1788]): Kritik der praktischen Vernunft, Köln.

Kay, J. J. (2008): Framing the situation, in: D. Waltner-Toews (Ed.): The ecosystem approach, New York, p.15–34.

Kay, J. J.; Boyle, M. (2008): Self-Organizing, Holarchic, Open Systems (SOHOs), in: D. Waltner-Toews (Ed.): The ecosystem approach, New York, p.51–78.

Kay, J. J.; Schneider, E. D. (1992): Thermodynamics and measures of ecosystem integrity, in: D. H. McKenzie; D. E. Hyatt; V. J. McDonals (Eds.): Ecological Indicators, Volume 1: Proceedings of the international Symposium on Ecological Indicators, Fort Launderdale.

Kern, K.; Knoll, C.; Schophaus, M. (2002): Die lokale Agenda 21. Ein innerdeutscher und internationaler Vergleich, in: Forschungsjournal Neue Soziale Bewegungen, 15(4), p.40–48.

Kerr, R. A. (2011): Peak oil production may already be here, in: Science, 331(6024), p.1510–1511.

Kindleberger, C. P.; Aliber, R. Z. (2011): Manias, panics and crashes – A history of financial crises, 6. Auflage, Hampshire.

Kivinen, O.; Piiroinen, T. (2018): The evolution of Homo discens: natural selection and human learning, in: Journal for the Theory of Social Behaviour, 48, p.117–133.

Klafki, W. (1991): Neue Studien zur Bildungstheorie und Didaktik. Zeitgemäße Allgemeinbildung und kritisch-konstruktive Didaktik, 2. erweiterte Auflage, Weinheim.

Kneer, R., et al. (2006): Entwicklung eines CO_2-emissionsfreien Kohleverbrennungsprozesses zur Stromerzeugung. Kongressbeitrag zum 22. Deutschen Flammentag, Verbrennung und Feuerungen, VDI-Gesellschaft Energietechnik (VDI-GET), Braunschweig, 21.–22. September 2005, in: VDI-Berichte, 1888, p.5–15.

Knoke, T., et al. (2008): Admixing broadleaved to coniferous tree species: a review on yield, ecological stability and economics, in: European Journal of Forest Research, 127(2), p. 89–101.

Knorr, K., et al. (Eds.) (2017): Energiewirtschaftliche Bedeutung der Offshore-Windenergie für die Energiewende, Update 2017. Eine Studie des Fraunhofer-Instituts für Windenergie und Energiesystemtechnik (IWES) im Auftrag der Stiftung Offshore Windenergie, Varel, Berlin.

Knutsson, B. (2009): The intellectual history of development – Towards a widening potential repertoire. Perpectives no. 13. School of Global Studies (Centre for African Studies) at Göteborg University, Göteborg.

Kocka, J. (2003): Zivilgesellschaft in historischer Perspektive, in: Forschungsjournal Neue Soziale Bewegungen, 16(2), p. 29–37.

Koestler, A. (1967): The ghost in the machine, London.

Koestler, A. (1970): Beyond atomism and holism – the concept of the holon, in: Perspectives in Biology and Medicine, 13(2), p. 131–154.

Koestler, A.; Smythies, J. R. (Eds.) (1969): Beyond reductionism: New perspectives in the life sciences, London.

Kooiman, J. (2003): Governing as governance, London.

Kordkheili, P. Q., et al. (2013): Energy input-output and economic analysis for soybean production in Mazandaran province of Iran, in: Elixir Agriculture, 56, p. 13246–13251.

Koska, I. (2001): Ökohydrologische Kennzeichnung. Standortskundliche Kennzeichnung und Bioindikation, in: M. Succow; H. Joosten (Eds.): Landschaftsökologische Moorkunde, Stuttgart, p. 92–111.

Kösters, W. (1993): Ökologische Zivilisierung. Verhalten in der Umweltkrise, Darmstadt.

Krausmann, F., et al. (2013): Global human appropriation of net primary production doubled in the 20th century, in: Proceedings of the National Academy of Sciences, 110(25), p. 10324–10329.

Kreft, S., et al. (2012): Warum gibt es Naturschutz?, in: P. L. Ibisch; S. Kreft; V. Luthardt (Eds.): Regionale Anpassung des Naturschutzes an den Klimawandel. Strategien und methodische Ansätze zur Erhaltung der Biodiversität und Ökosystemdienstleistungen in Brandenburg, Eberswalde, p. 32–37.

Krippner, R. (Ed.) (2016): Gebäudeintegrierte Solartechnik. Photovoltaik und Solarthermie – Schlüsseltechnologien für das zukunftsfähige Bauen. DETAIL Green Books, München.

Krol, G.-J. (2000): Ökonomische Verhaltenstheorie, in: H. May (Ed.): Handbuch zur ökonomischen Bildung, 4. unwesentlich veränderte Auflage, München, Wien.

Kross, E., et al. (2013): Facebook use predicts declines in subjective well-being in young adults, in: PLoS ONE, 8(8), e69841.

Krupenye, C.; Hare, B. (2018): Bonobos prefer individuals that hinder others over those that help, in: Current Biology, 28(2), p. 280–286.

Kuckshinrichs, W.; Hake, J.-F. (Eds.) (2012): CO_2-Abscheidung, -Speicherung und -Nutzung. Technische, wirtschaftliche, umweltseitige und gesellschaftliche Perspektive. Advances in Systems Analysis 2. Schriften des Forschungszentrums Jülich, Reihe Energie & Umwelt Band 164 [http://juser.fz-juelich.de/record/131987/files/FZJ-131987.pdf; 20. 03. 2018], Jülich.

Küppers, E. W. U. (2015): Systemische Bionik. Impulse für eine nachhaltige gesellschaftliche Weiterentwicklung, Wiesbaden.

Kurz, H. D. (2017): Geschichte des ökonomischen Denkens, 2. Auflage, München.

Landesbetrieb Hessen Forst (2016): Hessische Waldbaufibel – Grundsätze und Leitlinien zur naturnahen Wirtschaftsweise im hessischen Staatswald, Kassel.

Laszlo, A.; Krippner, S. (1998): Systems theories. Their origins, foundations, and development, in: J. S. Jordan (Ed.): Systems theories and a priori aspects of perception, Amsterdam, p. 47–74.

Laszlo, E. (1975): The meaning and significance of general sytem theory, in: Behavioral Science, 20(1), p. 9-24.

Le Monde diplomatique (Ed.) (2012): Atlas der Globalisierung. Die Welt von morgen, Berlin.

Lebensministerium (2012): Zukunftsdossier – Alternative Wirtschafts- und Gesellschaftskonzepte, in Österreichisches Bundesministerium für Land- und Forstwirtschaft, Umwelt und Wasserwirtschaft (Lebensministerium) (Ed.): Reihe Zukunftsdossier Nr. 3, Wien.

Lechner, A. M.; Chan, F. K. S.; Campos-Arceiz, A. (2018): Biodiversity conservation should be a core value of China's Belt and Road Initiative, in: Nature Ecology & Evolution, 2(3), p. 408–409.

Lehmann, A. (2010): Der deutsche Wald, in: Depenheuer, O.; Möhring, B. (Eds.): Waldeigentum – Dimensionen und Perspektiven. Bibliothek des Eigentums Band 8 (im Auftrag der Deutschen Stiftung Eigentum), Berlin, Heidelberg.

Lehner, M. (2009): Allgemeine Didaktik. UTB-basics, 3245, Bern.

Leopold, A. (1949): A Sand County Almanac, New York.

Leschke, M. (2015): Alternativen zur Marktwirtschaft. Ein kritischer Blick auf die Ansätze von Niko Paech und Christian Felber aus Sicht der konstitutionellen Ökonomie, in: Beiträge zur Jahrestagung des Ausschusses für Wirtschaftssysteme und Institutionenökonomik im Verein für Socialpolitik: "Marktwirtschaft im Lichte möglicher Alternativen", September 2015, Bayreuth [http://hdl.handle.net/10419/140887; 08.03.2018].

Lexikon der Nachhaltigkeit (2015): Schutz des Menschen und der Umwelt. Ziele und Rahmenbedingungen, in: IHK Nürnberg für Mittelfranken (Ed.): Lexikon der Nachhaltigkeit [https://www.nachhaltigkeit.info/artikel/13_bt_ek_mensch_umwelt_664.htm; 08.03.2018].

Li, H.; Sakamoto, Y. (2014): Social impacts in social media: An examination of perceived truthfulness and sharing of information, in: Computers in Human Behavior, 41, p. 278–287.

Liang, J., et al. (2016): Positive biodiversity-productivity relationship predominant in global forests, in: Science, 354(6309), aaf8957.

Libreria Editrice Vaticana (Ed.) (2015): Enzyklia Laudato Si von Papst Franziskus über die Sorge für das gemeinsame Haus. Herausgegeben vom Sekretariat der Deutschen Bischofskonferenz in der Reihe: Verlautbarungen des Apostolischen Stuhls Nr. 202, Bonn.

Lietaer, B., et al. (2013): Geld und Nachhaltigkeit. Von einem überholten Finanzsystem zu einem monetären Ökosystem, E-Book-Version, Berlin.

Lima Action Plan as endorsed by the 4th World Congress of Biosphere Reserves on 17 March 2016, and as adopted by the 28th MAB ICC on 19 March 2016, Lima, Peru.

Loorbach, D. (2010): Transition management for sustainable development: A prescriptive, complexity-based governance framework, in: Governance, 23(1), p. 161–183.

Lorenz, K. (1973): Die acht Todsünden der zivilisierten Menschheit, München.

Lorenzen, E. D., et al. (2011): Species-specific responses of Late Quaternary megafauna to climate and humans, in: Nature, 479(7373), p. 359–364.

Loske, R. (2016): Politik der Zukunftsfähigkeit. Konturen einer Nachhaltigkeitswende, Frankfurt/Main.

Loske, R.; Bleischwitz, R. (1997): Zukunftsfähiges Deutschland. Ein Beitrag zu einer global nachhaltigen Entwicklung, Basel.

Luhmann, N. (1986): Ökologische Kommunikation, 1. Auflage, Opladen.

Luhmann, N. (1987): Soziale Systeme. Grundriß einer allgemeinen Theorie, Frankfurt/Main.

Luthardt, V.; Wichmann, S. (2016): Ökosystemdienstleistungen von Mooren, in: W. Wichtmann; C. Schröder; H. Joosten (Eds.): Paludikultur – Bewirtschaftung nasser Moore für regionale Wertschöpfung, Klimaschutz und Biodiversität, Stuttgart, p. 15–20.

Malik, F. (2006): Strategie des Managements komplexer Systeme, 9. Auflage, Bern, Stuttgart, Wien.

Maloney, M. P.; Ward, M. P. (1973): Ecology: let's hear from the people. An objective scale for the measurement of ecological attitudes and knowledge, in: American Psychologist, 28(7), p. 583–586.

Malthus, T. R. (1798): An essay on the principle of population, London.

Martin, W.; Russell, M. J. (2007): On the origin of biochemistry at an alkaline hydrothermal vent, in: Philosophical Transactions of the Royal Society B – Biological Sciences, 362(1486), p. 1887–1925.

Marx, K. (1844): Ökonomisch-philosophische Manuskripte.

Marx, K.; Engels, F. (1964): Werke. Herausgegeben vom Institut für Marxismus-Leninismus beim ZK der SED, 43 Bände, Band 25, Berlin.

Maschkowski, G., et al. (2017): Transition-Initiativen. Vom Träumen, Planen, Machen und Feiern des Wandels, den wir selbst gestalten, in: Konzeptwerk Neue Ökonomie; DFG-Kolleg Postwachstumsgesellschaften (Ed.): Degrowth in Bewegung(en) – alternative Wege zur sozial-ökologischen Transformation, München, p. 368–379.

Maslow, A. H. (1943): A theory of human motivation. Psychological Review, 50(4), p. 370–396.

Maslow, A. H. (2016 [1954]): Motivation und Persönlichkeit, 14. Auflage, Reinbek.

Mathar, R. (2014): Der "Whole Institution Approach". Zentrale Elemente der Gestaltung des Weltaktionsprogramms Bildung für nachhaltige Entwicklung. Transformation und Neuorientierung von Bildung – Workshop auf der Nationalen Konferenz zum Abschluss der UN-Dekade, 29.–30. September 2014, Bonn [http://www.bnekonferenz2014.de/fileadmin/bnekonferenz/Dateien/Dokumente/Der_whole_institution_approach.pdf; 04.01.2018].

McCloskey, D. N. (2013): Tunzelmann, Schumpeter, and the Hockey Stick, in: Research Policy, 42(10), p. 1706–1715.

McHenry, H. M. (2004): Origin of human bipedality, in: Evolutionary Anthropology, 13(3), p. 116–119.

Meadows, D. (2010): Die Grenzen des Denkens – wie wir sie mit System erkennen und überwinden können, neu bearb. von Diana Wright unter Mitwirkung von Stephanie Weis-Gerhardt, München.

Meadows, D., et al. (1972): Die Grenzen des Wachstums. Bericht des Club of Rome zur Lage der Menschheit, Stuttgart.

Meadows, D.; Seiler, T. (2005): Jenseits der ökologischen Grenzen gibt es keine nachhaltige Entwicklung [http://www.umweltethik.at/wp/wp-content/uploads/MeadowsSeiler Gespraech.pdf; 06.12.17], in: Natur und Kultur: transdisziplinäre Zeitschrift für ökologische Nachhaltigkeit, 6(2), p. 19–37.

Meyer-Abich, K. (2012): Was hindert uns daran, nachhaltig zu wirtschaften?, in: N. Jung; H. Molitor; A. Schilling (Eds.): Auf dem Weg zu gutem Leben. Die Bedeutung der Natur für seelische Gesundheit und Werteentwicklung. Eberswalder Beiträge zu Bildung und Nachhaltigkeit 2, Opladen, p. 93–112.

Meyer-Abich, K. M. (1987): Naturphilosophie auf neuen Wegen, in: O. Schwemmer (Ed.): Über Natur. Philosophische Beiträge zum Naturverständnis, Frankfurt/Main, p. 63–74.

Michelsen, G. (2009): Kompetenzen und Bildung für nachhaltige Entwicklung, in: B. Overwien; H.-F. Rathenow (Eds.): Globalisierung fordert politische Bildung. Politisches Lernen im globalen Kontext, Opladen, Farmington Hills, p. 75–86.

Michelsen, G.; Siebert, H.; Lilje, J. (2011): Nachhaltigkeit lernen – Ein Lesebuch, Bad Homburg.

Miersch, M. (2009): Warum der Kapitalismus beibehalten werden muss [https://www.welt.de/wirtschaft/article4017664/Warum-der-Kapitalismus-beibehalten-werden-muss.html; 28.01.2018], in: Welt, 28.06.2009.

Miller, S. A.; Horvath, A.; Monteiro, P. J. M. (2018): Impacts of booming concrete production on water resources worldwide, in: Nature Sustainability, 1(1), p. 69–76.

Mohammed, M.; Thomas, K. (2014): Enabling community and trust: shared leadership for collective creativity, in: The Foundation Review, 6(4), p. 96–105.

Mohrs, T. (2002): Unfit für Nachhaltigkeit? "Bildung für nachhaltige Entwicklung" und die "Erblast der Gene", in: A. Beyer (Ed.): Fit für Nachhaltigkeit? Biologisch-anthropologische Grundlagen einer Bildung für nachhaltige Entwicklung, Opladen, p. 47–68.

Morriss-Kay, G. M. (2010): The evolution of human artistic creativity, in: Journal of Anatomy, 216(2), p. 158–176.

MUGV (Ministerium für Umwelt, Gesundheit und Verbraucherschutz des Landes Brandenburg) (Ed.) (2014): Nachhaltigkeitsstrategie des Landes Brandenburg. Natürlich. Nachhaltig, Brandenburg, Potsdam.

Müller-Christ, G. (2014): Nachhaltiges Management. Eine Einführung in Ressourcenorientierung und widersprüchliche Managementrationalitäten, 2., überarbeitete und erweiterte Auflage, Baden-Baden.

Müller, G. (1977): Zur Geschichte des Wortes Landschaft, in: A. von Wallthor; H. Quirin (Eds.): "Landschaft" als interdisziplinäres Forschungsproblem. Vorträge und Diskussionen des Kolloquiums am 7./8. November 1975 in Münster, Münster, p. 4–13.

Müller, R. A. E.; Clasen, M. (2008): Der Stern-Review. Klimapolitische Beratung im Grenzbereich von Ethik und Ökonomie, Kiel.

Münk, H. J. (2006): Von der Umweltproblematik zur nachhaltigen Entwicklung. Das Schöpfungsargument in der Umweltethik-Diskussion der (deutsch-sprachigen) Katholisch-Theologischen Ethik, in: M. Durst; H. J. Münk (Eds.): Schöpfung, Theologie und Wissenschaft. Theologische Berichte, Band 29, Freiburg, p. 115–194.

Murphy, J.; Roser, M. (2017): Internet. Published online by OurWorldInData.org [https://ourworldindata.org/internet; 15. 01. 2018].

NABU (Naturschutzbund Deutschland e.V.) (2010): Natürliche Waldentwicklung bis 2020 – Positionspapier [https://www.nabu.de/imperia/md/content/nabude/wald/nabu-positionspapier_nat__rliche_waldentwicklung_bis_2020.pdf; 29. 08. 2017].

Nanz, P.; Fritsche, M. (2012): Handbuch Bürgerbeteiligung. Bundeszentrale für politische Bildung. Schriftenreihe Band 1200, Bonn.

Nanz, P.; Leggewie, C. (2016): Die Konsultative. Mehr Demokratie durch Bürgerbeteiligung, Berlin.

Nationale Plattform Bildung für nachhaltige Entwicklung (2017): Nationaler Aktionsplan Bildung für nachhaltige Entwicklung. Der deutsche Beitrag zum UNESCO-Weltaktionsprogramm, Berlin.

Naturkapital Deutschland – TEEB DE (2015): Naturkapital und Klimapolitik – Synergien und Konflikte. Ed. von V. Hartje, H. Wüstemann und A. Bonn. Technische Universität Berlin, Helmholtz-Zentrum für Umweltforschung – UFZ, Berlin, Leipzig.

Netzwerk Plurale Ökonomik (2018): Ziele des Netzwerks [https://www.plurale-oekonomik.de/das-netzwerk/ziele-und-aktivitaeten/; 02. 01. 2018].

Neuberger, D. (1997): Mikroökonomie der Bank – eine industrieökonomische Perspektive, München.

Nicolis, G.; Prigogine, I. (1977): Self-organization in nonequilibrium systems. From dissipative structures to order through fluctuations, New York, London, Sydney, Toronto.

NOAA (National Oceanic and Atmospheric Administration) (2018): Ecosystem-Based Management [http://ecosystems.noaa.gov/Home.aspx; 02. 02. 2018].

NOAA National Centers for Environmental Information (2018): State of the Climate: Global Climate Report for Annual 2017 [https://www.ncdc.noaa.gov/sotc/global/201713; 25. 01. 2018].

Nussbaum, M. (2014 [1999]): Gerechtigkeit oder Das gute Leben, Frankfurt/Main.

Obergassel, W., et al. (2015): Phoenix from the ashes – an analysis of the Paris Agreement to the United Nations Framework Convention on Climate Change – Part I, in: Environmental Law & Management, 27(6), p. 243–262.

Obergassel, W., et al. (2016): Phoenix from the ashes – an analysis of the Paris Agreement to the United Nations Framework Convention on Climate Change – Part II, in: Environmental Law & Management, 28(1), p. 3–12.

Odum, E. P. (1964): The new ecology, in: Bioscience, 14(7), p. 14–16.

Odum, E. P. (1968): Energy flow in ecosystems: A historical review, in: American Zoologist, 8(1), p. 11–18.

OECD (Organisation für wirtschaftliche Zusammenarbeit und Entwicklung) (2016): Stromerzeugung, in: OECD (Ed.): Die OECD in Zahlen und Fakten 2015–2016: Wirtschaft, Umwelt, Gesellschaft, Paris, p. 102–103.

OECD (Organisation für wirtschaftliche Zusammenarbeit und Entwicklung) (2017a): OECD Data: Gross Domestic Product (GDP) [https://data.oecd.org/gdp/gross-domestic-product-gdp.htm; 22.05.2017].

OECD (Organisation für wirtschaftliche Zusammenarbeit und Entwicklung) (2017b): OECD Data: Household financial assets [https://data.oecd.org/hha/household-financial-assets.htm; 29.05.2017].

Olk, T.; Klein, A.; Hartnuß, B. (2010): Engagementpolitik. Die Entwicklung der Zivilgesellschaft als politische Aufgabe, Wiesbaden.

Olson, D. M., et al. (2001): Terrestrial ecoregions of the world: a new map of life on Earth, in: Bioscience, 51(11), p. 933–938.

Onishi, N.; Senguptajan, S. (2018): Dangerously low on water, Cape Town now faces 'Day Zero' [https://www.nytimes.com/2018/01/30/world/africa/cape-town-day-zero.html; 31.01.2018], in: The New York Times, 30.01.2018.

Ott, H. E. (2013): Theoretischer Nachhaltigkeitsdiskurs trifft Politik. Ein Erfahrungsbericht, in: H. Rogall, et al. (Eds.): 3. Jahrbuch Nachhaltige Ökonomie 2013/2014. Im Brennpunkt: Nachhaltigkeitsmanagement, Marburg, p. 115–134.

Ott, K.; Döring, R. (2001): Nachhaltigkeitskonzepte, in: Zeitschrift für Wirtschafts- und Unternehmensethik (zfwu), 2(3), p. 315–339.

Overwien, B.; Rathenow, H.-F. (Eds.) (2009): Globalisierung fordert politische Bildung. Politisches Lernen im globalen Kontext, Opladen, Farmington Hills.

Paech, N. (2011): Befreiung vom Überfluss. Auf dem Weg in die Postwachstumsökonomie, München.

Paech, N. (2015): Gesellschaft an der Wachstumswende. Vom Wachstumsdogma zur Postwachstumsökonomie, in: S. Elsen, et al. (Eds.): Die Kunst des Wandels, München, p. 25–44.

Pagel, L. (2013): Information ist Energie. Definition eines physikalisch begründeten Informationsbegriffs, Berlin, Heidelberg, New York.

Partzsch, L. (2015): Kein Wandel ohne Macht. Nachhaltigkeitsforschung braucht ein mehrdimensionales Machtverständnis, in: GAIA, 24(1), p. 48–56.

Patten, B. C. (2013): Systems ecology and environmentalism: Getting the science right, in: Ecological Engineering, 61(Part A), p. 446–455.

Patten, B. C. (2014): Systems ecology and environmentalism: getting the science right. Part I: Facets for a more holistic Nature Book of ecology, in: Ecological Modelling, 293(10), p. 4–21.

Pauli, G. (2017): The Blue Economy 3.0: The marriage of science, innovation and entrepreneurship creates a new business model that transforms society.

Pecl, G. T., et al. (2017): Biodiversity redistribution under climate change: Impacts on ecosystems and human well-being, in: Science, 355(6332), eaai9214.

Pfeiffer, D. A. (2006): Eating fossil fuels: Oil, food, and the coming crisis in agriculture, Gabriola.

Pfeil, W. (Ed.) (1822–1859): Kritische Blätter für Forst- und Jagdwissenschaft in Verbindung mit mehreren Forstmännern und Gelehrten, Leipzig.

Piantadosi, S. T.; Kidd, C. (2016): Extraordinary intelligence and the care of infants, in: Proceedings of the National Academy of Sciences of the United States of America, 113(25), p. 6874–6879.

Pimentel, D. (2009): Energy inputs in food crop production in developing and developed nations, in: Energies, 2(1), p. 1–24.

Pindyck, R. S.; Rubinfeld, D. L. (2015): Mikroökonomie, 8. Auflage, Hallbergmoos.

Plumpe, W. (2010): Wirtschaftskrisen. Geschichte und Gegenwart, München.

Pohl, H.-J. (1970): Kritik der Drei-Sektoren-Theorie, in: Mitteilungen aus der Arbeitsmarkt- und Berufsforschung, 4, p. 313–325.

Polanyi, K. (2001 [1944]): The Great Transformation, Boston.

Potapov, P., et al. (2017): The last frontiers of wilderness: Tracking loss of intact forest landscapes from 2000 to 2013, in: Science Advances, 3(1), e1600821.

Potts, R. (2013): Hominin evolution in settings of strong environmental variability, in: Quaternary Science Reviews, 73, p. 1–13.

Prammer, H. K. (2009): Integriertes Umweltkostenmanagement. Bezugsrahmen und Konzeption für eine ökologisch-nachhaltige Unternehmensführung, Wiesbaden.

Preisendörfer, P. (2000): Umwelteinstellungen und Umweltverhalten in Deutschland. Empirische Befunde und Analysen auf der Grundlage der Bevölkerungsumfragen "Umweltbewußtsein in Deutschland 1991–1998". Herausgegeben vom Umweltbundesamt, Opladen.

Pretzsch, H. (2003): Diversität und Produktivität von Wäldern, in: Allgemeine Forst- und Jagdzeitung, 174(5-6), p. 88–98.

Prognos AG; Öko-Institut; Ziesing, H.-J. (2017): Modell Deutschland – Klimaschutz bis 2050. Vom Ziel her denken. Eine Studie im Auftrag des WWF Deutschland [http://www.wwf.de/themen-projekte/klima-energie/modell-deutschland/klimaschutz-2050/; 21. 03. 2018].

Programm Transfer-21 (FU Berlin) (o.J.): Gestaltungskompetenz. Lernen für die Zukunft – Definition von Gestaltungskompetenz und ihrer Teilkompetenzen [www.institutfutur.de/transfer-21/index.php?p=222; 16. 03. 2018].

Projektwerkstättenrat an der HNEE (2016): Satzung des Projektwerkstättenrat vom 7. Januar 2016 [https://www.hnee.de/de/Studium/Infos-zum-Studium/Fuer-Studierende/Projektwerkstaetten/Projektwerkstaetten-Von-Studis-fuer-Studis-E8439.htm; 26. 01. 2018].

Prüfer, K., et al. (2012): The bonobo genome compared with the chimpanzee and human genomes, in: Nature, 486(7404), p. 527–531.

Quaschning, V. (2015): Regenerative Energiesysteme. Technologie – Berechnung – Simulation, 9. aktualisierte und erweiterte Auflage, München.

Randers, J.; Maxton, G. (2016): Ein Prozent ist genug. Mit wenig Wachstum soziale Ungleichheit, Arbeitslosigkeit und Klimawandel bekämpfen, München.

Rat der Europäischen Union (Ed.) (2006): Die neue EU-Strategie für nachhaltige Entwicklung, 10117/06, Brüssel.

Reif, A., et al. (2010): Waldbewirtschaftung in Zeiten des Klimawandels. Synergien und Konfliktpotenziale zwischen Forstwirtschaft und Naturschutz, in: Naturschutz und Landschaftsplanung, 42(9), p. 261–266.

Rieckmann, M. (2018a): Key themes in Education for Sustainable Development, in: UNESCO (Ed.): Issues and trends in Education for Sustainable Development, Paris, p. 61–84.

Rieckmann, M. (2018b): Learning to transform the world: key competencies in Education for Sustainable Development, in: UNESCO (Ed.): Issues and trends in education for sustainable Development, Paris, p. 39–59.

Ripl, W.; Wolter, K.-D. (2002): Ecosystem function and degradation: Carbon assimilation in marine and freshwater ecosystems, in: P. J. le B. Williams; D. N. Thomas; C. S. Reynolds (Eds.): Phytoplankton productivity, Oxford, p. 291–317.

Ripple, W. J., et al. (2015): Collapse of the world's largest herbivores, in: Science Advances, 1(4), e1400103.

Ripple, W. J., et al. (2017): World scientists' warning to humanity: a second notice, in: Bioscience, 67(12), p. 1026–1028.

Ritchie, H.; Roser, M. (2017): CO_2 and other Greenhouse Gas Emissions. Published online by OurWorldInData.org [https://ourworldindata.org/co2-and-other-greenhouse-gas-emissions; 15. 01. 2018].

Robertson, B. J. (2016): Holacracy. Ein revolutionäres Management-System für eine volatile Welt, München.

Röchert, R. (2008): Aufklären oder verführen? Zur Bedeutung der Faktoren "Emotionalität" und "Integrierte Kommunikation" für Besucherzentren und Ausstellungen, in: K.-H. Erdmann; T. Hopf; C. Schell (Eds.): Informieren und faszinieren – Kommunikation in Natur-Infozentren. Naturschutz und biologische Vielfalt Band 54, Bonn, p. 7–24.

Rockström, J. (2009): A safe operating space for humanity, in: Nature, 461, p. 472–475.

Rodrigues Mororó, R. (2014): Der demokratische Mythos Porto Alegre. Widersprüche und Wirklichkeit eines partizipativen "Planungsmodells", Wiesbaden.

Rogall, H. (2012): Nachhaltige Ökonomie. Ökonomische Theorie und Praxis einer nachhaltigen Entwicklung, 2. überarbeitete und stark erweiterte Auflage, Marburg.

Rogall, H. (2013): Volkswirtschaftslehre für Sozialwissenschaftler. Einführung in eine zukunftsfähige Wirtschaftslehre, 2. Auflage, Wiesbaden.

Rohr, von C. R., et al. (2015): Chimpanzees' bystander reactions to infanticide, in: Human Nature, 26(2), p. 143–160.

Rosch, M.; Frey, D. (1994): Soziale Einstellungen, in: D. Frey; S. Greif (Eds.): Sozialpsychologie. Ein Handbuch in Schlüsselbegriffen, 3. Auflage, Weinheim, p. 296–306.

Rosenberg, K.; Trevathan, W. (1995): Bipedalism and human birth: The obstetrical dilemma revisited, in: Evolutionary Anthropology, 4(5), p. 161–168.

Rosenkranz, L., et al. (2017): Verbundforschungsbericht WEHAM-Szenarien: Stakeholderbeteiligung bei der Entwicklung und Bewertung von Waldbehandlungs- und Holzverwendungsszenarien. Thünen Working Paper 73, Braunschweig.

Roser, M. (2018a): Economic growth. Published online by OurWorldInData.org [https://ourworldindata.org/economic-growth; 25.01.2018].

Roser, M. (2018b): Tourism. Published online by OurWorldInData.org [https://ourworldindata.org/tourism; 25.01.2018].

Roser, M.; Ortiz-Ospina, E. (2018): World Population Growth. Published online by Our WorldInData.org [https://ourworldindata.org/world-population-growth; 25.01.2018].

Rösler, C., et al. (1999): Lokale Agenda im europäischen Vergleich, herausgegeben vom Bundesministerium für Umwelt, Naturschutz und Reaktorsicherheit und vom Umweltbundesamt, Bonn.

Rostow, W. W. (1956): The take-off into self-sustained growth, in: The Economic Journal, 66(261), p. 25–48.

Roszak, T. (1994): Ökopsychologie. Der entwurzelte Mensch und der Ruf der Erde, Stuttgart.

Rotmans, J.; Kemp, R.; van Asselt, M. (2001): More evolution than revolution: transition management in public policy, in: Foresight, 3(1), p. 15–31.

Sachs, W. (1997): Sustainable Development. Zur politischen Anatomie eines internationalen Leitbilds, in: K.-W. Brandt (Ed.): Nachhaltige Entwicklung. Eine Herausforderung an die Soziologie, Opladen, p. 93–110.

Salafsky, N., et al. (2008): A standard lexicon for biodiversity conservation: unified classifications of threats and actions, in: Conservation Biology, 22(4), p. 897–911.

Sandström, U. F. (2002): Green Infrastructure planning in urban Sweden, in: Planning Practice and Research, 17(4), p. 373–338.

Schahn, J. (1993): Die Kluft zwischen Einstellung und Verhalten beim individuellen Umweltschutz, in: J. Schahn; T. Giesinger (Eds.): Psychologie für den Umweltschutz, München, p. 29–49.

Schahn, J.; Matthies, E. (2008): Moral, Umweltbewusstsein und umweltbewusstes Handeln, in: E.-D. Lantermann; V. Linneweber (Eds.): Grundlagen, Paradigmen und Methoden der Umweltpsychologie, Göttingen, Bern, Toronto, Seattle, p. 663–689.

Schäpke, N., et al. (2017): Reallabore im Kontext transformativer Forschung. Ansatzpunkte zur Konzeption und Einbettung in den internationalen Forschungsstand. IETSR discussion papers in Transdisciplinary Sustainability Research Nr. 1/2017, Lüneburg.

Scheer, H. (2010): Der energethische Imperativ. 100 Prozent jetzt. Wie der vollständige Wechsel zu erneuerbaren Energien zu realisieren ist, München.

Scheler, M. (1978): Die Stellung des Menschen im Kosmos (Erstveröffentlichung 1927; published online by http://gutenberg.spiegel.de/buch/die-stellung-des-menschen-im-kosmos-8160/1), Bern, München.

Scheuthle, H.; Frick, J.; Kaiser, F. G. (2010): Personenzentrierte Intervention zur Veränderung von Umweltverhalten, in: V. Linneweber; E.-D. Lantermann; E. Kals (Eds.): Spezifische Umwelten und umweltbezogenes Handeln, Göttingen, Bern, Toronto, Seattle, p. 643–667.

Schmidt, M. (2008): Die Bedeutung der Effizienz für Nachhaltigkeit. Chancen und Grenzen, in: S. Hartard; A. Schaffer; J. Giegrich (Eds.): Ressourceneffizienz im Kontext der Nachhaltigkeitsdebatte, Baden-Baden, p. 31–46.

Schneider, E. D.; Kay, J. J. (1994): Complexity and thermodynamics: Towards a new ecology, in: Futures, 26(6), p. 626–647.

Schneidewind, U.; Scheck, H. (2012): Zur Transformation des Energiesektors. Ein Blick aus der Perspektive der Transition-Forschung, in: U. Schneidewind, H.-G. Servatius, D. Rohlfing (Eds.): Smart Energy, Berlin, Heidelberg, p. 45–61.

Schneidewind, U.; Singer-Brodowski, M. (2013): Transformative Wissenschaft. Klimawandel im deutschen Wissenschafts- und Hochschulsystem, Marburg.

Schraml, U. (2016): Peter und der Wald oder: Woher kommt die Begeisterung für die Geheimnisse von Bäumen?, in: Holz-Zentralblatt, 17, p. 437–438.

Schröder, C., et al. (2015): Steckbriefe für Niedermoorbewirtschaftung bei unterschiedlichen Wasserverhältnissen, Müncheberg.

Schumpeter, J. A. (1942): Capitalism, socialism and democracy, New York.

Schwartz, S. H.; Howard, J. A. (1981): A normative decision-making model of altruism, in: J. P. Rushton; R. M. Sorrentino (Eds.): Altruism and helping behavior, Hillsdale, p. 189–211.

Shultz, S.; Nelson, E.; Dunbar, R. I. M. (2012): Hominin cognitive evolution: identifying patterns and processes in the fossil and archaeological record, in: Philosophical Transactions of the Royal Society – Biological Sciences, 367(1599), p. 2130–2140.

Sik, O. (1987): Wirtschaftssysteme: Vergleiche – Theorie – Kritik, Berlin, Heidelberg.

Simonis, U. (1989): Wir müssen anders wirtschaften. Ansatzpunkte einer ökologischen Umorientierung der Industriegesellschaft, Frankfurt/Main.

Sintomer, Y., et al. (Eds.) (2013): Participatory Budgeting in Asia and Europe. Key Challenges of Participation, Basingstoke.

Skidelsky, R.; Skidelsky, E. (2013): Wie viel ist genug? Vom Wachstumswahn zu einer Ökonomie des guten Lebens, München.

Smith, A. (1776): Inquiry into the nature and causes of the wealth of nations, Oxford.

Smuts, J. (1926): Holism and evolution, London.

Solow, R. M. (1974): The economics of resources or the resources of economics, in: American Economic Review, 64(2), p. 1–14.

Sommer, J. (2015a): Die vier Dimensionen gelingender Bürgerbeteiligung, in: J. Sommer (Ed.): Kursbuch Bürgerbeteiligung, Berlin, p. 11–21.

Sommer, J. (Ed.) (2015b): Kursbuch Bürgerbeteiligung, Berlin.

Spada, H. (1990): Umweltbewußtsein. Einstellung und Verhalten, in: L. Kruse; C.-F. Graumann; E.-D. Lantermann (Eds.): Ökologische Psychologie. Ein Handbuch in Schlüsselbegriffen, München, p. 623–631.

Spanier, H. (2008): Was kann die Naturschutzkommunikation von der Werbung lernen?, in: K.-H. Erdmann; T. Hopf; C. Schell (Eds.): Informieren und faszinieren – Kommunikation in Natur-Infozentren. Naturschutz und biologische Vielfalt, Band 54, p. 25–48.

Spathelf, P. (1997): Seminatural silviculture in southwest Germany, in: Forestry Chronicle, 73(6), p. 715–722.

Spathelf, P.; Bolte, A.; van der Maaten, E. (2015): Is Close-to-Nature Silviculture (CNS) an adequate concept to adapt forests to climate change?, in: Landbauforschung, 65(3/4), p. 161–170.

Spathelf, P., et al. (2014): Climate change impacts in European forests: the expert-views of local observers, in: Annals of Forest Science, 71(2), p. 131–137.

Spitzer, M. (2009): Lernen. Gehirnforschung und die Schule des Lebens, Heidelberg.

SRU (Sachverständigenrat für Umweltfragen/German Advisory Council on the Environment) (2002): Für eine neue Vorreiterrolle, Umweltgutachten 2002, Deutscher Bundestag Drucksache 14/8792.

SRU (Sachverständigenrat für Umweltfragen/German Advisory Council on the Environment) (2008): Umweltschutz im Zeichen des Klimawandels. Umweltgutachten 2008, Berlin.

SRU (Sachverständigenrat für Umweltfragen/German Advisory Council on the Environment) (2012): Umweltgerechte Waldnutzung (Kapitel 6), in: SRU (Ed.): Umweltgutachten 2012: Verantwortung in einer begrenzten Welt, Berlin, p. 211–239.

SRU (Sachverständigenrat für Umweltfragen/German Advisory Council on the Environment) (2016): Zur Neuauflage der deutschen Nachhaltigkeitsstrategie, Stellungnahme Nr. 21, Berlin.

Stadt Freiburg i. B. (Ed.) (2016): 2. Freiburger Nachhaltigkeitsbericht 2016. Beispielhafter Ausschnitt zur Darstellung des Nachhaltigkeitsprozesses, Freiburg i. B.

Statista (2017a): Bruttoinlandsprodukt (BIP) in Deutschland von 1991 bis 2016 [https://de.statista.com/statistik/daten/studie/1251/umfrage/entwicklung-des-bruttoinlands produkts-seit-dem-jahr-1991/; 22.05.2017].

Statista (2017b): Entwicklung des durchschnittlichen Umsatzes pro Handelstag am weltweiten Devisenmarkt von 1995 bis 2016 (in Milliarden US-Dollar) [https://de.statista.com/statistik/daten/studie/239512/umfrage/umsaetze-pro-handelstag-am-weltweiten-devi senmarkt/; 29.05.2017].

Statista (2018): Facebook: number of monthly active users worldwide 2008- 2017 [https://www.statista.com/statistics/264810/number-of-monthly-active-facebook-users-world wide/; 25.01.2018].

Steffen, W., et al. (2015): Planetary boundaries: Guiding human development on a changing planet, in: Science, 347(6223), 1259855.

Stern, N. (2007): The economics of climate change. The Stern Review, Cambridge.

Stevens, M. (2015): Drought update: Dry wells, debate over water cutbacks [http://www.latimes.com/local/lanow/la-me-ln-drought-hearing-20150512-story.html; 31.01.2018], in: Los Angeles Times, 12.05.2015.

Stiglitz, J. (2010): Freefall: America, Free Markets, and the Sinking of the World Economy, New York.

Stiner, M. C. (2002): Carnivory, coevolution, and the geographic spread of the genus Homo, in: Journal of Archaeological Research, 10(1), p. 1–63.

Stober, I.; Bucher, K. (2014): Geothermie. 2. Auflage, Berlin, Heidelberg.

Stokstad, E. (2014): More than twice as much mercury in environment as thought [http://www.sciencemag.org/news/2014/09/more-twice-much-mercury-environment-thought; 25.01.2018], in: Science, 04.09.2014.

Stoll-Kleemann, S.; Welp, M. (Eds.) (2006): Stakeholder dialogues in natural resources management – Theory and practice, Berlin, Heidelberg.

Stoltenberg, U. (2009): Mensch und Wald. Theorie und Praxis einer Bildung für eine nachhaltige Entwicklung am Beispiel des Themenfelds Wald, München.

Stroeve, J., et al. (2007): Arctic sea ice decline: Faster than forecast, in: Geophysical Research Letters, 34(9), L09501.

Sumser, E. (2016): Evolution der Ethik: Der menschliche Sinn für Moral im Licht der modernen Evolutionsbiologie, Berlin.

Swanson, A. (2015): How China used more cement in 3 years than the U.S. did in the entire 20th Century [https://www.washingtonpost.com/news/wonk/wp/2015/03/24/how-china-used-more-cement-in-3-years-than-the-u-s-did-in-the-entire-20th-century/?utm_term=.a5e062fc998c; 25.01.2018], in: The Washington Post, 24.03.2015.

Tansley, A. G. (1935): The use and abuse of vegetational concepts and terms, in: Ecology, 16(3), p. 284–307.

Taylor, B. (2001): Earth and nature-based spirituality (Part II): From Earth First! and Bioregionalism to scientific paganism and the New Age, in: Religion, 31(3), p. 225–245.

TEEB (2010): The Economics of Ecosystems and Biodiversity – Ecological and Economic Foundations, London.

The World Bank (2017a): Agricultural land (sq km). World Bank Open Data [https://data.worldbank.org/; 25.01.2018].

The World Bank (2017b): Land under cereal production (hectares). World Bank Open Data [https://data.worldbank.org/; 25.01.2018].

The World Bank (2017c): Population density (people per sq. km of land area). Food and Agriculture Organization and World Bank population estimates. World Bank Open Data [https://data.worldbank.org/; 26.01.2018].

The World Bank (2017d): Population in urban agglomerations of more than 1 million. World Bank Open Data [https://data.worldbank.org/; 26.01.2018].

The World Bank (2017e): Urban population. World Bank staff estimates based on the United Nations Population Division's World Urbanization Prospects. World Bank Open Data [https://data.worldbank.org/; 26.01.2018].

The World Bank (2017f): World Development Report 2017: Governance and the law [http://www.worldbank.org/en/publication/wdr2017#; 25.01.2018], Licence: Creative Commons Attribution CC BY 3.0 IGO, Washington, DC.

Thrash, J. C., et al. (2011): Phylogenomic evidence for a common ancestor of mitochondria and the SAR11 clade, in: Scientific Reports, 1, Art. 13.

Tingley, D.; Wagner, G. (2017): Solar geoengineering and the chemtrails conspiracy on social media, in: Palgrave Communications, 3(1), Art. 12.

Torres, A., et al. (2017): A looming tragedy of the sand commons, in: Science, 357(6355), p. 970–971.

Trevathan, W. R. (1996): The evolution of bipedalism and assisted birth, in: Medical Anthropology Quarterly, 10(2), p. 287–290.

Tzoulas, K., et al. (2007): Promoting ecosystem and human health in urban areas using Green Infrastructure: A literature review, in: Landscape and Urban Planning, 81(3), p. 167–178.

Uchiyama, K. (2016): Environmental Kuznets Curve hypothesis and carbon dioxide emissions, Springer Briefs in Economics, Development Bank of Japan Research Series, Tokyo.

UIG (Umweltinformationsgesetz in der Fassung der Bekanntmachung vom 27. Oktober 2014 [BGBl. I p. 1643], das zuletzt durch Artikel 2 Absatz 17 des Gesetzes vom 20. Juli 2017 [BGBl. I p. 2808] geändert wurde).

Umweltbundesamt (2010): Energieziel 2050. 100% Strom aus erneuerbaren Quellen, Dessau.

Umweltbundesamt (2014): Rebound-Effekte [https://www.umweltbundesamt.de/themen/abfall-ressourcen/oekonomische-rechtliche-aspekte-der/rebound-effekte; 02.01.2018].

Umweltbundesamt (Ed.) (2015a): Umweltbewusstsein in Deutschland 2014. Ergebnisse einer repräsentativen Bevölkerungsumfrage, Reinheim.

Umweltbundesamt (2015b): Vorsorgeprinzip [https://www.umweltbundesamt.de/themen/nachhaltigkeit-strategien-internationales/umweltrecht/umweltverfassungsrecht/vorsorgeprinzip; 12.03.2018].

Umweltbundesamt (Ed.) (2015c): Wie Transformationen und gesellschaftliche Innovationen gelingen können, Dessau.

Umweltbundesamt (2016a): Berichterstattung unter der Klimarahmenkonvention der Vereinten Nationen und dem Kyoto-Protokoll 2016. Nationaler Inventarbericht zum Deutschen Treibhausgasinventar 1990–2014. Climate Change 23/2016.

Umweltbundesamt (2016b): Faktor X – Die Idee Faktor X [https://www.umweltbundesamt.de/themen/abfall-ressourcen/ressourcenschonung-in-produktion-konsum/faktor-x; 02.01.2018].

Umweltbundesamt (2017a): Abfallaufkommen der Kategorie Siedlungsabfälle [http://www.umweltbundesamt.de/indikator-abfallmenge-siedlungsabfaelle; 22.05.2017].

Umweltbundesamt (2017b): Luftqualität in Ballungsräumen [http://www.umweltbundesamt.de/indikator-luftqualitaet-in-ballungsraeumen; 22.05.2017].

Umweltbundesamt (Ed.) (2017c): Umweltbewusstsein in Deutschland 2016. Ergebnisse einer repräsentativen Bevölkerungsumfrage, Reinheim.

Umweltbundesamt (2018): Treibhausgas-Emissionen [https://www.umweltbundesamt.de/themen/klima-energie/treibhausgas-emissionen; 01.02.2018].

Umweltgutachterausschuss (2015): In 10 Schritten zu EMAS. Ein Leitfaden für Umweltmanagementbeauftragte, 2. Auflage [http://www.emas.de/fileadmin/user_upload/06_service/PDF-Dateien/EMAS-Leitfaden-Umweltmangementbeauftragte.pdf; 29.01.2018].

UN-Dekade Biologische Vielfalt Geschäftsstelle (2017): UN-Dekade Botschafter/innen [https://www.undekade-biologischevielfalt.de/un-dekade/die-botschafterinnen-der-un-dekade/; 07.06.2017].

UNCTADSTAT (2017): GDP at current prices [http://unctadstat.unctad.org/wds/TableViewer/tableView.aspx?ReportId=96; 29.05.2017].

UNECE (United Nations Economic Commission for Europe) (1998): Convention on Access to Information, Public Participation in Decision-making and Access to Justice in Environmental Matters (Aarhus Convention), Aarhus.

UNESCO (United Nations Educational, Scientific and Cultural Organization) (2006): Framework for the UN DESD International Implementation Scheme. ED/DESD/2006/PI/1, Paris.

UNESCO (United Nations Educational, Scientific and Cultural Organization) (2014): Shaping the future we want. UN Decade of Education für Sustainable Development (2005–2014). Final Report, Paris.

United Nations (1948): Universal Declaration of Human Rights. Resolution 217 A (III) of the General Assemby of the United Nations on 10 December 1948.

United Nations (1962): The United Nations Development Decade, Proposals for Action: Report of the Secretary-General. Part 49, New York.

United Nations (1992a): Agenda 21: Earth Summit: the United Nations Programme of Action from Rio [http://www.un-documents.net/a21-36.htm; 20.02.2019].

United Nations (1992b): Convention on Biological Diversity [https://www.cbd.int/doc/legal/cbd-en.pdf; 07.03.2018].

United Nations (2000): United Nations Millennium Declaration. Resolution adopted by the General Assembly 55/2 [http://www.un.org/millennium/declaration/ares552e.pdf; 08.11.2017].

United Nations (2002): General Framework of the Islamic agenda for sustainable development. Islamic Declaration on Sustainable Development: Background Paper No. 5 submitted by the First Islamic Conference of Environment Ministers to the World Summit on Sustainable Development 2002 in Johannesburg.

United Nations (2012): The future we want. Outcome document of the United Nations Conference on Sustainable Development in Rio de Janeiro, Brazil [https://sustainabledevelopment.un.org/content/documents/733FutureWeWant.pdf; 08.11.2017].

United Nations (2013): Intergenerational solidarity and the needs of future generations. Report of the Secretary-General. A/68/100 [https://sustainabledevelopment.un.org/content/documents/2006future.pdf; 17.12.2017].

United Nations (2015): Transforming our World: The 2030 Agenda for Sustainable Development A/RES/70/1 [https://sustainabledevelopment.un.org/post2015/transformingourworld/publication; 08.11.2017].

United Nations (2017): Sustainable Development Goals. 17 goals to transform our world [http://www.un.org/sustainabledevelopment/news/communications-material/; 08.11.2017].

United Nations Statistics Division (2015): Slum population as percentage of urban. United Nations Millennium Development Goals database [http://mdgs.un.org; 26.01.2018].

Ura, K., et al. (2012): A short guide to Gross National Happiness Index, Centre for Bhutan Studies [http://www.grossnationalhappiness.com/wp-content/uploads/2012/04/Short-GNH-Index-edited.pdf; 02.01.2018].

Vallée, R. (1993): Systems theory, a historical presentation, in: R. Rodriguez Delgado; B. Banathy (Eds.): International Systems Science Handbook, Madrid, p. 84–104.

Vincent, K.; Davidson, C. (2015): The toxicity of glyphosate alone and glyphosate-surfactant mixtures to western toad (Anaxyrus boreas) tadpoles, in: Environmental Toxicology and Chemistry, 34(12), p. 2791–2795.

Voland, E. (2006): Anthropologische Hürden auf dem Weg zu einer erfolgreichen Umweltbildung, in: B. Hiller; M. Lange (Eds.): Bildung für nachhaltige Entwicklung – Perspektiven für die Umweltbildung. Vorträge und Studien, Heft 16, Münster, p. 45–54.

Voland, E. (2007): Die Natur des Menschen. Grundkurs Soziobiologie, München.

Vor, T., et al. (Eds.) (2015): Potenziale und Risiken eingeführter Baumarten. Baumartenportraits mit naturschutzfachlicher Bewertung. Göttinger Forstwissenschaften Band 7, Göttingen.

Vörösmarty, C. J., et al. (2010): Global threats to human water security and river biodiversity, in: Nature, 467, p. 555–561.

Voß, J.-P.; Bauknecht, D.; Kemp, R. (Eds.) (2006): Reflexive governance for sustainable development, Cheltenham.

Voß, J.-P., et al. (2007): Steering for sustainable development: A typology of problems and strategies with respect to ambivalence, uncertainty and distributed power, in: Journal of Environmental Policy & Planning, 9(3/4), p. 193–212.

Voss, K. (2014): Internet und Partizipation. Bottom-up oder Top-down? Politische Beteiligungsmöglichkeiten im Internet, Wiesbaden.

Wagner, M.; Oehlmann, J. (2009): Endocrine disruptors in bottled mineral water: total estrogenic burden and migration from plastic bottles, in: Environmental Science and Pollution Research, 16(3), p. 278–286.

Walk, H. (2008): Partizipative Governance. Beteiligungsrechte und Beteiligungsformen im Mehrebenensystem der Klimapolitik, Wiesbaden.

Wand, G. (2008): The influence of stress on the transition from drug use to addiction, in: Alcohol Research and Health, 31(2), p. 119–136.

wandelBar (2015): Wandelbar, Transition Charta 1.0 vom 04. 09. 2015 [http://stadt-und-land-im-wandel.de/transition-charta/; 02. 01. 2018].

WBGU (German Advisory Council on Social Change) (2011): World in Transition. A Social Contract for Sustainability, Berlin. [https://www.wbgu.de/fileadmin/user_upload/wbgu.de/templates/dateien/veroeffentlichungen/hauptgutachten/jg2011/wbgu_jg2011_en.pdf].

WBGU (Wissenschaftlicher Beirat der Bundesregierung Globale Umweltveränderungen) (Ed.) (1993). Welt im Wandel: Grundstruktur globaler Mensch-Umwelt-Beziehungen. Jahresgutachten 1993, Bonn.

Weber, M. (1919): Politik als Beruf, München, Leipzig.

WEC (World Energy Council) (2013): World Energy Scenarios – Composing energy futures to 2050, London.

Weizsäcker, E. U. von (1994): Erdpolitik. Ökologische Realpolitik an der Schwelle zum Jahrhundert der Umwelt. 4. aktualisierte Ausgabe, Darmstadt.

Welp, M.; Kasemir, B.; Jaeger, C. C. (2009): Citizens' voices in environmental policy: The contribution of integrated assessment focus groups to accountable decision-making, in: F. H. J. M. Coenen (Ed.): Public participation and better environmental decisions: The promise and limits of participatory processes for the quality of environmentally related decision-making, Dordrecht, p. 21–34.

Weltkommission für Umwelt und Entwicklung (1988): Unsere gemeinsame Zukunft – Bericht der Weltkommission für Umwelt und Entwicklung. Deutsche Übersetzung vom Staatsverlag der DDR, Berlin.

Welzer, H. (2008): Klimakriege. Wofür im 21. Jahrhundert getötet wird, Frankfurt/Main.

Wessel, K.-F. (2015): Der ganze Mensch. Eine Einführung in die Humanontogenetik oder Die biopsychosoziale Einheit Mensch von der Konzeption bis zum Tode, Berlin.

Westman, W. E. (1977): How much are nature's services worth?, in: Science, 197(4307), p. 960–964.

Wichtmann, W.; Schröder, C.; Joosten, H. (Eds.) (2016): Paludikultur – Bewirtschaftung nasser Moore für regionale Wertschöpfung, Klimaschutz und Biodiversität, Stuttgart.

Williams, H. T. P., et al. (2015): Network analysis reveals open forums and echo chambers in social media discussions of climate change, in: Global Environmental Change, 32, p. 126–138.

Williams, J. M., et al. (2008): Causes of death in the Kasekela chimpanzees of Gombe National Park, Tanzania, in: American Journal of Primatology, 70(8), p. 766–777.

Williams, M. (2006): Deforesting the earth: From prehistory to global crisis. An abridgment, Chicago, London.

Willner, S. N., et al. (2018): Adaptation required to preserve future high-end river flood risk at present levels, in: Science Advances, 4(1), eaao1914.

Wilson, M. L., et al. (2014): Lethal aggression in Pan is better explained by adaptive strategies than human impacts, in: Nature, 513(7518), p. 414–417.

Winkelmann, R., et al. (2015): Combustion of available fossil fuel resources sufficient to eliminate the Antarctic Ice Sheet, in: Science Advances, 1(8), e1500589.

Wittchen, H. U., et al. (2011): The size and burden of mental disorders and other disorders of the brain in Europe 2010, in: European Neuropsychopharmacology, 21(9), p. 655–679.

Wöhe, G.; Döhring, U. (2013): Einführung in die Allgemeine Betriebswirtschaftslehre, 25. Auflage, München.

Wohlleben, P. (2015): Das geheime Leben der Bäume, München.

World Commission on Environment and Development (1987): Our Common Future, Oxford.

WWF (World Wildlife Fund) (2018): Ecoregions [https://www.worldwildlife.org/biomes; 18.01.2018].

WWF Deutschland; CDP (Carbon Disclosure Project) (2016). Vom Emissionsbericht zur Klimastrategie. Grundlagen für ein einheitliches Emissions- und Klimastrategieberichtswesen [http://klimareporting.de/sites/default/files/klimareporting.de_leitfaden_0.pdf; 15.05.2017].

Yang, C. Z., et al. (2011): Most plastic products release estrogenic chemicals: a potential health problem that can be solved, in: Environmental Health Perspectives, 119(7), p. 989–996.

Zarfl, C., et al. (2015): A global boom in hydropower dam construction, in: Aquatic Sciences, 77(1), p. 161–170.

ZEIT ONLINE (2017): Klimaschutz-Index 2018. Staaten halten sich kaum an Pariser Klimavertrag [http://www.zeit.de/wirtschaft/2017-11/klimaschutz-index-2018-klimaziele-pariser-abkommen; 21.03.2018], in: ZEIT ONLINE, 15.11.2017.

Zeitz, J.; Luthardt, V. (2018): DSS-TORBOS. Ein Entscheidungsunterstützungssystem zur torfschonenden Bewirtschaftung organischer Böden [http://www.dss-torbos.de/index.html; 02.02.2018].

Ziaei, S. M.; Mazloumzadeh, S. M.; Jabbary, M. (2015): A comparison of energy use and productivity of wheat and barley (case study), in: Journal of the Saudi Society of Agricultural Sciences, 14(1), p. 19–25.

Zuckerman, B., et al. (2008): Planetary systems around close binary stars: The case of the very dusty, sun-like, spectroscopic Binary BD+20 307, in: The Astrophysical Journal, 688(2), p. 1345–1351.

Index

List of editors, authors and specialist advisers

Editors and co-authors

Pierre L. Ibisch holds a higher doctorate in biology, and, after many years working in development cooperation, in 2004 became Professor of Nature Conservation at Eberswalde University for Sustainable Development, where he heads the Centre for Econics and Ecosystem Management. He was founding head of the international Master's programme Global Change Management. Since 2009, he has held various research professorships (in biodiversity and natural resource management in global change, until 2015; since 2015, in ecosystem-based sustainable development) and has been involved in worldwide research and consulting activities. He is a member of the board of the Deutsche Umweltstiftung and co-editor of the *Jahrbuch Ökologie*.

Heike Molitor has been Professor for Environmental Education and Education for Sustainable Development at Eberswalde University for Sustainable Development since 2009, and is currently a Research Professor there (2018–2020). She has been a member of various advisory boards (Advisory Council for Sustainable Development of the State of Brandenburg 2010–2014; Expert Advisory Council for Nature Conservation and Landscape Conservation Berlin 2017–2021; specialist higher education forum of the Global Action Programme in Education for Sustainable Development 2015–2019).

Alexander Conrad has been Professor of Economics at Eberswalde University for Sustainable Development since 2014, specializing in Sustainable Regional Management and Local Government Finance. He is head of the Regional Management degree programme and focuses on sustainable supply systems in rural areas.

Heike Walk holds a higher doctorate in political science and is Professor of Transformation Governance and Vice-President for Curriculum and Teaching at Eberswalde University for Sustainable Development. Since 2003 she has been co-editor of the book series 'Civil Society and Democracy', published by Springer Fachmedien. In addition, she is spokesperson for the working group on 'Civil Society Research' of the Federal German Network for Civic Engagement.

Vanja Mihotovic (now Vanja Spoo) is Professor of Engineering Mechanics, Physics and Electrical Engineering at Eberswalde University for Sustainable Development. He holds a PhD in production technology/production engineering and was employed for many years at Siemens Energy, Gas Turbine Division.

Juliane Geyer was awarded a doctorate for her work on the adaptation of nature conservation management to climate change, and since 2009 has worked on various projects at the Centre for Econics and Ecosystem Management at Eberswalde University for Sustainable Development, especially on management of protected areas. She is also involved in local initiatives for sustainable development.

Co-authors and specialist advisers for individual chapters

Kerstin Kräusche is a graduate chemist who has been managing the environmental management process at Eberswalde University for Sustainable Development since 2007. Since 2014, she has been responsible for the establishment, further development and networking of internal sustainability management, and links aspects of sustainable development with management, teaching and research processes.

Benjamin Nölting is a political scientist and Professor of Governance of Regional Sustainability Transformation at Eberswalde University for Sustainable Development. He teaches and researches on the sustainable development of regions and organizations, leads the further education course on 'Strategic Sustainability Management', and works on developing concepts for the transfer of sustainability in higher education.

Vera Luthardt has a doctorate in biology and is Professor of Vegetation Science and Applied Plant Ecology at Eberswalde University for Sustainable Development. Between 2009 and 2015, and again since 2018, she has held the post of Research Professor with a focus on the management of moorlands. Since 2011, she has headed the State Nature Conservation Advisory Council at the Brandenburg Ministry of the Environment, and has been an appointed member of the MAB National Committee for Germany since 2018.

Martin Welp is Professor of Socioeconomics and Communication at Eberswalde University for Sustainable Development. He received his doctorate from the TU Berlin, and was employed for many years at the Potsdam Institute for Climate Impact Research (PIK). As a Research Professor, he focused on Transformation and Change Management in urban and periurban areas (2015–2018). He is a member of the Centre for Econics and Ecosystem Management.

Other authors

Hans-Peter Benedikt is Professor of Entrepreneurship and Human Resource Management at Eberswalde University for Sustainable Development, was Chair of the Advisory Council Network for Study Quality of the Universities and Higher Education Colleges of Brandenburg (2008–2010), a member of the Advisory Board of the Institute for Entrepreneurship and SMEs in Brandenburg (2007–2011), and higher education adviser to SIFE (Students in Free Enterprises) (2009–2012).

Martin Guericke is a forest scientist and, since 2006, Professor of Forest Growth Science at Eberswalde University for Sustainable Development. In addition to his teaching and applied research, he is Director of the study programme on forestry.

Norbert Jung is a biologist, psychotherapist, group trainer and supervisor (inpatient psychotherapy, depth psychology), and was Professor of Environmental Education at Eberswalde University of Applied Sciences from 2000–2008. He is co-founder of the interdisciplinary 'Eberswalde Symposium for Environmental Education' and co-editor of the book series 'Eberswalder Beitrage für Bildung und Nachhaltigkeit' ('Eberswalde Articles on Education and Sustainability'). In 1990, he was a member of the first freely elected East Berlin city council for the New Forum, which later became part of Alliance 90/The Greens.

Jan König has a doctorate in economics and is head of the department of social affairs in the city of Eberswalde and Honorary Professor for municipal economic development at Eberswalde University of Applied Sciences.

Hermann E. Ott has a doctorate in law, is Head of the Germany office of Client-Earth and Honorary Professor for sustainability strategies and governance at Eberswalde University of Sustainable Development. He was previously Senior Advisor at the Wuppertal Institute for Climate, Environment and Energy, and from 2009–2013 was a Member of the German Bundestag. He is the author of numerous books and articles on environmental and climate protection, sustainability and global governance. He is an honorary member of the Presidium of the Deutscher Naturschutzring (DNR, German Nature Conservation Alliance).

Peter Spathelf is a forest scientist and Professor of Applied Silviculture at Eberswalde University for Sustainable Development. His work focuses on close-to-nature silviculture, forest adaptation to climate change, and forest restoration. He is the representative responsible for climate change at the German Forestry Association.

Wilhelm-Günther Vahrson, who holds a higher doctorate in geography, is Professor for Landscape Science and, since 1998, head of the University of Applied Sciences Eberswalde and then Eberswalde University for Sustainable Development, first as Rector, then as President. He has been Speaker of the Brandenburg Rectors' Conference several times. He is also a member of various committees, such as the Nature Conservation Council of the State of Brandenburg (Chair 2005–2010), the Advisory Council for Sustainable Development of the State of Brandenburg (2010–2014), and the National Platform of the Global Action Programme for Education for Sustainable Development (2015–2019).

Other advisers and reviewers

Hartmut Ihne, a political scientist and philosopher, has been Honorary Professor for Ethics and Politics at Eberswalde University for Sustainable Development since 2008 and President of the University of Applied Sciences Bonn-Rhein-Sieg. Author of a standard work on development policy, he was previously (among other things) Managing Director of the Centre for Development Research (ZEF) and the Centre for European Integration Research (ZEI) at the University of Bonn.

Carsten Mann is Professor for Sustainable Forest Resource Economics at Eberswalde University for Sustainable Development. He wrote his doctoral thesis on conflicts in protected areas at the Institute for Forestry and Environmental Policy Freiburg, and his higher doctorate on the adaptation of institutions to socio-ecological systems at Humboldt University of Berlin. His work focuses on the analysis of political and economic influences on natural resource management as well as on inter- and transdisciplinary sustainability research.

Dörte Martens has a doctorate in psychology. She is a research assistant at the Eberswalde University for Sustainable Development, Faculty of Landscape Use and Nature Conservation, working on the project 'Nature Experience Spaces in Large Cities: the example of Berlin'. Her research focus is on the impact of nature and nature experiences on child development and health.

Udo E. Simonis was formerly Professor of Economics at the Technical University of Berlin, and is now Emeritus Professor of Environmental Policy and Director of the International Institute for Environment and Society at the Social Science Research Centre Berlin. A member of numerous committees (including WBGU, Advisory Board of the Global Change Management degree programme) and editor and co-editor of the *Jahrbuch Ökologie*, he was awarded the Order of Merit of the Federal Republic of Germany in 2017.

Fabian Wulf is a graduate engineer in wood technology and works in the Faculty of Wood Science and Technology at Eberswalde University for Sustainable Development, in the working group 'Chemistry and Physics of Wood in Material and Process Development for Wood and Composite Materials'. On a voluntary basis — for the local Transition Town group 'wandelBar', among others — he works on sustainable mobility and the energy transition and their implementation in small-scale local projects.